Reef Evolution

by

RACHEL WOOD

Royal Society University Research Fellow,
Department of Earth Sciences,
University of Cambridge
and Fellow, Gonville and Caius College

OXFORD

UNIVERSITY PRESS

OXFORD

UNIVERSITY PRESS

Great Clarendon Street, Oxford OX2 6DP

Oxford University Press is a department of the University of Oxford.
It furthers the University's objective of excellence in research, scholarship,
and education by publishing worldwide in

Oxford New York

Athens Auckland Bangkok Bogotá Buenos Aires Calcutta
Cape Town Chennai Dar es Salaam Delhi Florence Hong Kong Istanbul
Karachi Kuala Lumpur Madrid Melbourne Mexico City Mumbai
Nairobi Paris São Paulo Singapore Taipei Tokyo Toronto Warsaw

with associated companies in Berlin Ibadan

Oxford is a registered trade mark of Oxford University Press
in the UK and in certain other countries

Published in the United States
by Oxford University Press, Inc., New York

A catalogue record for this book is available from the British Library

Library of Congress Cataloging in Publication Data
Wood, Rachel.
Reef evolution / by Rachel Wood.
Includes bibliographical references (p. 354) and index.
1. Reef ecology. 2. Reefs. 3. Animal communities.
4. Palaeontology. I. Title.
QH541.5.C7W662   1999   577.7′89—dc21   98–45212

ISBN 0 19 854999 7 (Hbk)
0 19 857784 2 (Pbk)

Typeset by EXPO, Holdings, Malaysia

Printed in Great Britain by
on acid-free paper by
Biddles Ltd, Guildford & Klag's Lynn

Preface

Reefs are an extraordinary natural phenomenon. Visible from outer space, coral reefs form the largest biologically constructed features known; in close proximity their spectacular beauty and abundance of life is dazzling. Such reefs are usually restricted to shallow, tropical seas and although estimates suggest that they occupy only 0.2% of the world's ocean area, their influence is many-faceted and global. As long-lived topographic structures, coral reefs control sediment transport processes in that they protect coastlines from erosion, help create sheltered harbours, and allow the development of shallow basins with associated mangrove and seagrass communities. As major producers of carbonate sediment, reefs are important not only as storehouses of carbon but also as regulators of atmospheric CO_2, which in turn may contribute to the control of climate and sea-level fluctuations. To society, the unique species found in coral reef communities are the source of many raw materials for commercial and subsistence food species (most notably fish), as well as increasing numbers of medically useful compounds. Ancient, subsurface reefs and their associated sediments can provide extensive reservoirs for oil and gas. But from a biological standpoint, the great significance of reefs lies in the fact that they generate and maintain a substantial proportion of tropical marine biodiversity.

Modern coral reefs are renowned for their diversity, not only in terms of numbers of species but particularly because of the occurrence of representatives of most extant phyla. But coral reefs are in some ways paradoxical in that they appear to present oases of biological growth in seas of otherwise low productivity. They are highly specialized communities with intricate inter-relationships characterized by widespread symbioses, and are regulated by intense predation. In particular, it is the photosymbiotic association of corals with algae that not only provides the driving energy and physical structure for the whole reef community, but which may also aid their survival in the highly dynamic, tropical marine environment. Yet the fossil record of reef-building shows that the acquisition of photosymbiosis and the appearance of modern predator groups are relatively recent geological occurrences: many ancient reefs clearly grew under profoundly different environmental and ecological controls from those which govern the functioning of modern coral reefs. It is clear then that the origins of the ecosystems which dominate Earth today cannot be fully understood without a historical perspective, and it is the fossil record which offers the only documentation of their evolutionary development. This book aims to explore the role of environmental change and biological innovation in the history of reef-building as evidenced in that record.

Marine organisms have aggregated to form reefs for at least 3.5 billion years and the durability of these structures lends them particular significance. But the common expression of reefs in the geological record tends to obscure many of the profound differences in both the ecological structure and environmental setting of ancient reefs. Ancient reefs present a great variety of communities—so varied that all attempts to define a reef have been thwarted even though a plethora of terminology exists to describe the shape and size of individual reefs, as well as ancient reef fabrics. Yet being the result of *in-situ* growth, ancient reefs frequently preserve exquisite details of past ecological interactions lost from most other fossil communities. Such ecological interactions are of profound importance and interest, as not only are they the ultimate determinants of reef growth but also offer an opportunity to document evolutionary processes at a level capable of generating testable hypotheses. Indeed, reefs offer some of the most persuasive evidence for past natural selection that the fossil record can provide, but this rich resource has been barely explored. It is therefore surprising that little emphasis has been given to either the careful description of ancient reef communities, or to those biological processes that have governed their ecology and evolution.

It is not my intention here to detail the myriad of fossil communities that have constructed reefs over geological time. Indeed, this would be an impossible and futile task as there are as many different communities as there are individual reefs. Rather, I wish to emphasize fundamental processes and trends by documenting those biological innovations and environmental controls that have moulded the evolution of reef ecosystems and given rise to the highly complex communities found today. The theme developed in this book is that reef construction is ultimately controlled by the biological occupation of space, but that the demands on reef organisms to achieve this have changed over the course of geological time. This approach reflects my belief that consideration of the processes common to the formation of all reefs can provide a new perspective on reef evolution.

The evolution of reefs in the context of the ancient carbon cycle, the appearance of clonality and the modular habit, the radiation of predator and herbivore groups, and the acquisition of photosymbiosis are treated as a series of topics. To this end, an interdisciplinary approach is adopted that attempts to avoid cumbersome specialist nomenclature. Necessary terms and phrases are highlighted in italics, and definitions (according to my usage) are given where they first appear; they are also collected in a glossary at the end of the text. To improve the flow and keep the references within bounds, I have not always substantiated every assertion with a reference. This may have led to the generalization and oversimplification of some issues for which I apologise. Where fuller explanations or detailed descriptions are deemed necessary, they have been placed in boxes. The resulting book expresses a personal, and possibly somewhat idiosyncratic, viewpoint, but my aim has been to provide an analytical text which will be of value not only to advanced undergraduates and postgraduates, but also to researchers in ancient reef ecology. I have therefore assumed some basic knowledge of geology and biology.

Reefs are highly susceptible to environmental change, and it has been estimated that up to 70% of living coral reefs are currently under threat. In the past few years, we have learnt much as to how reefs respond to global change and catastrophe on a human timescale, but the origins of many characters we seek to explain are in the past, and so only study of the palaeontological record can provide the necessary time-perspective to reveal the dynamics of long-term, evolutionary change.

Ancient reefs are too often studied solely as geological phenomena with little regard to the detailed biological interactions within their constructional communities. Here, I hope to demonstrate that if one does not understand the biology of a reef, one does not understand the reef at all.

Many have greatly assisted with the writing of this book. I am particularly grateful to those colleagues and friends who made detailed and constructive comments on earlier drafts: D.R. Bellwood, J.A.D. Dickson, A.E. Douglas, E. Insalaco, F.K. McKinney, C. Oppenheimer, B.R. Rosen, and P.D. Taylor. A large number of colleagues have been very generous in providing illustrations: D.R. Bellwood, C. Birkeland, D. Bosence, P. Copper, R. Cuffey, J.A.D. Dickson, A.E. Douglas, E. Flügel, G. Forsythe, G. Galletly, G.M. Grammer, Glenat Publishers, P. Hallock-Muller, P. Hoffman, E. Insalaco, N.P. James, S. Jensen, P.D. Kruse, J.D. Lang, M.M. and D.S. Littler, D.J.C. Mundy, C. Oppenheimer, B.R. Pratt, J. Reitner, D. Schumann, L. Sorbini, B.R. Spincer, R. Steene, R. S. Steneck, F. Talbot, P.D. Taylor, J. Warme, G.E. Webb, J. Wendt, and A.Yu. Zhuravlev. Dudley Simons undertook much photographic work, and Hilary Alberti drafted many figures. John Sibbick drew the reconstructions of ancient reef communities with considerable patience and artistic flair. I am most appreciative of the continued support of the Royal Society, the Department of Earth Sciences, and Gonville and Caius College, Cambridge, and for the professional advice given by Oxford University Press. Finally, I would like to thank my parents, John and Joyce Wood, and especially Clive Oppenheimer for his unwavering support: this time, the book really is finished.

Cambridge   R.W.
1998

Contents

Part I

Introduction to reefs both ancient and modern

1 An introduction to reefs

In an old-standing reef, the corals, which greatly differ in kind on different parts of it, are probably all adapted to the stations they occupy, and hold their places, like other organic beings, by a struggle one with another and with external nature…

Charles Darwin (1872). The structure and distribution of coral reefs.

1.1 What is a reef?

The term 'reef' derives from the Old Norse nautical term *rif*, meaning a hazardous 'rib' of rock, sand, or biological material that lies close to the surface of the sea. But as organic constructions of calcium carbonate, the term reef has evolved different scientific meanings from its original nautical usage, such that now no definition is universally accepted by ships' captains, biologists, and geologists. Consider the diversity of the following four examples.

1. Living coral reefs grow to sea level in clear, tropical waters and may cover many square kilometres. They are constructed by large, heavily calcified organisms (mainly corals and coralline algae) whose growth is driven by photosynthesis. Such reefs show mechanical resistance to all but the most vigorous currents and storms, and often form steep walls in the face of severe wave impact. Their cavernous, complex structure provides habitats for thousands of species—many of which are unique to coral reef ecosystems. With time, the reef is reduced to mainly sediment and rubble by both physical abrasion and a host of rasping, boring, and excavating organisms.

2. In the eastern Java Sea, on the Nicaraguan Rise, and off the Great Barrier Reef, long, narrow mounds are constructed by the calcified codiacean alga *Halimeda*. These structures usually grow at depths greater than 30 m under the influence of moderate currents, but can reach up to 12 m in relief. Sponges may grow among the algae. After death, *Halimeda* readily disarticulates to form a chaotic accumulation of loose plates and carbonate sediment, which becomes bound and lithified through the rapid growth of cements.

3. Extinct bivalved molluscs, known as *rudists*, formed extensive communities of closely packed individuals in shallow, Late Cretaceous seas (~ 100–68 million years ago). Rudists usually grew rooted in—and were stabilized by—soft sediment and were oriented downstream of the prevailing currents. Such communities were often dominated by only one or a few species, were poorly cemented, and had minimal topographic relief.

4. Substantial mounds of calcium carbonate mud (*micrite*) and cement are well known from the geological record, especially from the Lower Carboniferous, some 350–330 million years ago. These mounds were poorly bedded, had steep sides, and could grow as deep as 200 m. They often contain large volumes of inorganic cement, but there is little obvious evidence for any biological contribution except abundant *peloids* (structureless micritic

grains of multiple origin) and the scattered debris of relatively fragile suspension-feeding organisms.

All these communities have at some time been described as reefs. Yet many palaeontologists do not consider that the deep-water Carboniferous mud-mounds, with their abundant micrite and cement but few skeletal organisms, constitute a reef (e.g. Bosence and Bridges 1995). That rudists formed reefs in the Cretaceous is also not accepted by all, as there is little evidence to suggest that such communities achieved sufficient relief to modify their surrounding environment (e.g. Gili *et al.* 1995). And while the alga *Halimeda* can form structures with considerable relief, these are generally not referred to as reefs since they grow at depth and are not formed by organisms that remain in their growth position after death. Indeed, the most widespread usage of the term is also the most restrictive: that reefs consist of a rigid, wave-resistant framework constructed by large skeletal organisms (e.g. Ladd 1944; Lowenstam 1950; Newell *et al.* 1953; Dunham 1970). Using such a definition, of the four examples described above only the coral community represents a true reef.

There are many problems, however, with the application of this strict definition. If only the characteristics of living coral reefs are taken as diagnostic, then very few fossil examples would qualify as reefs. This is due to two interrelated reasons: modern coral reefs possess unique ecological features and environmental requirements, but these are also the very features that are notoriously difficult to detect in the geological record.

1.1.1 Wave-resistant framework

To assess whether a fossil marine organism or community formed a wave-resistant framework is problematic for two reasons.

First, wave-resistance is a relative term because the force of breaking waves is highly variable, being dependent upon local environmental factors such as water depth, fetch (the distance travelled by wind or waves across open water), and wind speed. The presence of an *in-situ* marine community which has formed relief upon the sea floor is therefore evidence only that the constructional organisms are able to withstand the ambient hydrodynamic regime, whether it is one of high-impact surf, moderate waves, or gentle currents. Growth into surface waters clearly demands a gradational spectrum of wave-resistance.

Second, reef formation is highly dynamic, involving both constructional processes of skeletal organism growth and those of physical and biological destruction. While a living coral reef community is demonstrably wave-resistant, boreholes that have penetrated beneath the growing surface of the reef show that the original framework can be almost completely obliterated, with between 40–90% of the rock volume consisting of rubble, sediment, and voids (Hubbard *et al.* 1990). A community with such preservation in the geological record would therefore be excluded from the most strict definition of a reef. Moreover, those complex processes that transform a sediment into a rock known as *diagenesis*—which include dissolution, compaction, and the precipitation of cements—will further modify the buried reef sediments, sometimes to render the ecology of the original living reef community almost unrecognizable.

Such factors can make detection of the former presence of a framework, determination of the depth at which a reef grew, and even reconstruction of original relief difficult in ancient examples. It is not surprising then, that the expression of a reef in the geological record may bear little resemblance to the ecology of the living community.

1.1.2 The importance of large skeletal organisms

The origin of many ancient reefs is not clear, as for significant periods of geological time large skeletal organisms are not conspicuous components of reef communities. Ancient reefs, which consist of significant amounts of cement and peloidal micrite, nonetheless formed substantial topographic barriers that separated deep basins from shallow lagoons behind. These two sources of carbonate can alone account for up to 75% of the reef rock volume. Some of these micrites do not represent reworked sedimentary grains, but probably had an organic origin precipitated by the growth and decay of microbial communities.

These observations show that biological processes other than the growth of large, skeletal organisms can be responsible for the *in-situ* production of calcium carbonate that resists the ambient hydrodynamic regime to form a reef. Moreover, they also demonstrate that the ecological properties of modern coral reef communities cannot automatically be transferred to ancient examples. Modern and ancient reefs clearly encompass a whole spectrum of structures, with reef formation being dependent upon a variety of both inorganic and organic *in-situ* phenomena. These include:

(1) biomineralization to form calcareous skeletons;

(2) the formation of sediment grains by skeletal disintegration and destruction;

(3) the baffling, binding, or trapping of loose sediment by organisms; and

(4) the precipitation of carbonate cement and micrite.

This diversity of process not only makes comparison between recent and ancient reefs fraught with difficulty, it also necessitates the formulation of universally applicable criteria for the recognition of reefs.

So returning to the definition of a reef, while we have seen that it is not appropriate to include wave-resistance and the presence of large skeletal organisms as essential characteristics, all reefs are in some way laterally restricted—be they small, isolated reefs or extensive linear reef complexes—and all show evidence of biological influence upon *in-situ* carbonate production to form a rigid structure. This distinguishes reefs from the surrounding unconsolidated sediments whose distribution is governed by processes such as winnowing and transport. I here favour the broad definition that *a reef is a discrete carbonate structure formed by in-situ or bound organic components that develops topographic relief upon the sea floor.* This simple definition not only allows illuminating exploration of the processes common to the formation of all reefs, but also highlights the ways in which reef communities can differ and how they have changed through geological time.

1.2 The diversity of modern reefs

Some understanding of the biological processes common to all reef formation can be gained through an appreciation of the diversity of living reef communities. In addition to corals, a great variety of microbes, algae, and many other invertebrates form reefs in an apparently wide range of environmental settings (Table 1.1). Modern reefs are found in Arctic, temperate, and tropical habitats, and range from virtually monospecific communities (those formed by a single species) to those that harbour thousands of species.

Table 1.1 Modern reef-building communities with their environmental requirements and global distribution

Main reef-builders	Environmental requirements	Global distribution
Photosymbiotic corals + coralline algae	Shallow, photic zone, hard-substrates, low nutrient levels	Tropical to subtropical
Non-photosymbiotic corals	Deep, cold waters (> 200 m, up to 6200 m)	Tropical, temperate, and Arctic
Sabellariid polychaetes	Shallow (< 10 m), turbulent and turbid waters	Subtropical to temperate
Oysters	Intertidal to subtidal, hard-substrates, often 15–25 ppt salinity, low energy	Subtropical to temperate
Vermetid gastropods	Intertidal to shallow subtidal, often 25–35 ppt salinity	Subtropical
Serpulid polychaetes	< 30 m, brackish to hypersaline, extreme temperatures	Tropical, temperate, and Arctic
Bryozoans	Intertidal (< 2 m), brackish	Tropical to subtropical
Calcified algae:		
— *Halimeda*	> 30 m, active currents, elevated nutrient levels	Tropical to subtropical
— Coralline algae (including algal cups and ridges)	0–150 m, often high energy in tropics, moderate to intense grazing	Tropical, temperate, and Arctic
Sponges:		
— Lithistids	Subtidal-deep (30–200 m)	Tropical to subtropical
— Hexactinellids	Deep (>150 m)	Sub-Arctic
— Demosponges	Subtidal to deep waters (30–150 m)	Subtropical, Arctic
— Calcified demosponges	Deep (>70 m)	Subtropical
— Sponges + bryozoans	Deep (>100 m) nutrient upwelling	Subtropical
Non-calcified organisms, especially microbial (including cyanobacteria, chlorophytes, and diatoms) and sponge biofilms:		
— Stromatolites (laminated; cyanobacterially derived)	Shallow (< 10 m), high carbonate supersaturation levels or degassing rates; low nutrient levels or high sedimentation rates; brackish, hypersaline, or normal marine	Tropical to temperate
— Microbialites/thrombolites (clotted textures; microbial or sponge-mediated	Cryptic areas and deep forereef of coral reefs; low sedimentation rates, locally high carbonate supersaturation levels	Tropical to subtropical

Table 1.1 *Continued*

Main reef-builders	Environmental requirements	Global distribution
Mud banks—probably baffled by sea grasses/mangroves	Shallow, protected	Subtropical (Florida)
Bivalves and vestimentiferan worms	Deep waters around methane, petroleum, or hypersaline brine seeps and hydrothermal vents	Various
Non-photosymbiotic corals (*Lophelia*) and codiacean algae (*Halimeda*)	Deep (100–550 m); near methane-seeping faults	Temperate and tropical (W. Ireland and Timor Sea)
Skeletal debris ('lithoherms')	Very deep (400–600 m), intense marine cementation	Subtropical (Florida)

Reef formation is contingent upon the successful growth of closely packed immobile (*sessile*) organisms, and it is not surprising to note that all modern reef communities develop under specific environmental conditions that allow such growth. When we consider the conditions under which any given species will form a reef—that is show proliferative growth at the expense of other sessile biota—we see that these are far more restricted than the potential maximum environmental range of that species. For example, living tropical corals are not restricted to reefs—they are common in a broad range of habitats that include those with lower mean annual temperatures than reefs (14–16 °C, compared to 18 °C), and those characterized by muddy, rather than hard, substrates.

The distribution of modern reefs shows that their development is dictated by avoiding competition from other organisms and predation, or by adapting to physical or biological disturbance.

1.2.1 Coral reefs

Modern coral reefs are specialized communities that develop in clear, well-lit tropical and subtropical waters (that is, within the *photic zone*) characterized by high levels of aragonite *supersaturation* (i.e. where concentrations of calcium and carbonate exceed the thermodynamic mineral solubility product). They also often form in seas where nutrient levels are low. Most corals and coralline algae are capable of permanent and secure encrustation to hard substrates (Fig. 1.1(a), but see Fig. 1.1(f)), and so preferentially colonize elevated, stable substrates. Many of the corals have entered into a symbiosis with algae (*photosymbiosis*). These organisms form a photosynthetic community whose ecology is regulated to a considerable degree by predation, and these are the processes that are responsible for many of the unique characteristics of coral reefs.

1.2.2 Other reefs

Cyanobacterial communities can form *stromatolite* reefs—finely layered constructions of micrite (Fig. 1.1(b))—where environmental conditions (such as high sedimentation rates or low nutrients) exclude potential competitors for substrate space and where water chemistry favours the rapid precipitation of carbonate (Reid *et al.* 1995). Reefs formed by tropical coralline algae (Fig. 1.1(c)), develop where non-skeletal algae and corals are excluded by intense grazing pressure and wave impact (Adey and Steneck 1985), and temperate and

(a)   20 cm

(b)   20 cm

(c)

(d)   10 cm

(e)

(f)

Fig. 1.1 The diversity of modern reef communities. (a) An exposed coral reef flat at low tide, Nananu-i-ra Island, Fiji (Photograph: C. Oppenheimer). (b) Intertidal stromatolites, formed mainly by cyanobacterial and diatom communities, Hamelin Pool, Shark Bay, western Australia. (c) Coralline algal reef at low tide, constructed by *Lithophyllum congestum*, Isaacs Reef, St Croix (Photograph: D. Bosence). Hammer = 40 cm long (d) A reef formed by the calcified alga *Halimeda* growing at a depth of 30 m in the Great Barrier Reef lagoon, Australia (Photograph: P. Hallock-Muller). (e) A small vermetid gastropod reef, Sarasota, Florida (this example is approximately 4 million years old). Hammer = 32 cm long. (f) A unique reef constructed solely by solitary fungiid corals, Great Astrolabe Reef, Fiji (Photograph: M.M. and D.S. Littler).

Arctic coralline algal reefs form in response to the presence of specialist grazers (Freiwald and Henrich 1994). Likewise, although *Halimeda* is found throughout shallow coral reef complexes, this alga forms reefs only at depths below about 30 m (Fig. 1.1(d))—at light intensities too low and nutrient levels too high for most reef coral growth (Roberts *et al.* 1987; Hallock *et al.* 1988; Hine *et al.* 1988).

Many other modern non-coral reefs are constructed by organisms which possess little inherent stability under agitated conditions and, being non-photosynthetic, are not dependent upon light. Consequently, they often develop in low-energy settings, either in relatively deep waters or in marginal, shallow-water environments. These include serpulid polychaete, vermetid gastropod (Fig. 1.1(e)), and bryozoan communities. Deep-water habitats, such as the area supporting bryozoan-sponge reef growth off the shelf of south-eastern Australia, receive minimal or no light but offer abundant nutrients (Boreen and James 1993). These environments are relatively benign in that predation pressure and turbulence are often low. Shallow-water non-coral reefs are usually constructed by specialized organisms which are able to grow under conditions (such as non-marine salinity) that exclude other normal marine competitors or predators. Such reefs often grow in embayments which are protected from wave destruction. There are exceptions however: sabellariid polychaete reefs thrive in areas of vigorous wave- and current-action which causes the suspension and transport of sand particles needed for rigid tube construction.

1.3 What is a reef community?

Many ecologists now consider that biological communities are not fixed entities with precise boundaries, but are chance associations of species with similar requirements. This notion is supported by several models that adequately predict community composition on the basis of immigration and extinction, the spatial distribution of environments, and the size of the species pool alone (e.g. Cornell and Karlson 1996; Hubbell 1997). The fossil record shows that communities have constantly changed through local extinction and recruitment of their component species, and that new communities which have developed in previously unoccupied habitats are composed of species from the available population which have had geographic access to the new area (Buzas and Culver 1994). Indeed, it has been well documented that formerly co-occurring species are now found in entirely different associations to those which they occupied in the past. For example, in Indo-Pacific reefs, as the sea level fluctuated during the Pleistocene corals responded to the changing availability of the shallow marine habitat by switching their membership of different communities (Potts 1984). Associations within local reef communities have altered as new species have evolved whilst older taxa persist. Clearly, community traits are not heritable: natural selection cannot operate upon communities but only upon their constituent species.

Such observations undermine the notion of a strong cohesion within communities, and support the view that communities are chance associations. But while it is now widely accepted that many marine communities are not discrete, this has not yet been universally accepted for coral reefs (for example see Jackson 1992; Pandolfi 1996). This is because the sheer abundance of specialized interactions and symbioses, as well as the diversity of ecological niches, seem to indicate a long-lived coexistence between organisms. However, it is now apparent that even seemingly specific interactions can in fact be modified as the species membership of a reef community changes—which occurs frequently—so that, like the constituent species, the ways in which organisms interact are not fixed either. For example, the highly territorial three-spot damselfish (*Eupomacentrus planifrons*) was noted by Kaufman (1977) to graze preferentially upon the algae growing on loosely branching corals such as the staghorn coral *Acropora cervicornis*. But after the near-complete destruction of the fragile *Acropora* thickets in Jamaica by Hurricane Allen in 1980, three-spot damselfish territories were found to have re-established on a variety of massive, hemispherical corals, which being more robust had preferentially survived the catastrophe (Knowlton *et al.* 1981). This suggests that there is an enormous ecological redundancy of species in reefs, i.e. many species can occupy a broadly similar niche.

As organisms vary continuously in distribution and abundance both spatially and temporally, local communities can be regarded simply as representing arbitrary subdivisions along environmental gradients (Whittaker 1975). Individual species are not limited to membership of only one type of community: they will become part of any community where environmental conditions enable their survival. However, in order to maintain membership of that community they must be able to accommodate constant long-term fluctuations, such as climate and sea-level changes. Environments are not constant: they have changed continually on all timescales throughout Earth's history, and species distributions have shifted in response to these changes as well as being exposed to new biotic environments in the process.

1.4   Reefs as the biological occupation of space

By definition, all reefs have a biological origin, and develop due to the close packing (*aggregation*) and successful growth of sessile marine organisms that grow upon the sea floor (an *epibenthic* mode of life): reef communities are associations of species with similar ecological and environmental requirements.

Possession of a sessile mode of life places strict demands on an organism. First, it must compete with other organisms to acquire and maintain a suitable substrate for growth until it reaches reproductive maturity. Second, a sessile habit makes an organism highly vulnerable to disturbance—both physical abrasion and biological attack by predation—such that the risk of mortality is probably the major control on its morphology and distribution (Coates and Jackson 1985). An understanding of how reef-building organisms respond to the demands of a sessile lifestyle, and the mechanisms responsible for their aggregating behaviour, are therefore fundamental to the study of reef ecology and evolution.

The theme of this book, then, is to reinterpret the evolution of reefs from a biological perspective—as a history of the changing way in which reef organisms have successfully occupied the sea floor. Such an approach provides a unified theme that allows consideration of the processes common to the formation of all reefs. It also places an emphasis upon the relationship between the ecology of the reef community and its final geological expression.

We have seen that organisms can contribute to an increased rate of carbonate production within reefs by a number of processes. Which process of *in-situ* carbonate production prevails is determined largely by environmental factors such as the hydrodynamic regime, sedimentation rate, light, and temperature that control both the precipitation of calcium carbonate and the distribution of organisms. In particular, the relative stability of organisms, their competitive abilities, and their defences against predation, i.e. their ability to secure and maintain substrate, will determine under what degree of disturbance they are able to form a reef. Ambient disturbance levels—both physical and biological—can thus largely determine the style of reef formation.

1.5   The historical perspective

Attempts to understand ancient ecologies by direct reference to modern communities are hampered by several obstacles. First, the incomplete nature of the fossil record and the impossibility of dating and correlating geological strata (known as *stratigraphy*) at the fine scale appropriate to modern ecological analysis makes the direct transfer of information and techniques inappropriate. This problem is compounded by the general coarsening of temporal resolution with increased time back from the present. While hypotheses that attempt to explain the distribution of organisms can be tested experimentally, such experiments have the weakness of excluding virtually all factors that operate over periods longer than human timescales. Second, there are no modern analogues for many ancient ecologies, not only because many of the constituent organisms are extinct, but also because the environmental conditions under which they grew have changed radically during the history of life. The mechanisms employed and the conditions under which organisms aggregate to form reefs are varied, and it is evident that ancient reef communities not only grew in environments now unoccupied by modern coral reefs but that they also show profound differences in substrate preference and *trophic structure* (the feeding relationships within a community). Many evolutionary innovations, as well as extinction events, have exerted great changes on reef communities through geological time. As predominantly shallow marine communities, reefs are highly susceptible to changes in sea level, climate, and sea-water chemistry, all of which are driven by processes such as climatic oscillations—paced by variations in the Earth's orbit—and plate *tectonics* (forces involved in the Earth's crust). A full appreciation of ancient reef ecology and evolution clearly calls for a non-uniformitarian approach.

Reef formation concerns the biological utilization of calcium carbonate for the maintenance of stability. In principal, any sessile organism has the ability to form a reef. Indeed, the geological record shows that a multitude of different groups of algae and skeletal metazoans (most now extinct) have formed reefs since the early Archaean, some 3.5 billion years ago (Fig. 1.2). In this historical context, modern coral reefs are very recent phenomena: scleractinian corals did not appear until the mid-Triassic (~240 million years ago) and coralline algae, although known since the Carboniferous, did not become diverse until the late Mesozoic to early Tertiary, some 70–60 million years ago. Likewise, the fish that characterize modern coral reef communities and are so central to their functioning appeared only about 50 million years ago.

1.6   The approach adopted in this book

One of the central debates within palaeobiological studies is the extent to which the evolution of organisms is shaped through their adaptation by natural selection, through constraints imposed by their basic organization, through environmental change, or simply by chance events. The processes considered to be important for reef evolution occur on a variety of

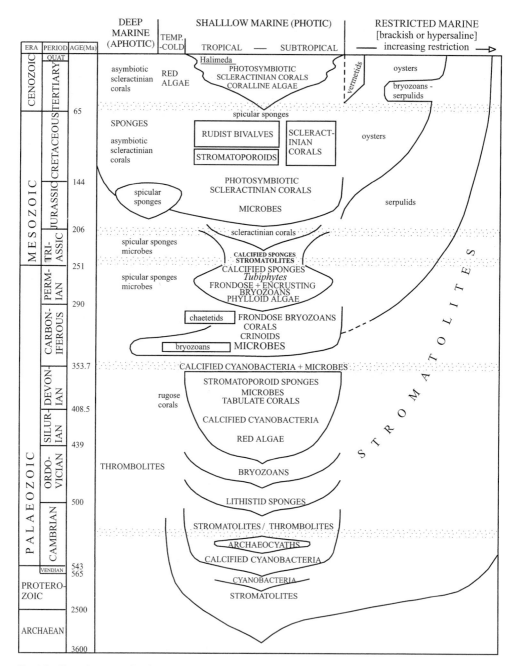

Fig. 1.2 The main groups of reef-builders through geological history. Stippled areas mark major extinction events for reef communities. Temp., temperate; Quat., Quaternary. (Modified from Heckel 1974, by kind permission of SEPM Society for Sedimentary Geology.)

timescales (Table 1.2), and one of the main aims of this book is to gain some understanding of their relative importance. Ecological interactions between organisms are generally assumed to be intense selective forces in the evolution of all biotas, and some consider it likely that interspecific interactions alone, especially those between predators and prey, would enable evolution to continue even in the absence of any change in the physicochemical environment. Reef distribution is clearly controlled by biotic factors as well as more traditionally cited physical controls, and this book is also concerned with how an understanding of these processes might translate into the historical record of reefs we see on an evolutionary timescale.

We recognize three main types of interspecific interaction: competition, predator or herbivore/prey or parasite/host relationships, and mutualism (Table 1.3). These interactions are not always mutually exclusive and may even be interrelated: for example, an originally

Table 1.2 Timescales relevant to the evolution of reef communities. To what extent can the ecological timescales of individual interactions be reconciled with long-term environmental change to produce evolutionary change?

Process	Frequency
Ecological	
Ecological interactions	Days–weeks–months
Organism life span	Months–centuries
Reef community longevity	Months–centuries–millennia
Epidemic disease	Decades–centuries
Evolutionary	
Speciation	Millennia–millions of years
Environmental	
Hurricanes	Months–decades
Sea-level or climatic change (orbital oscillations)	10 000–100 000 years
Mass extinctions	10–100 million years

Table 1.3 The major classes of interspecific interaction. (Modified from Majerus *et al.* 1996, by kind permission of Addison Wesley Longman Ltd.)

Term	Type of interaction
Competition	Fitness of individuals of species A decreased by interaction with individuals of species B; an individual of species A also suffers decreased reproductive success as a result of the interaction
Predation and parasitism	Reproductive success of an individual of species A increased by interaction with B; reproductive success of an individual of species B decreased by interaction
Mutualism	Reproductive success of an individual of species A increased by interaction with B and *vice versa*. Removal of either species will result in a reduction in the reproductive success of the other

parasitic relationship can evolve into a mutualism. Morphological and life-history attributes are considered to be the products of past interactions, and so the approach adopted here is to reconstruct ancient reef ecologies using inferences as to the life histories of their constructing organisms, together with information gained from the observation and experimentation of modern biotas. The evolutionary consequences of ecologically important parameters, particularly with regard to environmental preference, are then used to explain the distribution of modern and ancient reefs. As the detailed evolutionary ancestry (*phylogeny*) of many ancient reef organisms is difficult to assess, I hope to demonstrate that such an approach overcomes some of the difficulties inherent in reconstructing the ecology of ancient reef communities.

This book is divided into three parts:

Part I   Introduction to reefs both ancient and modern;

Part II   Analysis of the role of physicochemical change on reef growth; and

Part III   Documentation of the appearance and development of key biological innovations in the history of reef communities.

This chapter has emphasized that, notwithstanding the diversity of reef communities and their environmental settings, there are fundamental processes common to all reefs. Reef formation concerns the successful occupation of space and the biological control of *in-situ* carbonate production.

Chapter 2 presents criteria for the recognition of reefs in the geological record, and their environmental setting. The record of ancient reef ecology over the past 3.5 billion years is outlined in Chapter 3, although much of the detail remains inadequately documented. This serves as a foundation for the study of the relative role of environmental change in reef evolution (Part II, Chapters 4 and 5) and evolutionary innovations documented in Part III (Chapters 6–8).

Chapter 4 considers how physicochemical controls such as the temperature, hydrodynamic regime, and carbonate saturation levels govern the distribution of carbonate production and the growth of different reef communities on a variety of timescales. Have the geological processes responsible for reef formation and preservation remained constant throughout geological time? Reefs often grow at or near sea level, and so are susceptible to rapid sea-level changes. This, and longer term processes, such as changing tectonic regime and global climatic conditions, are shown to have been major determinants of the resultant record of reef-building.

Extreme perturbations (catastrophic events) play a highly important restructuring role in reef communities, and the potentially devastating effect of mass extinctions has led to an apportioning of special status within the study of evolution. Whether a species occupies a habitat until it is displaced by a competitively superior species, or remains in place until severe disturbance causes its removal, is of fundamental importance. Reefs are generally assumed to be particularly susceptible to mass extinction events, and details of their recovery are only just becoming clear (Chapter 5).

Part III concerns the development of evolutionary innovations in the history of reef ecosystems. Here, particular emphasis is placed upon the origins and diversification of those ecological features unique to modern coral reefs. The ability of organisms to form reefs has changed through time, in response to a changing biological environment. How do organisms acquire sufficient stability to form a reef, and how might these mechanisms vary according to environmental predictability? Acquisition of a modular organization, which may lead to the ability to encrust and so gain firm attachment to a substrate is vital for the stability and long-term growth of any shallow-water reef. Not surprisingly, modularity has appeared independently in many different *clades* (groups of related species derived from a common ancestor) of reef-associated organisms through the *Phanerozoic* (the Eon of 'visible life', from the Cambrian to the Recent) (Chapter 6).

Intensity of predation and herbivory have not remained constant throughout the Phanerozoic: in particular, the ability to rasp, scrap, and excavate calcareous substrates has increased in a long-lived evolutionary radiation that started some 220 million years ago. As a result, shallow marine environments have become progressively more disturbed, and the processes employed by reef organisms to achieve the necessary occupation of space have changed. The effect of this radiation on the evolution of reef communities is examined in Chapter 7.

Without doubt one of the most significant evolutionary innovations in the history of reef-building has been the acquisition of algal symbionts by metazoans. But how important is photosynthesis for reef-building? The timing of the acquisition of photosymbionts by reef-builders, and the ecological and environmental effects of this innovation, are explored in Chapter 8.

2 The recognition of ancient reefs

All reefs form as the direct or indirect result of organic activity that has promoted a higher rate of *in-situ* carbonate production than in surrounding sediments (Fig. 2.1). Reefs develop only where the rate of carbonate production exceeds any loss due to erosion and transport away from the reef. In the Ancient, this produces massive, often unbedded, accumulations within otherwise normally bedded strata (Figs 2.1(c) and (d)) indicating that the processes operating within the reef differed from those beyond it.

The study of ancient reefs has been dogged by an elaborate terminology abounding with synonyms but with little consensus as to exact meaning. Many of these terms have limited ecological utility, but some of the basic terminology necessary for understanding ancient reefs is introduced in the following section. Here, the fundamental biological and geological characteristics of reefs are outlined, together with the processes necessary for their preservation. These characteristics serve as criteria for recognizing reefs in the geological record.

2.1 Fundamental processes: criteria for the recognition of ancient reefs

2.1.1 Organic origin

Reef-builders are generally invertebrates (especially cnidarians, sponges, and bryozoans) or plants with calcareous skeletons, although some sponges produce siliceous skeletons. Non-skeletal organisms including microbes (proteobacteria and cyanobacteria) and eukaryotic algae (protists) may also induce carbonate precipitation so as to form reefs under particular environmental conditions, although their importance—or indeed presence—may not always be readily apparent.

2.1.2 Topographic relief

All epibenthic organisms produce relief on the sea floor, but many sessile organisms grow significantly raised above a substrate in order to increase their feeding efficiency, and to avoid competition or smothering by sediment. Aggregating sessile communities grow preferentially on topographic highs or patches of otherwise stable substrates. Some organisms prefer to settle upon the same type of substrate as their own skeletons form, i.e. elevated hard surfaces. For others, the *recruitment* of larvae (those that settle from the plankton onto a substrate and subsequently metamorphose into adults) may be density-dependent, i.e. the next generation is more likely to grow around existing members of the same species. In time, all such behaviours will result in the formation of a structure with contemporary elevation upon the sea floor, known as *topographic relief.*

Fig. 2.1   Modern (a,b) and ancient (c,d) reef geometries show a higher rate of carbonate production than surrounding sediments. (a) Patch reefs, Belize. (b) Barrier reef complex, Belize. (c) Two Upper Devonian patch reefs expressed as bioherms, Flathead Range, Canadian Rocky Mountains. Note the unbedded, reef fabrics and slightly compacted, bedded flanking beds (Photograph: B.R. Pratt). (d) View looking south over an exhumed Upper Devonian reef complex, Laidlaw Range and Glenister Knolls, Canning Basin, western Australia.

The existence of topographic relief is therefore implicit evidence that the constructional organisms have sufficient stability to withstand the ambient hydrodynamic regime. The degree of relief is controlled by:

(1) the elevation of the reef organisms and their durability;

(2) the length of time over which the community has grown; and

(3) the rate of sedimentation.

Relief will only be present if the rate of reef growth exceeds that of sedimentation in the surrounding environment.

Modern reefs are relatively thin veneers of growth, and their apparent relief is often a reflection of the complex morphology of underlying sediments. A great deal of relief on many modern reefs is attributable to such inherited (often Pleistocene) topography, known as *relict topography*.

If relief cannot be detected within the reef itself, it may be apparent within surrounding sediments:

1. The onlap of the surrounding *flanking beds* onto the reef will indicate the position of centres of constructional relief (Fig. 2.1(c)). This relationship can be exaggerated by *post-depositional compaction*, where the relatively weakly cemented flanking beds are preferentially compacted during burial around the dense, cement-rich reef.

2. The presence of organic debris derived from the reef (known as reef *talus*) within flanking beds or otherwise reef-associated sediments (Fig. 2.1(c)) is good evidence for the contemporaneous existence of a topographic high that was also a centre of carbonate sediment production. (These must be proven, however, to be derived from the reef in question rather than residual talus from pre-existing deposits.) Much reef talus is produced by postmortem disintegration of rapidly growing, short-lived, reef-dwelling epibenthos, as well as by physical and biological destruction (*bioerosion*) of carbonate skeletons or rock. The presence of eroded pieces of reef rock (*talus blocks*) within surrounding sediments also provides evidence for both the contemporaneous lithification and positive relief of the reef. The type and quantity of talus formation, and subsequent export, if any, is often related to local physiography, and the tectonic or hydrodynamic regime which, in turn, may be controlled by the larger scale forces of sea-level dynamics and basin *subsidence* rates (the rate of progressive depression of the Earth's crust that allows sediments to accumulate).

3. Quantitative measurement of the original relief can be derived from *geopetals* (partially sediment-filled voids capped by cement) within the reef or flanking sediment. The differential attitude of geopetals to the reef surface, as with the orientation of any flat-based reef organisms, will indicate the slope angle of the reef. Disorientated geopetals allow the distinction of dislodged reef blocks from *in-situ* reef.

4. Ecological and sedimentological differences within a reef body can indicate the contemporaneous presence of an environmental gradient due to topographic relief. For example, this may include indications that a reef possessed leeward and windward sides.

Apparent topographic relief of a reef in a geological outcrop can be misleading, as many ancient reef sequences are not single depositional units but composites constructed by the superimposition of many generations of reef-building communities. This may be expressed by subtle *discontinuity surfaces* caused by the temporary cessation of reef growth, which can be traced into the flanking beds. The succession of communities bounded by such hiatuses may also show compositional changes in response to changing environmental parameters, such as upward growth into more shallow and energetic waters which may also involve a change in the consistency of the sea floor. Indeed, the accumulation of the hard parts of one organism which change a substrate so facilitating the colonization of another, is a common occurrence on reefs (Scoffin 1992). Also, apparent or exaggerated contemporaneous relief can be caused by the emergence and subaerial erosion of the lithified reef body.

The environmental setting of a reef, as well as its relative hydrodynamic resistance, topographic relief, lateral extent, and depth of formation, will control the degree to which it will modify the local depositional environment and processes of sedimentation. Large, wave-resistant constructions growing in very shallow waters will exert an influence considerably beyond their actual dimensions; deep-water reefs with minimal relief generate little sediment or influence over the local hydrodynamic regime.

2.1.3 Reef morphology

Many terms have been proposed to express the myriad of reef form and type. Modern reefs are classified according to their three-dimensional shape, their position relative to land, and the dominant constructional community. The availability of a shallow substrate for colonization determines the resultant shape of a coral reef: *patch reefs* are small, isolated reefs that usually form in low-energy settings, typically on shallow shelves; *fringing reefs* are linear structures in subtidal waters near land; *barrier reefs* are also linear, but form some distance from land separated by a lagoon of considerable depth and width. *Atolls* form horseshoe or ring-shaped reefs that surround volcanic islands. These reef types can form a continuum; for as Darwin rightly surmised, as a volcano subsides a fringing reef would evolve into a barrier reef and then eventually into an atoll.

Ancient-reef terminology describes geometry as presented in either a two-dimensional geological outcrop or as reconstructed from subsurface data, and limestone or dolomite texture (*fabric*). A *bioherm* refers to a relatively small, discrete, lens-shaped reef (Fig. 2.1(c)); *biostrome* to a bed of often *in-situ* skeletal organisms without significant relief. The term *reef mound* is widely used for constructions of relatively delicate organisms, often not in life-position, and *mud-mound* for build-ups dominated by fine-grained sediment, usually micrite, with relatively few sessile skeletal biota. *Reef complex* refers to laterally extensive reefs that have developed sufficient topographic relief such that distinct zones are created (see Figs 2.1(b) and (d)), including:

- the *reef margin* (the reef community itself that forms a break in slope at the seaward margin);
- the *back-reef*—the landward side of a reef, including the area behind the *reef crest* (a sharp break in slope at the seaward margin of a reef, or edge of a reef flat), the *reef flat* (any flat area behind a reef crest) and carbonate sand passing into a shelf lagoon;
- the *fore-reef* (reef slope and talus deposits).

Any localized, thick carbonate deposit built by organisms but not always with an apparent bound fabric is termed a *build-up*.

These terms confer little ecological information, but they do, in part, reflect the varying abilities of different groups of organisms to build reefs under differing environmental conditions, so influencing the style of reef fabric and sedimentary association.

2.1.4 Framework and community differentiation

In some reefs, sufficient stability to withstand the ambient hydrodynamic regime can be achieved solely by the dense aggregation of organisms. Such communities will be preserved in life-position unless their skeletons are prone to disintegration (Fig. 2.2(a)), or have

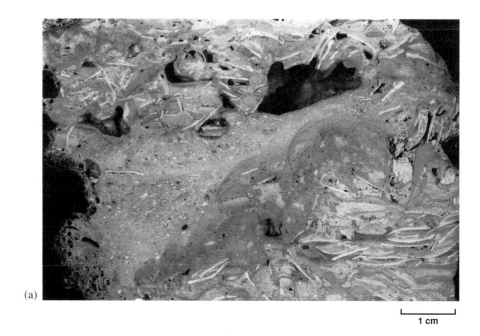

(a)

1 cm

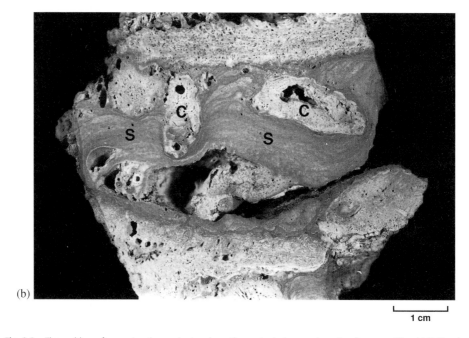

(b)

1 cm

Fig. 2.2 The problem of reconstructing ancient ecology: the geological expression of reef communities. (a) *Halimeda* readily disarticulates to form abundant loose plates, such that in the geological record this alga is usually preserved as a chaotic packstone (see Fig. 2.6). Compare with the living community shown in Fig. 1.1(d). (b) Reworked coral fragments (C) which have been heavily bored, and then encrusted by stromatolites (S). Compare with the living community shown in Fig. 1.1(a). (Photographs: B.R. Pratt.)

suffered biological or physical destruction and reworking (Fig. 2.2(b)). Many organisms, however, cannot generally withstand energetic environments without some further means of imparting rigidity. These include:

(1) mutual interconnection, *encrustation* (the laterally expanding growth form of an organism that attaches to a hard substrate) or successive overgrowth by the principal or *primary reef-building* community (such as corals and coralline algae in modern coral reefs); or

(2) the binding of the primary community by *secondary encrusters* (organisms that encrust or bind the primary reef-building community) and otherwise attached organisms (such as foraminifera and encrusting bryozoans); or

(3) the precipitation of inorganic cement around and between the skeletons of the reef community.

All these phenomena can produce a bound, organic carbonate structure or *framework* (Fig. 2.3). The presence of a reef framework can be demonstrated by either the bound or interconnected growth of principal reef builders (be they skeletal metazoans or algae, or microbial precipitations), or demonstration that open spaces or cavities were present within the original reef structure implying growth, with rigidity and physical support, above a substrate (Fig. 2.4(a)). The growth form and the degree of elevation of the framework builders

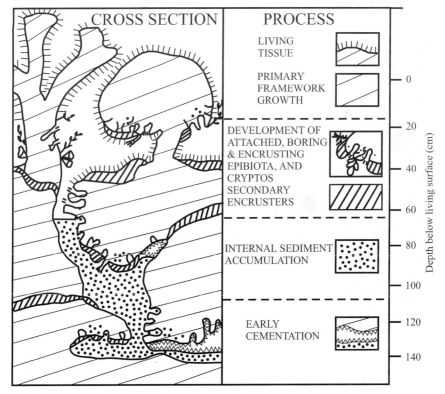

Fig. 2.3 Development of a reef framework showing the main constructional and destructional processes. (Modified from Scoffin 1972a, by kind permission of the American Association for the Advancement of Science.)

above the substrate surface will determine the shape, size, and extent of resultant *primary cavities*, with reefs constructed by small organisms often possessing a higher volume of cavity space than those formed by large forms. *Secondary cavities* can also be formed by subsequent bioerosion, or sometimes by biological disturbance of unconsolidated sediment (*bioturbation*) within the reef (Fig. 2.4(b)).

The reef framework may be colonized by secondary encrusters on exposed substrates, but protected, hidden surfaces—under overhangs, in cavities, or even within rubble—present particularly attractive areas for settlement. These sites are collectively known as *crypts* (Fig. 2.4). The presence of an attached *cryptic* community on a primary framework as well as a crypt-filling sediment of clearly marine origin is further evidence for the original existence of an open-growth framework which allowed the free flow of water. Secondary encrusters have varying demands and tolerances for light, food and sediment supply, and water circulation, such that those which colonize substrates close to the reef surface may differ markedly from those which inhabit the dark recesses of an abandoned reef framework near the sediment–water interface. Crypts, especially large caves and grottoes, have traditionally been considered to be *refugia*—safe havens—where formerly open-surface dwellers have retreated in the face of new competition. However, the differentiation of reefs into distinct open-surface and cryptic communities has a long geological history, being known since the Palaeoproterozoic some 1.5 billion years ago, and their status as refugia is now open to question.

Reef framework formation is thus dependent upon a variety of *in-situ* inorganic and organic processes. But these processes operate at different timescales in the development of a reef: mutual encrustation occurs during the growth of the primary reef framework; secondary encrusters can only colonize once the primary framework has been established; and although cement precipitation may be contemporary with primary framework growth, it can take centuries, millennia, or even millions of years to completely occlude crypts. Therefore, the processes responsible for the preservation of a reef in the geological record are often quite independent from the original ecology.

2.1.5 Internal sediment

Reef crypts often contain *internal sediment* formed by either *in-situ* processes of disintegration and bioerosion, or washed-in material derived from the physical breakage of the reef biota (Fig. 2.4(a)). Some internal sediment may also contain organisms that live preferentially within crypts. Internal sediment is commonly composed of many generations, which may be graded from coarse at the base to fine-grained at the top. These are known as *fining-upwards cycles*. As particle size is determined by the width of crypt openings and hydrodynamic regime, so a decrease in grain size may indicate progressive restriction and sealing of the reef framework during sedimentation. Internal sediment can also show *cross-lamination* (bedding inclined at an angle to the horizontal) or bioturbation (Fig. 2.4(a)), and may be cemented and subsequently bored. Internal sediments are often associated with marine cements: partially filled cavities will form geopetal structures; other cavities may be totally filled with either sediment or cement.

2.1.6 Early lithification

Reefs in tropical and subtropical settings are highly susceptible to early lithification, which is a near-surface phenomenon. This is due to:

(1) their open, porous construction;

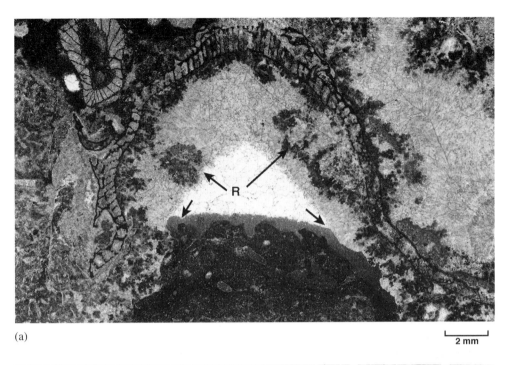

(a)

2 mm

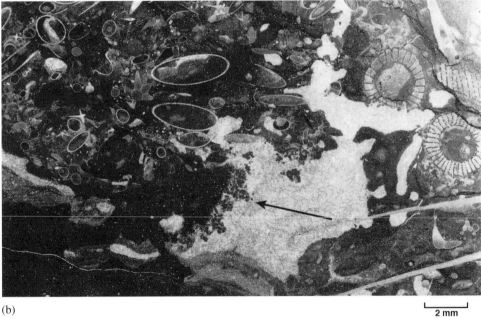

(b)

2 mm

Fig. 2.4 Ancient reef crypts from the Lower Cambrian (~525 million years old). (a) Geopetal structure showing early cementation. A domal cup of an archaeocyath sponge has formed a primary crypt whose undersurface has been colonized by the alga *Renalcis* (R). The micrite-fill indicates horizontal at the time of deposition: the first generation is burrowed and is followed by a second layer. Note how all internal sediment postdates both the *Renalcis* growth and cement precipitation (small arrows). (b) Secondary crypt formed by a lithified burrow system (perhaps enhanced by scouring) within micrite. Parts of the crypt walls have been colonized by the alga *Renalcis* (arrowed). (Photomicrographs: P.D. Kruse).

(2) the immobility of both the reef framework and sediment trapped therein (Fig. 2.3), combined with a slow rate of sedimentation;

(3) their growth in areas where there is a high flux of warm, oxygenated sea water supersaturated with respect to $CaCO_3$ that is pumped through the reef framework by tidal wave or current action. Seaward, steep-walled reefs are more heavily cemented than reefs with low-angle profiles growing in more sheltered settings.

Early lithification is manifested by the precipitation of cements within either open crypts or internal sediment. Evidence for such early cementation includes:

(1) a demonstration of the contemporaneous nature of the internal sediment and cement, such as interlayered growth (Fig. 2.4(a));

(2) a demonstration of the contemporaneous growth of biota and cement, such as the biological encrustation upon, or destruction (microboring) of, cements or lithified internal sediment (Fig. 2.4(b));

(3) the incorporation of fragments of reworked cement or lithified reef sediment (*intraclasts*) into surrounding strata.

Carbon-dating of cements found in modern coral reefs reveals them to have grown remarkably rapidly, but the growth rate of cements in ancient reefs is more difficult to quantify.

Early lithification will protect a reef from compaction, and rapid burial will retard reworking and so aid preservation of the reef community in the geological record. Final burial terminates both the physical and biological destruction of a reef, such that its timing will determine the extent of preservation of both the intact framework and original topographic relief. The rate of burial will be determined by comparing the relative growth of the reef (as expressed by topographic relief) to the rate of flanking bed sedimentation and internal sediment accumulation.

The preferential cementation of reefs at the margins of shelves may lead to the formation of fractures (*fissures*). These result in instability of the reef edge, causing rotational failure and the eventual dislodgement of huge blocks of lithified reef down the fore-slope. Fissures are often filled by many generations of marine sediment and cement, and those that form parallel to the margin are known as *neptunian dykes* (Fig. 2.5). Some contain organisms that dwelt preferentially within these submarine habitats. Cementation can thus influence the morphology of shelf margins, forming precipitous scarps elevated high above the basin floor.

Pervasive early cementation appears to have been more prevalent in pre-Cenozoic (older than 65 million years) reefs than in younger examples. Indeed, some Palaeozoic reefs (those older than 250 million years) contain over 75% by volume of cement, which may be a reflection of differences in both sea-water chemistry and the proportion of intact reef framework, and, hence, the volumetric abundance of crypts.

Many parts of a reef move during settling and compaction subsequent to deposition, thus fracturing and disturbing the depositional fabric. This may involve the formation of cavities by dissolution, fracturing of skeletons, squeezing and splitting of micrite, and the mobilization and migration of clay minerals. Dissolution of carbonate is often most active on the walls and ceilings of open framework cavities.

Fig. 2.5 Neptunian dyke developed within a well-cemented Upper Devonian reef margin. Tunnel Creek, Canning Basin, western Australia.

2.2 Classification of ancient reef carbonate

I shall follow the fabric-based terminology proposed to describe the carbonate rock (limestone and dolomite) associated with reefs by Dunham (1962) and Embry and Klovan (1971) (Fig. 2.6). This classification is fully described in many standard texts (e.g. Tucker and Wright 1990). There are three main divisions which in part, reflect the interaction of hydraulic energy and biological production.

(1) matrix-supported (micrite and *wackestone*, where micrite is defined as grains <20 μm, and *floatstone* where more than 10% of the grains are >2 mm);

(2) grain-supported (*packstone* and *grainstone*, and *rudstone* where more than 10% of the grains are >2 mm);

(3) carbonate which was biologically bound during deposition—including that formed by *in-situ* micrite (*boundstone*), i.e. a reef.

A variety of terms have been proposed to describe the different fabrics of boundstones. The most widely used are framestone, bafflestone, and bindstone (Embry and Klovan 1971). Such terminology is explicitly abandoned here because it can often serve to obscure the ecology of an ancient reef. First, these terms simply describe the morphology of the visually or volumetrically dominant organisms within a reef community often without direct reference (but often inference) as to their actual constructional or other ecological role. Second, the characteristic differentiation of reefs into distinct open surface and cryptic communities, as well as the tremendous variety of reef-builder morphologies, cannot be reasonably accommodated within any fabric-based scheme: reef growth fabrics are highly complex and

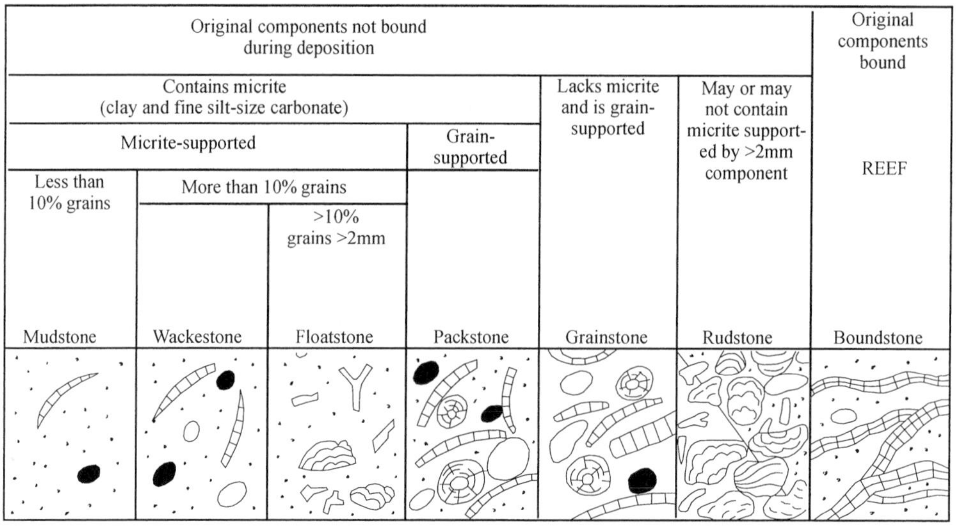

Original components not bound during deposition						Original components bound
Contains micrite (clay and fine silt-size carbonate)				Lacks micrite and is grain-supported	May or may not contain micrite support-ed by >2mm component	REEF
Micrite-supported			Grain-supported			
Less than 10% grains	More than 10% grains					
		>10% grains >2mm				
Mudstone	Wackestone	Floatstone	Packstone	Grainstone	Rudstone	Boundstone

Fig. 2.6 Limestone classification. (Modified from Dunham 1962, by kind permission of the American Association of Petroleum Geologists, and Embry and Klovan 1971.)

variable even on a small scale, and reefs are, by their nature, patchy and heterogeneous communities. Third, such a classification does not adequately describe the microbially precipitated, boundstone fabrics that are common in ancient reefs.

2.3 Reconstructing ancient reef ecology: taphonomic bias

Being the result of *in-situ* growth, ancient reefs frequently preserve exquisite details of past ecological interactions that are lost from other fossil communities. But many processes occur on a reef that compound to ensure that the composition of the biotic community as finally preserved in the geological record may be markedly different from that of the original living reef. Notwithstanding the almost total loss of soft-bodied and motile organisms (which may travel to feed but not actually live on the reef), skeletal organisms are also susceptible to *taphonomic bias*. Taphonomy is the study of all the processes of preservation.

We have already noted that a considerable proportion of modern coral reefs can be preserved in the geological record as rubble, sediment, and voids as a result of physical and biological destruction. Moreover, much of the reef sediment may be demonstrably removed from its original location, primarily during storms which, although intermittent, have a disproportionate influence: in some cases this may account for up to 25% of the total carbonate produced, with the remainder being redistributed and reincorporated within the reef interior (Hubbard 1992). Likewise, the rate of bioerosion can be substantial: for example, Hubbard *et al.* (1990) calculated that up to 60% of the carbonate produced on reefs growing on the north coast reefs of St Croix is reduced to sediment by bioerosion. However, bioerosion did not become a major destructive force in reefs until the mid-Jurassic (~170 Ma), suggesting that we might discount the effects of significant bioerosion in older reefs and predict that a greater proportion of reef framework will be preserved *in situ*.

The composition of skeletal debris associated with a reef may not be a reliable indicator of the original reef community composition, as organisms vary greatly in their relative rates of skeletal material production as well as their durability in the face of destructive forces (compare Figs 1.1(a) and (d) with Figs 2.2(b) and (a), respectively). The way in which an organism died will also control the final preservation of the skeleton. Rapidly growing skeletal organisms which are in some way protected from destructive processes (such as by rapid burial, overgrowth by secondary encrusters, and reduced vulnerability to bioerosion) will often be disproportionately represented in the fossil community. While *Halimeda* may only cover 10% of a living coral reef framework, the high rate of carbonate production by this alga will result in its disarticulated plates contributing to loose reef sediment in excess of that of corals (Scoffin 1992). Being prone to storm damage and bioerosion, delicate, branching forms will have a lower *preservation potential* even though they may have been more important volumetrically on the living reef. For example, although the relatively fragile coral *Agaricia* represents 54% of the living scleractinian genera on the shelf-edge reefs of St Croix, US Virgin Islands, it is completely absent from cores taken through the underlying abandoned reef and so would not be preserved in the geological record (Hubbard *et al*. 1986). By contrast, in a comparative study of similar Pleistocene reefs in the Bahamas and living reefs in Florida dominated by relatively resilient massive corals, Greenstein and Curran (1997) found that the fossil community preserved zonation and provided a relatively accurate reflection of the living community composition. However, in a similar study such massive forms were noted to represent less than one-quarter of the primary-framework carbonate on the equivalent living reefs nearby (Scoffin *et al*. 1980).

The importance of physical and biological reworking on reefs is illustrated by the fate of thousands of individuals of the echinoid *Diadema* which died *en masse* in a disease that swept the Caribbean in 1983. Pre-1983 surveys showed their exceptionally high abundances on reefs (for example, Scoffin *et al*. (1980) reported 23 individuals per square metre in Barbados fringing reefs), but even though substrates associated with the reefs were littered with the disarticulated spines and tests of this echinoid several weeks after the mass mortality event, less than 1 year later the sediments on neighbouring reefs betrayed no evidence of any increase in the remains of *Diadema* (Greenstein 1989). As such, this profound and catastrophic event would have gone unrecorded in the geological record.

Cryptic encrusting organisms are relatively protected, not only from physical and biological destruction but also because crypts can be sites of rapid early cementation. Those that inhabit the deepest recesses of the reef framework will be the most protected. However, taphonomic loss is probably also not inconsiderable in crypts due to the likely overwhelming proportion of soft-bodied biota (85% of surface area accounting for 62% of total species' diversity was recorded in one study of modern reef caves), and to the corrosive effects of some encrusters (Rasmussen and Brett 1985). Moreover, cryptic organisms attaching to the roofs of cavities will suffer the effects of corroding solutions that may be flushed through the reef framework (Scoffin 1972*b*).

To summarize, the internal form of a reef—that which is recorded in the geological record—is determined by many factors. These include:

(1) the ecology and environmental demands of the reef community;

(2) the rate of skeletal carbonate production;

(3) the degree of physical and biological destruction;

(4) the pattern of sediment removal from or storage within the reef;

(5) the degree of early cementation.

These factors compound such that ancient reef complexes may show a disproportionately large volume of intact framework preserved in lagoonal or deep fore-reef settings rather than on the reef crest. So while the growth of the living reef organisms in response to local, short-term processes can be studied with precision, the growth and development of the reef itself is the end result of complex, longer term processes, many of which remain poorly documented (Scoffin 1992).

2.4   The setting of ancient reefs

Before describing the diversity of ancient reef ecologies, it is necessary to understand variations in the environmental setting of reefs through geological time. This is introduced in the following section.

   As with all aspects of Earth's history, plate tectonics provides a unifying theory that explains the global distribution of reefs. The constant motion of oceanic and continental crust creates zones of rifting, sea-floor spreading, and crustal collision. This controls the formation of seas, the movement of landmasses into equatorial waters, and influences climate as well as creating or removing barriers for the dispersal of the larvae of marine organisms. The local tectonic regime also controls local terrestrial drainage patterns and determines the rate of subsidence. Tectonics can also control local topography and bathymetry.

   Modern and ancient reefs are usually found within shallow marine carbonate sequences that develop on horizontal shelfs known as *carbonate platforms*. Shelf architecture is a major control on the development of carbonate platforms. Figure 2.7 shows the four types of platform that are recognized: rimmed shelf, ramp, epeiric platform (intracontinental sea), and isolated platforms. Each of these carbonate-platform types produces a distinctive pattern of sediments, whose distribution is further modified by tectonic effects, sea-level changes, and the influence of reefs on platform bathymetry.

2.4.1   Carbonate platforms

A *rimmed shelf* is a carbonate platform with a pronounced break of slope at shallow depth that extends into deep water. The shelf margin is a high-energy area subject to considerable wave activity, storms, and often tidal currents, and the warming and CO_2-degassing of oceanic water as it passes onto the shallow shelf creates an environment that promotes rapid rates of *in-situ* carbonate production. As a result, rimmed shelves are characterized by the development of a series of often well-cemented barrier reefs, or skeletal or oolitic sand shoals. [Ooids are coated carbonate grains produced by the inorganic precipitation of laminae around a nucleus.] The presence of these barriers creates further turbulence along the steep shelf margin, so promoting the constant erosion and the reworking of sediment. The barriers also restrict circulation to varying degrees in the landward shelf lagoon or basin, which may be extend from a few to up to 100 km. Where continuous and wave-resistant barriers form, a very low-energy lagoon with highly restricted circulation and often seasonally high salinity will develop dominated by packstones and wackestones. These

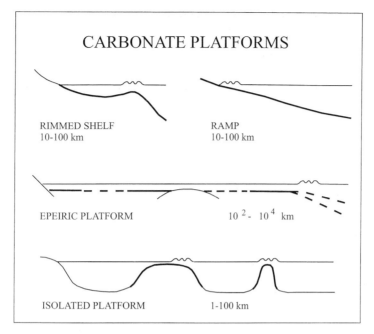

Fig. 2.7   The major categories of carbonate platforms. (Modified from Tucker and Wright 1990, by kind permission of Blackwell Science Ltd.)

back-reef sediments are often cyclic—showing repeating metre-scale, shallowing-upwards units reflecting a decrease in turbulence and water depth, as well as an increase in restriction as tidal flats are established, followed by periodic flooding through a rise in sea level (*transgression*). Modern examples of rimmed shelves include the Belize Shelf and Queensland Shelf, eastern Australia, both of which support barrier reefs. The distribution of carbonate sediments on these rimmed shelves is strongly controlled by relict topographies, formed during the Pleistocene.

Rimmed shelves can be divided further into those that are:

- *accretionary*—which show a lateral migration (*progradation*) of shallow shelf-margin reefs and carbonate sand bodies over fore-reef and reef slope deposits;
- *bypass* rimmed *margins* that deposit little sediment on the shelf slope, because shelf margin sedimentation is able to keep pace with rising sea level but insufficient sediment is deposited on the slope for any significant lateral accretion;
- *erosional* rimmed margins which occur in areas with strong tides or ocean currents such that cliff or escarpments characterize the shelf slope.

In erosional rimmed margins, debris may often accumulate at the toe of the slope. Where progradation occurs, characteristic *clinoform* geometries form, which consist of large-scale dipping surfaces, formed largely by shallow-water debris shed from the shelf margin (Fig. 2.8).

A *ramp* is a gently sloping surface (generally less than 1°) on which shallow-water carbonate sediments pass gradually offshore to deeper waters dominated by muddy, basinal sediments. Although there are no major breaks in slope, there may be many local shoal

Fig. 2.8 Windjana Gorge, Canning Basin, western Australia showing well-developed clinoforms of Late Devonian reef-slope sediments prograding into the basin towards the left.

areas. Wave energy is not as high as along a shelf margin, but the gradual shoaling of the ramp results in strong wave action in shallow areas so that isolated carbonate structures such as skeletal or ooid grainstone shoals can form. Barrier reefs are absent, but patch reefs are common seaward of the beach-barrier in the lagoon, and mud-mound mounds and *pinnacle reefs* (those on topographic highs within a basin) can develop on the outer, deep ramp, especially on topographic highs. The Trucial Coast of the Arabian Gulf and Shark Bay in western Australia are modern examples of carbonate ramps. A ramp may develop into a rimmed shelf as a result of long-term prograding reef growth, or a shelf may become a ramp due to differential subsidence or drowning.

Epeiric—or *epicontinental*—*platforms* are very extensive (between 100 and 10 000 km wide) generally flat, *cratonic* areas (the stable central parts of continents) which are covered by a very shallow sea, generally less than 10 m deep. Although the oceanward edges of epeiric platforms may be bounded by ramp or shelf margins, often formed by contemporaneous fault movement, the epeiric platform itself shows distinct sequences dominated by low-energy, subtidal to intertidal sediments. Movement of sediment by storms may have been important in such environments, although it is not clear whether epeiric seas were subject to a tidal regime. Epeiric platforms were far more common in the geological past when the sea level was much higher than it is today and large expanses of continents were flooded. The closest modern analogues to ancient epeiric seas are the interiors of the Great Bahama Bank and Florida Bay.

Isolated platforms are shallow-water carbonate platforms which are surrounded by deep water, often with steep margins. The sediments that develop on these platforms are controlled, in large part, by prevailing winds and storms, and reefs often develop—especially on open, windward margins. If the rate of carbonate production at the margins is high, combined with a steady rate of subsidence, then a shallow reef rim will grow around the platform forming a deeper central lagoon with patch reefs. Such isolated platforms are known as *atolls*, but true oceanic examples form on extinct, subsiding volcanoes.

2.4.2 Tectonic setting

Carbonate platforms develop in a range of tectonic settings, but particularly along passive continental margins, intracratonic basins, rifted continental margins, and tectonically active areas (Fig. 2.9).

1. *Passive continental margins*: At times of relatively high sea level when sediment input from continents (*terrigenous sediment*) into shallow marine areas is reduced, carbonate platforms often develop on passive continental margins. All types of carbonate platform can form on open, ocean-facing shelves, which may develop on top of earlier rift sequences. Subsidence is often slow (although decreasing exponentially) and prolonged in such settings, so that the growth of substantial, thick, complex reefs is possible. Rapid subsidence rates also produce thick reef sequences if reefs are able to maintain growth. Isolated platforms may be created where faulting has occurred. Examples include the reefs which grew around the margins of the Iapetus ocean in the Cambrian–Ordovician, and the Tethys ocean during the Jurassic–Cretaceous.

2. *Intracratonic basins*: Many ancient reefs developed within shallow basins formed by downwarping of the craton. In such situations, subsidence is often relatively slow and the reef sequences are interrupted by many hiatuses. Reefs may develop around the basin margins, on fringing platforms, or shelves, or on topographic highs within the basin-forming pinnacle reefs. A well-known example is the Permian Capitan reef which developed around the partially isolated Delaware Basin of Texas and New Mexico. Such basins may develop stratified seas with anoxic bottom waters.

INTRACRATONIC BASINS

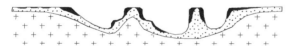

RIFTED CONTINENTAL MARGINS

PASSIVE CONTINENTAL MARGINS

Fig. 2.9 Major tectonic settings in which ancient carbonate platforms have developed. Black areas represent reefs. (Modified from James and Macintyre 1985.)

3. *Rifted continental margins*: Continental margins may be actively rifting such that subsidence rates are highly variable. Elevated topographies can form on newly rifted margins allowing the development of marginal platforms, or on the top of horst blocks enabling the growth of isolated reefs (atolls and platforms) in the axis of the rift. Examples include the Miocene Gulf of Suez and the Eocene Sirte Basin.

4. *Tectonically active areas*: Reefs can develop in tectonically active regions such as thrust belts, volcanic islands and arcs, and open-ocean volcanoes. Such settings are locally important, and examples include modern Indonesia and the Cenozoic of the Middle East.

3 The record: ancient reef ecologies

Reefs have existed since the appearance of microbial mat communities over 3.5 billion years ago. Since that time, many groups of extinct and extant skeletal algae and metazoans as well as microbes have aggregated to form a tremendous variety of reef ecologies in widely differing environmental settings.

To describe this history in full is not only an impossible but futile undertaking as there has never been a single global-evolving reef ecosystem: all reef communities are individual in their ecology and development, and each geographical area has a unique history of reef growth and distribution closely linked to local environments, biogeography, and tectonic history. Notwithstanding this diversity, however, evolutionary innovation and long-term global environmental change have moulded the persistence and development of reef ecosystems, each of which had a limited temporal distribution. I have therefore divided the record of reef-building into a series of episodes that are bounded by either the radiation, or the eclipse, of major groups of reef-builders, although it should be noted that the history of reefs does not always parallel that of these groups. Essential characteristics of major reef biota or reef-associated structures are described separately, and the palaeoecology of some representative reef communities is highlighted as a series of case studies. (A detailed analysis of mass extinctions and the subsequent recovery of reef ecosystems is given in Chapter 5.)

The descriptive, chronological record presented in this chapter is of necessity a selective compilation, but it does serve to underline the diversity and complexity of ancient reef communities, as well as providing a source of reference for later chapters that examine the processes responsible for the evolution of reef ecologies. The geological time scale used here is given in Appendix 1, where names and ages of stratigraphic units follow Harland *et al.* (1990). Geological time is divided into three Eons: the Archaean, from 3.6 to 2.5 billion years ago; the Proterozoic from 2.5 billion years ago to ~543 million years ago, and the Phanerozoic (the Eon of 'visible life', from the Cambrian to Recent) from 543 million years ago to the present day. The units 'Ga', 'Ma', and 'ka' are used to refer to dates before the present in billions (10^9), millions (10^6), and thousands (10^3) of years, respectively. 'Gyr', 'Myr', and 'kyr' refer to duration of time in billions, millions, and thousands of years, respectively.

3.1 Archaean to Lower Cambrian (Nemakit-Daldynian): 3.5 Ga–530 Ma

The oldest known reefs were constructed by *microbialites*—calcareous deposits formed by the interaction of benthic microbial communities and detrital or chemical sediments (see Box 3.1). From their first appearance ~3.5 Ga and for the next 2.5 Gyr, these microbialites were expressed solely as finely laminated structures known as stromatolites (Figs 3.1(a,b)).

(a)

(b)

Fig. 3.1 Modern and ancient stromatolites. (a) Living stromatolites, Hamelin Pool, Shark Bay, western Australia. Stromatolites are approximately 30–50 cm in diameter. (b) Elongate columnar stromatolites from the Palaeoproterozoic Taltheilei Formation, Great Slave Lake, North-West Territories, Canada. Hammer = 40 cm long. (Photograph: P. Hoffman and Geological Survey of Canada.)

Box 3.1: **Microbialites, stromatolites, and thrombolites**

Microbial carbonates—or microbialites—are fine-grained carbonates precipitated *in situ* either directly or indirectly by the physiological activity of benthic micro-organisms. Microbialites accrete due to the sediment-trapping, binding, or precipitating activities of varied mat or biofilm communities, often supplemented by the direct inorganic precipitation of cements. These mat communities are usually composed of bacteria, blue-green algae (single-celled photosynthetic prokaryotes called cyanobacteria), and green algae (chlorophytes), or biofilms, sometimes associated with sponges (Reitner 1993).

Although their recognition in the geological record can be problematic and controversial, microbialites may be characterized by the presence of clotted textures with abundant peloids, dendrolitic structures, or a weakly laminated internal structure due to incremental movement of the mat community, and small cement-filled cavities known as *fenestrae* together with evidence for early lithification such as bioerosion or encrustation. Most microbialites are encrusting or free-standing, and their growth and coalescence can result in the formation of a reef framework.

Presumed microbialites are common as primary framebuilders and secondary encrusters within many Phanerozoic reefs: indeed, it is probable that they were more far important than currently recognized. Microbialites were probably also widespread in the Proterozoic, but they are difficult to identify unequivocally. Although the relative importance of microbialites in modern coral reefs is not clear, the increasing numbers of examples recognized also suggest a significant role.

Stromatolites are finely layered (laminated) microbialites produced by photosynthetic communities (mainly cyanobacteria) that form a range of morphologies including domes, columns, and mounds. Stromatolite growth-form changes with the taxonomic composition of the mat community, in response to underlying topography and environmental factors such as wave energy and currents. Preservation of the original mat community, however, is rare in both recent and ancient examples: the vast majority of stromatolites leave no trace of their original biologies, and although stromatolites 3.5 Ga are known, identifiable cyanobacteria have not been recognized before ~2.5 Ga (Klein *et al.* 1987). Recently, even the biological nature of some Proterozoic stromatolites has been questioned as the microscopic textures and morphological form of these stromatolite growth surfaces can be adequately explained by a purely inorganic model of chemical precipitation, fallout, and diffuse accretion of suspended sediment (Grotzinger and Rothman 1996).

Stromatolites form today in both marine and non-marine environments, often under conditions of non-marine salinity such as in Shark Bay, western Australia (Fig. 3.1(a)). However, the finding of abundant stromatolites in a variety of intertidal and subtidal settings around Exuma Cays, in the Bahamas (Reid *et al.* 1995), suggests that normal open marine examples may be more common than previously supposed.

Modern stromatolites grow very slowly, with accretion rates of only 0.4–1 mm year^{-1}. Most cyanobacteria associated with these stromatolites, e.g. *Schizothrix*, are unable to produce a calcareous skeleton directly. Their lithification is due either to indirect postmortem inorganic cementation processes where cements are precipitated either on or within microbial sheaths induced by the decay or indirect metabolic processes prevalent in hypersaline or brackish waters, or to direct carbonate particle entrapment (see Section 4.1.3). As such they are poor analogues for fossil counterparts, since many ancient stromatolites show abundant evidence for the direct precipitation of laminae in normal marine waters, especially those known from the Palaeo- and Mesoproterozoic. In the

Neoproterozoic after ~850 Ma, trapped and bound clay-sized carbonate crystals (largely precipitated in the water column) predominate within stromatolitic laminae (Grotzinger 1989).

Modern stromatolites appear to form only where two criteria are satisfied:

(1) where environmental conditions such as high sedimentation rates (e.g. Exuma Cays) or low nutrient levels (e.g. Shark Bay) excludes the growth of other faster growing algal competitors for substrate space. Unlike most seaweeds, some modern cyanobacteria are able to fix nitrogen and so are not nitrogen-limited (Hay 1991).

(2) where oceanographic conditions create a water chemistry that is favourable for carbonate precipitation, such as high levels of supersaturation of carbonate, rapid degassing (loss of CO_2) rates, or local elevations of sea-water temperature.

Thrombolites are non-laminated microbial structures with clotted micritic textures, which often present a mottled or bioturbated appearance. Thrombolites commonly form free-standing mound- or loaf-shaped structures, and some may have relatively rapid growth rates (Webb 1996). While most modern stromatolites are formed predominantly by cyanobacterial mats, thrombolites may be formed by algal turf, other eukaryotic algae (Reid *et al.* 1995), or by coccoid cyanobacteria (McNamara 1992).

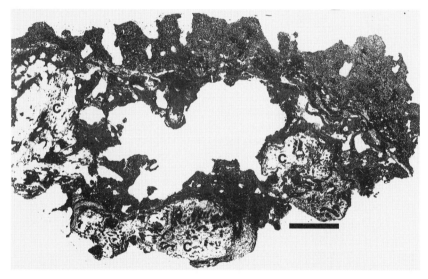

Fig. Box 3.1 Thin-section through a modern thrombolite that has encrusted the branches of a scleractinian coral (C). Both the corals and thrombolites have been heavily bioeroded. A large void is present in the centre of this specimen. Heron Island, Great Barrier Reef, Australia. Scale = 5 mm. (Photomicrograph: G.E. Webb, reproduced by permission of *Sedimentology* and Blackwells Science Ltd.)

The oldest recognizable, extensive stromatolite reefs date from the late Archaean (2.7–2.5 Ga), coincident with the large-scale formation of cratons that allowed the development of extensive, shallow marine platforms along continental margins. Stromatolites formed atoll-like reefs around subsiding volcanoes and as linear belts on early cratonic crust, although non-marine stromatolites are also known from this time. The presence of any earlier reef structures would have been rendered unrecognizable by metamorphism.

Prior to 2.3 Ga, the Earth was essentially anoxic, and atmospheric oxygen levels probably continued to be very low during the Palaeoproterozoic (2–1.65 Ga). Carbonate platforms with most of the essential features of those found during the Phanerozoic were well developed by 2.6–2.3 Ga (Grotzinger 1989), and stromatolites became abundant and diverse from 2.5 Ga. Archaean and Proterozoic platforms include both ramps and rimmed shelves which show strong environmental zonation as reflected by the distribution of stromatolites which grew in basinal, subtidal to supratidal environments in both carbonate and terrigenous-dominated settings (Fig. 3.2).

Reefs from the Palaeoproterozoic were particularly widespread and are among the largest known in the history of reef-building, extending over thousands of kilometres (Hoffman 1974; Grotzinger 1989); these reefs also coincide with the best-developed Proterozoic carbonate platforms. Indeed, the presence of wave-resistant stromatolite reefs was responsible for the strong palaeoenvironmental zonation of many carbonate platforms at this time (Grotzinger 1990). Examples include the Palaeoproterozoic Monteville Formation and the Mesoproterozoic Dismal Lakes Group in Canada. These early reefs were morphologically highly variable, with barrier and fringing systems cut by tidal channels, as well as pinnacle and patch reefs surrounded by oolites. Some of these growth forms are distinctive and appear to have had restricted stratigraphic distribution.

Palaeoproterozoic stromatolites, such as those from the Pethi Group, east of the Great Slave Lake, North-West territories, Canada (1.9 Ga), show that a variety of morphologies clearly formed in direct response to hydrodynamic conditions: laminar forms formed in peritidal areas, very large columnar forms dominated the more energetic platform margin, and weakly columnar forms grew on the basin floor (Hoffman 1974). The platform margin stromatolites achieved up to 3 m relief, and they show belts of transverse stromatolitic mounds and channels which may be analogous to modern spur-and-groove structures; these were certainly wave- and current-resistant structures. In this example, they also show striking elongation presumably perpendicular to the Proterozoic shoreline (Fig. 3.1(b)).

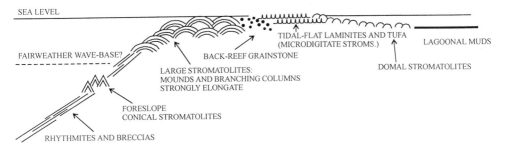

Fig. 3.2 Distribution of stromatolite types across a high-energy (windward), Proterozoic, rimmed carbonate shelf from lagoon to deep foreslope. The barrier-reef complex is characterized by strongly elongate stromatolite mounds and columns. Conical stromatolites form below wave base; domal stromatolites grew in the low-energy, back-reef lagoon; tufas, including microdigitate stromatolites, formed by precipitation on tidal flats. (Modified from Grotzinger 1989, by kind permission of SEPM (Society for Sedimentary Geology).)

Description of exceptionally well-preserved algal mat communities preserved in silicified stromatolites indicates that all modern classes and orders of cyanobacteria had evolved by 2 Ga.

Crypts are known to have been exploited in these earliest reefs. For example, organic-walled microfossils (*Huroniospera* sp. and *Gunflintia* sp.) and hematitic problematica (organisms of unknown affinity) such as *Frutexites* sp. have been described within lithified algal mat sequences from the Palaeoproterozoic Odjick Formation, Canada (Hofmann and Grotzinger 1985). Reef biotas clearly differentiated into open surface and cryptic communities as soon as cryptic niches became available.

After the late Palaeoproterozoic, stromatolites suffered three periods of decline in terms of both diversity and abundance (Fig. 3.3): at about 1 Ga, at the end of the Proterozoic (~545 Ma), and a final decline after the Early Ordovician (~490 Ma), following a slight resurgence in the Late Cambrian when they became locally abundant. Various microbialites in conjunction with skeletal forms continued, however, to be important components of reef ecology and preservation, especially in the late Palaeozoic to early Mesozoic.

Coincident with the decline of stromatolites at the end of the Proterozoic was the rise in abundance and diversity of thrombolites (Box 3.1) and calcified cyanobacteria (Box 3.2) within reefs. Thrombolites often developed in low-energy settings, such as on ramps and deep shelves. Well-preserved examples of Neoproterozoic thrombolite reefs are known from the Little Dal Group, north-western Canada (Case study 3.1). The presence of these reefs in the Neoproterozoic suggests that the subsequent appearance of abundant well-calcified cyanobacteria at the Precambrian–Cambrian boundary may record a taphonomic or environmental event that allowed the radiation of an already established flora (Knoll *et al.* 1993).

Microbial communities also built large pinnacle reefs up to 40 m high with exposed depositional slopes of 30°, within the terminal Neoproterozoic Huns Member of the Nama Group of southern Namibia. Here, the reef cores are constructed by stromatolites/thrombolites. Original topographic relief would have been undulating and in the order of several metres, with topographic lows being infilled by mud and talus.

Non-skeletal multicellular algae are known from the late Neoproterozoic (Xiao *et al.* 1997), but terminal Neoproterozoic carbonates also contain the first possible evidence for the skeletonization of algae, as well as for metazoans (Knoll and Walter 1992). Remnants of what appear to be thin platy or sheet-like aragonitic algae similar to late Palaeozoic phylloid ('leaf-like') algae are known from the Nama Group (S.W.F. Grant *et al.* 1991). These algae formed small patch reefs (1 m high, 2 m wide) that intergrew with thrombolites surrounded by cross-bedded grainstones.

The finding of exceptionally well-preserved biotas of late Neoproterozoic age (570 ± 20 Ma) supports the view that animals originated long before they became conspicuous elements of the fossil record (Xiao *et al.* 1997). The early evolution of animals probably proceeded as a single, protracted evolutionary radiation lasting some 55 Myr before culminating in the Cambrian explosion that records widespread skeletonization and the expansion of behavioural repertoires (Grotzinger *et al.* 1995), although there is evidence to support the presence of a terminal Proterozoic extinction event (Knoll 1996). Recently, it has been proposed that metazoan diversification only became possible after the end of the Varanger ice age, now thought to have terminated some 590 Ma (A.J. Kaufman, A.H Knoll, and

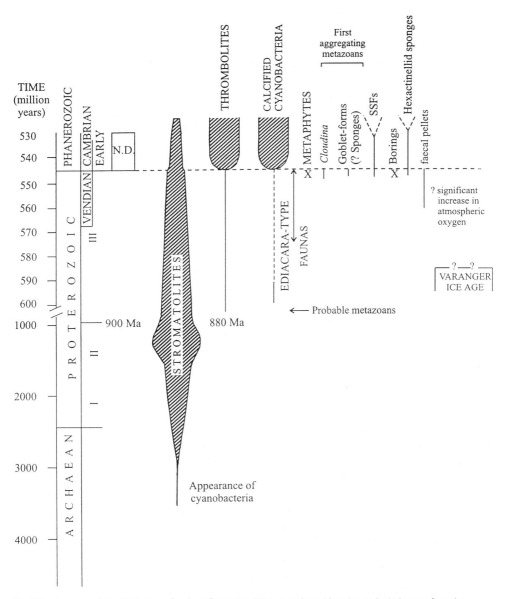

Fig. 3.3 Summary of the distribution of early reef-associated biota together with major geological events from the Archaean to Early Cambrian (Nemakit-Daldynian). N.D.: Nemakit-Daldynian; SSFs: Small shelly fossils.

G.M. Narbonne, unpublished data), and that an increase in atmospheric oxygen levels in the late Neoproterozoic may also have removed a barrier to the evolution of large body size in metazoans (e.g. Knoll 1992). Simple, centimetre-sized soft-bodied fossils of probable metazoans are known from rocks older than ~600 Ma (Hofmann *et al.* 1990), but the diverse and characteristic body fossils and traces of the Ediacara fauna appear about 570 Ma. Metazoans capable of forming calcitic skeletons appeared as early as the youngest Ediacaran

Box 3.2: **Calcified cyanobacteria (calcimicrobes)**

Calcified microbes (known as calcimicrobes) are the partial remains of microbes (probably the extracellular sheaths of cyanobacteria) preserved by carbonate precipitation. Although known from the latest Archaean, they became abundant on Cambrian and Later Palaeozoic carbonate platforms. All forms may have trapped sediment, and it has been suggested that the micrite present between colonies was originally a cement similar to modern sea-floor cements (James and Gravestock 1990).

Calcified cyanobacteria show significant morphological variation, therefore making secure taxonomy difficult; some have even suggested that the many established taxa may represent diagenetic end-members of a calcification series of coccoid forms (Pratt 1984). Three main groupings of calcified cyanobacteria are typified by:

(1) *Renalcis*—with a globular or cellular structure formed by clumps of micrite-walled lunnules or by clots of featureless micrite;

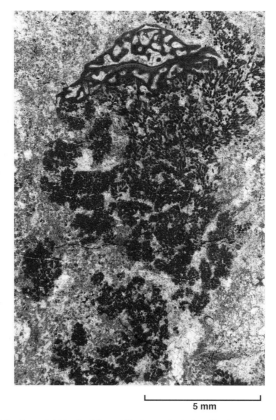

5 mm

Fig. Box 3.2 (a)   An upright bush of the calcified cyanobacterium *Gordonophyton* colonized by a juvenile archaeocyath sponge. Note the continued growth of *Gordonophyton* after the attachment of the archaeocyath, suggesting that lithification occurred either during life, or very shortly after localized death of the colony. Lower Cambrian (Atdabanian), western Mongolia.

(2) *Epiphyton*—shrub-like colonies with finely bifurcating branches and a dense, irregular microstructure; and

(3) *Girvanella*—sheets or crusts of numerous intertwined tubules.

Most calcified cyanobacteria formed substantial upright bushes or rafts, or encrustations around other reef components. Many were common and facultative inhabitants of crypts in the Lower Cambrian: *Korilophyton*, *Angulocellularia*, *Renalcis*, *Chabakovia*, *Gordonophyton*, and *Epiphyton* were especially abundant in crypts, but only *Chabakovia* was a preferential crypt dweller. *Renalcis*- and *Epiphyton*-group forms exhibited phenotypy, that is, showing globular and compact morphologies when growing upright on open surfaces, but delicate, arborescent branched forms in crypts (Kobluk and James 1979; Zhuravlev and Wood 1995).

Modern cyanobacteria are found at a depth of >200 m in poorly illuminated conditions, but the maximal depth for prolific growth appears to be 50–70 m in clear water (Saffo 1987). No living cyanobacteria precipitate carbonate obligately (Pentecost and Riding 1986), and so carbonate formation is more dependent upon environmental factors than physiology. Microbial calcification does not occur in Tertiary or Recent marine environments of normal salinity: modern cyanobacteria can calcify either during life (in freshwater) or postmortem, where heterotrophic bacteria may facilitate early diagenetic calcification of their mucopolysaccharide sheaths. Calcification of cyanobacteria is promoted locally by high concentrations of either Ca^{2+} or carbonate supersaturation.

L_____J 2 mm

Fig. Box 3.2 (b) Crypt ceilings formed by rafts of the calcified cyanobacterium *Razumovskia uralica*, which have been colonized by pendent colonies of *Epiphyton fruticosum*. Lower Cambrian (Botoman), South Urals, Russia.

Case study 3.1: Little Dal Group, Mackenzie Mountains, north-western Canada (Neoproterozoic: 1100–780 Ma)

The Neoproterozoic Little Dal group, in north-western Canada contains reefs 1–8 m thick that coalesced to form composite reefs up to 500 m in thickness and 3 km in diameter. These reefs show steep depositional slopes and topographic relief of at least several tens of metres. They grew 150 km seaward of the shelf edge on a flat basin floor below storm-wave base, initiating in deep waters within the photic zone but growing progressively into shallow water near sea level. Reef architecture closely reflects sea-level fluctuations as the reefs tracked sea level, with boulder-sized talus forming at periods of relatively low sea level in response to increased erosion and/or progradation over unconsolidated substrates.

The reefs were composed of tabular stromatolites in the deeper parts, and thrombolites and calcified cyanobacteria as the reef grew into shallower waters. The cyanobacteria formed a spectrum of digitate, laminar, and clotted structures. Geopetal structures several centimetres wide within the primary cavities are common, and early marine cement is abundant. The thrombolites formed bioherms over 8 m in diameter composed of columns, sometimes with pronounced lateral or downwards growth. These are composed of laminated and clotted micrite with abundant fenestrae, constructed by laminar 'cellular crusts' and a tubular cyanobacteria resembling *Girvanella*. Fine micritic laminae occur locally, which resemble the Mesozoic microbe *Bacinella* (Pratt 1995). Cavity roofs are encrusted by micrite, sheets of *Girvanella*, and pendent, clotted microbial structures similar to *Renalcis* (from Aitken 1989; Turner *et al.* 1993, 1997; Pratt 1995).

Fig. CS 3.1 Lacy crusts composed of intertwined filaments resembling *Girvanella*, which formed open framework crypts. These crypts are lined by microspar and infilled with micritic internal cement. Little Dal Group, Mackenzie Mountains, north-western Canada. (Photomicrograph: B.R. Pratt.)

communities, and several small shelly taxa of a Cambrian caste are now recorded to extend well into the terminal Neoproterozoic. Phytoplankton diversified during the latest Proterozoic–Early Cambrian (Moczydlowska 1991) which may have driven the radiation of suspension and filter-feeders (Butterfield 1994).

The first metazoan reefs are known from the latest Neoproterozoic, and particularly diverse examples are known from the Nama Group. These are composed of dense, monospecific aggregations of weakly skeletonized tubular, cup- or goblet-like organisms, which include the globally distributed *Cloudina*, a suspension-feeder—possibly solitary cnidarian or worm tube (S.W.F. Grant 1990; Fig. 3.4(a))—as well as seven other as yet undescribed forms (A.H. Knoll, personal communication). The range of these sessile organisms overlaps with the most diverse Ediacaran soft-bodied fossil assemblages (Grotzinger *et al.* 1995): *Cloudina*, extends from strata dated as 550.5 to 542.5 Ma, and an undescribed goblet-shaped form from 550 to 543 Ma.

The undescribed goblet-shaped forms of unknown affinity, but possible poriferan, formed small aggregations (Fig. 3.4(b)) which grew in sheltered depressions between thrombolite heads of the Huns Member pinnacle reefs (A.H. Knoll, personal communication). These forms are preserved as moulds, and were either non-skeletal or only weakly mineralized. Their aggregations show no framework development and little relief.

Endolithic (rock-boring) 'algal' microborings have been documented from ooids beginning in the Vendian (Green *et al.* 1988), and probable boreholes have been noted in *Cloudina* (Bengtson and Zhao 1992). The progenitor of these boreholes is unknown, and it is difficult to assess whether they are the result of postmortem boring or predatory attack.

Substantial pinnacle reefs in the Omkyk Member were constructed by microbialites and *Cloudina* (which becomes abundant especially towards the upper parts), and reach up to 100 m high and 3 km wide (Fig. 3.5). Bioclasts were locally sufficiently abundant to form grainstones (Grotzinger *et al.* 1995). These reefs contain a framework with cavities filled by fibrous cements, interpreted as formerly aragonite.

By late 'Nemakit-Daldynian' (= Manykaian; lowermost Cambrian; some 544–530 Ma according to Bowring *et al.* (1993)), calcified cyanobacterial reefs were widespread on the Siberian platform (Luchinina 1985) and in Mongolia (Drozdova 1980). These were built by domal heads of *Korilophyton* with minor *Tarthinia*, and are sometimes encrusted by fine, digitate stromatolites (Fig. 3.6). Many of these bioherms are succeeded by domal stromatolites, interpreted as recording growth into more shallow waters. Interbiohermal and cryptinfilling micrite is often burrowed by discrete, unbranched and meandering, millimetre-wide tubes which have a microspar infill with no backfilling. This provides evidence of abundant deposit-feeding (detritivory) by small metazoans (Zhuravlev and Wood 1995; Kruse *et al.* 1996), and the burrow style suggests a certain firmness of the sediment. It has also been suggested that tube-builders of unknown affinity such as *Anabarites* formed mounded aggregations during the Nemakit-Daldynian (Luchinina 1985).

Sponges were present as early as the late Neoproterozoic (Ediacaran), known from biomarkers (24-isopropylcholestane; Moldowan *et al.* 1994) and controversial medusiform impressions suggested to be sponges are associated with the Ediacara biota (Gehling and Rigby 1996). The first undisputed sponge spicules are triacts known from Doushantuo cherts dated at 570 ± 20 Ma (Zhao *et al.* 1985), and clusters of multiple spicule type including oxeas, hexactines, and pentactines referable to the Hexactinellidae from the late Ediacarian of south-western Mongolia (Brasier *et al.* 1997). Some of these soft-bodied sponges probably inhabited deep and cold-water settings. Cnidarians also have a fossil record extending back to the late Neoproterozoic, with soft-bodied representatives of all four extant classes known from this time (Conway Morris 1993).

In summary, by the Palaeoproterozoic widespread and substantial stromatolitic reefs were present in a variety of marine environments, and wave-resistant structures had evolved

(a)

1 cm

(b)

1 cm

Fig. 3.4 The earliest aggregating sessile metazoans (of problematic affinity) from the late Neoproterozoic Nama Group, southern Namibia. (a) Thickets of *Cloudina*; (b) an assemblage of goblet-shaped forms (?sponges) which grew in sheltered depressions between thrombolite heads of the pinnacle reefs;. (Photographs: S. Jensen.)

Fig. 3.5 Pinnacle reefs from the late Neoproterozoic Omkyk Member, lowermost Nama Group, Namibia. (Photograph: S. Jensen.)

Fig. 3.6 Small reefs built by calcified cyanobacteria. Lowermost Cambrian (Nemakit-Daldynian), Bayan Gol, western Mongolia. Small burrowing metazoans lived in the sheltered areas of sediment within these reefs. Lens cap = 52 mm. (Photograph: P.D. Kruse.)

which controlled sedimentation across extensive carbonate platforms. Reefs show differentiation into distinct open surface and cryptic communities as soon as crypts formed in the Palaeoproterozoic. Calcified cyanobacteria became increasingly important towards the end of the Proterozoic, and also built substantial wave-resistant reefs. Some of the earliest skeletal metazoans known from the late Proterozoic (e.g. *Cloudina*) were sessile, gregarious, probable suspension-feeding organisms that were capable of forming very limited topographic relief. Borings have been recorded from this time and a detritivorous fauna associated with reef cavities had developed by the Nemakit-Daldynian.

3.2 Lower Cambrian (Tommotian–Toyonian): 530–520 Ma

The latest Neoproterozoic to Early Cambrian was a time of plate fragmentation, and sediments were deposited in response to the gradual inundation of extensive, relatively shallow cratonic seas. These seas were bounded by narrow continental margins of carbonate sand shoals, and many reefs appear to have developed in the protected areas behind these shoals or downslope on the shelf margins (James and Macintyre 1985).

Although stromatolites remained common in intertidal–supratidal environments, the first biotically diverse metazoan–algal reefs formed subtidally with the appearance of the highly gregarious archaeocyath sponges (Box 3.3) within cyanobacterial–thrombolitic communities at the base of the Tommotian (530 Ma; Bowring *et al.* (1993)) on the Siberian Platform. By the late Atdabanian, archaeocyaths were widespread in subtropical and tropical environments, and although they were present in non-reef settings, they reached their highest diversity and abundances in reefs. These reef communities persisted until the virtual demise of the archaeocyaths at the end of the Toyonian, some 520 Ma (Bowring *et al.* 1993), although various calcified cyanobacteria continued to build substantial reefs for the remainder of the Cambrian. Calcified cyanobacterial diversity also rose considerably during the Lower Cambrian.

Contrary to earlier reports, the earliest metazoan reefs at the base of the Tommotian as exposed on the Aldan River, Siberia, were already ecologically complex (Riding and Zhuravlev 1995). Although of low diversity, the framework was built by branching archaeocyath sponges (*Cambrocyathellus tschuranicus*) with abundant secondary skeletal growth and *Renalcis*, with a cryptic biota of archaeocyaths (*Archaeolynthus polaris*) and pendent *Renalcis* (Fig. 3.7). These reefs are also associated with the skeletal debris of organisms characteristic of the Lower Cambrian: orthothecimorph hyoliths, molluscs, brachiopods, coleolids, sponge spicules, and various small shelly fossils. Microburrowing deposit-feeders continued to proliferate within the sheltered areas of the framework.

By the middle Tommotian, archaeocyath–cyanobacterial reefs became more diverse and ecologically complex (Kruse *et al.* 1995), with the appearance of other sessile, calcified organisms inferred to have been suspension-feeders (Boxes 3.4 and 3.5). These include radiocyaths, a variety of simple cup-shaped, often cryptic forms known as 'coralomorphs' and the problematic tube-like cribricyaths, and rare but locally abundant large skeletal cnidarians (Lafuste *et al.* 1991; Savarese *et al.* 1993) and stromatoporoid sponges (Pratt 1990). The first known tabulate corals are associated with Botoman archaeocyath–cyanobacterial reefs, e.g. *Moorowipora* (Sorauf and Savarese 1995). Several other large colonial, cnidarian-like skeletal organisms are known from this time, such as *Flindersipora* (Lafuste *et al.* 1991), but their affinity remains uncertain. Possible calcarean sponges (members of the class Calcarea)

Box 3.3: **Archaeocyatha (Lower Cambrian: ~ 530–520 Ma)**

Archaeocyaths were sponges and the first large, sessile, skeletal metazoans to inhabit reefs. These gregarious forms appeared during the Early Cambrian (Tommotian), reached their acme in the Late–Early Cambrian (mid-Botoman) and became virtually extinct at the end of the Lower Cambrian, a period of some 10 Myr. Only two genera are known from the post-Lower Cambrian (Debrenne *et al.* 1984; Wood *et al.* 1992*b*).

Archaeocyaths were probably a monophyletic group (that is all representatives were derived from a single ancestor) without spicules but with morphologically complex, calcified skeletons which probably formed by direct calcitization of an organic template (Wood l990). Their affinity within the Porifera is uncertain, but they appear to be most closely related to the Demospongiae (Debrenne and Zhuravlev 1994). The typical archaeocyath formed a double-walled inverted conical calcareous skeleton, where the area between the two walls is known as the *intervallum* which was filled with soft tissue. The intervallum may bear many skeletal structures, including septa, taenia and *tabulae*.

The Archaeocyatha has been traditionally subdivided into two subclasses, the Regulars and Irregulars, according to differences in the juvenile development of the skeleton. However, the absence of these supposed differences has now been demonstrated, and it has also been shown that these skeletal characters are a function of differences in soft-tissue distribution which are independent of systematic placing (Debrenne *et al.* 1990; Zhuravlev 1990; Debrenne and Zhuravlev 1992). Regulars (orders Monocyathida, Ajacicyathida, and Tabulacyathida), generally show no secondary skeleton, and are inferred to have had completely soft tissue-filled intervallums, whereas Irregulars (orders Archaeocyathida and Kazachstanicyathida) are inferred to have borne soft tissue in their upper parts only. Abundant skeletal structures such as tabulae are present which served to section-off abandoned areas of the skeleton as the soft tissue migrated upwards.

Archaeocyaths rarely reached more than 20 cm in height, with most being 2–10 cm in height. But they show a tremendous diversity of growth forms, including solitary tubular and open dish-shaped cups, and multiple branching, with few achieving more stable, encrusting morphologies. Many archaeocyath species were mutually able to encrust by

2 mm

Fig. Box 3.3 (a) Grainstone formed by a variety of small, reworked archaeocyath cups, showing the characteristic double-walled construction with radiating septae within the intervallum. Lower Cambrian (Atdabanian), western Mongolia.

forming skeleton that extended beyond the boundaries of the main cup—in the form of structures known as *stereoplasm* and *exothecae*. Generally however, archaeocyaths formed a reef framework only with the aid of calcified cyanobacteria.

2 mm

Fig. Box 3.3 (b)   Small branching archaeocyath sponges. Lower Cambrian (Atdabanian), western Mongolia.

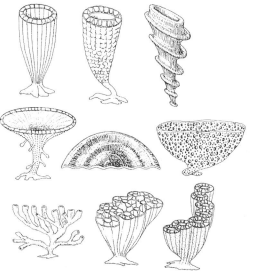

Fig. Box 3.3 (c)   Some of the great morphological variety shown by archaeocyath sponges. (From Hill 1972, by kind permission of the Geological Society of America, Boulder, Colorado, USA. Copyright © 1972 Geological Society of America.)

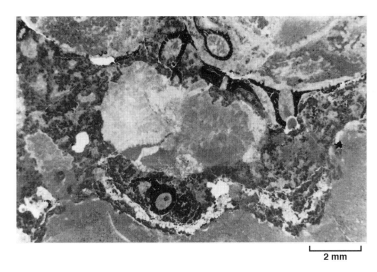

2 mm

Fig. 3.7 The earliest metazoan reefs—from the Aldan River, Siberia. Lower Cambrian (Tommotian). Although possessing low diversity, these archaeocyath sponge–calcified cyanobacterial reefs were already differentiated into framework-forming and cryptic biotas. (Photomicrograph: A. Yu. Zhuravlev.)

appeared in the early mid-Tommotian (Kruse *et al.* 1995), non-calcified demosponges in the late Atdabanian (Reitner 1992), and lithistid sponges are first known from the Toyonian archaeocyath reefs of Australia (Kruse 1991). Hollow, star-shaped chancelloriid debris is also abundant in Early Cambrian reefs—derived from a sessile, sclerite-bearing organism of problematic affinity. An associated fauna of ostracods, echinoderms, trilobites, and brachiopods was also locally present. Probable borings, possibly made by sponges, have been noted within coralomorph skeletons from the Canadian Rocky Mountains (Pratt 1991); indeed, silt-sized microspar grains resembling 'chips' from clionid-type sponges have been identified within Lower Cambrian reef cavities (Kobluk 1981*a*). (The makers of *Trypanites*—probably the work of sipunculid worms and reported to be associated with Toyonian reefs in Labrador—developed in hardgrounds which postdated reef growth.) Surprisingly, the endolithic 'algal' microborings documented from the Vendian are rare in Cambrian reef-associated bioclasts. All these elements produced one of the most diverse and ecologically complex reef-ecosystems known from the Palaeozoic.

Box 3.4: **Reef-associated sponges, and radiocyaths and receptaculitids**

Sponges can possess three types of skeleton:

(1) organic strands of poriferan collagen (known as *spongin*);

(2) siliceous or calcareous needle-like bodies (*spicules*); or

(3) massive calcareous skeletons that resemble those of corals.

Apart from the poorly documented role of sponge-associated biofilms, sponges have been associated with reefs since the late Proterozoic and those of demonstrable importance to ancient reef-building possessed a rigid skeleton. This skeleton is in the form of either fused robust spicules (as found in lithistid demosponges, where they are known as desmas), or a calcareous skeleton (calcified sponges). Although uncommon and of low diversity in the Recent, calcified sponges were diverse and abundant members of the pre-Cenozoic benthos, and reef-associated forms include the Lower Cambrian archaeocyaths (see Box 3.3), Mid-Palaeozoic and mid–Late Mesozoic 'stromatoporoids' (with well-layered skeletons, and often with abundant star-shaped canal structures known as astrorhizae) and 'chaetetids' (with skeletons composed of small tubules), and 'sphinctozoans' (sponges with a chambered organization).

Although sponges with calcareous skeletons or fused spicules form the most prominent aspect of the sponge fossil record, such features are of little value when assessing evolutionary ancestries. Lithistid sponges are demonstrably polyphyletic (Reitner 1992), and the spicules incorporated within the calcareous skeletons of sphinctozoans, and Mesozoic chaetetids and stromatoporoids also reveal a polyphyletic origin, with representatives from different orders of both the classes Demospongiae and Calcarea (Vacelet 1985; Wood 1987). Some Cambrian sphinctozoan sponges are derived from the Archaeocyatha (Debrenne and Wood 1990). All such sponges have independently acquired calcareous skeletons at different times in unrelated clades, and so these formerly discrete systematic groups can be better regarded as 'grades of organization': reflections of non-systematically related soft-tissue distribution and the arrangement of the filtration system (Wood 1987; Wood *et al.* 1992*a*).

Radiocyaths (Late Tommotian to Toyonian) and receptaculitids (Lower Ordovician–Devonian) were stalked, cup-shaped, sessile benthic organisms. They formed solitary or more often branching individuals of similar dimensions to the largest archaeocyaths.

Radiocyath cups are composed of loosely fused meromes, where each merome consists of a proximal and distal nesaster connected by a shaft. Meromes may separate on the

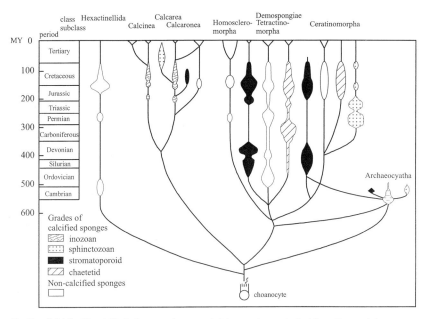

Fig. Box 3.4 (a) The ability to form a calcareous skeleton analogous to that found in corals has evolved independently in many groups of sponges, especially within the classes Demospongiae and Calcarea. (Modified from Vacelet 1983)

death of an individual and are often reworked where they further disintegrate to produce isolated nesasters. Nesasters may be abundant components both within bioherms, and in interbiohermal packstones and grainstones. Their skeletons are always totally recrystallized, indicating a probable original aragonitic mineralogy (Wood *et al.* 1993).

The affinity of radiocyaths and receptaculitids is still disputed: their morphology indicates a suspension-feeding mode of life, and, as result, they have been allied to the sponges (Termier and Termier 1979). Others have proposed a dascyclad algal affinity (Toomey and Nitecki 1979). Radiocyaths and receptaculids are typical of relatively low-energy settings, and appear to have been able to colonize soft substrates. They commonly provided sites for the colonization of encrusting skeletal metazoans and algae.

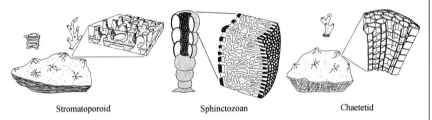

Stromatoporoid   Sphinctozoan   Chaetetid

Fig. Box 3.4 (b) Three groups of calcified sponges, whose living representatives can be placed within various clades of modern sponge classes. The groups reflect different distributions of soft tissues and organizations of the filter-feeding canal systems. (Modified from Stearn 1966; Zeigler and Rietschel 1970; West and Clarke 1983.)

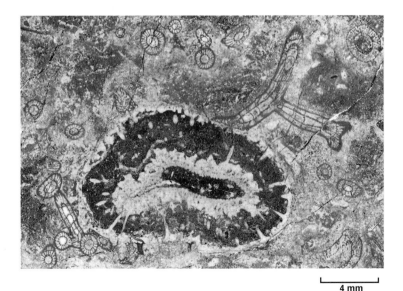

4 mm

Fig. Box 3.4 (c) Transverse section through a radiocyath cup, showing the arrangement of meromes. This cup has been colonized by a branching archaeocyath sponge.

Box 3.5: **Cambrian corals and coralomorphs**

Corals are anothozoan cnidarians with calcareous basal skeletons. Extinct and living corals do not have a skeletonized common ancestor, but probably arose as separate skeletonization events from the same groups of anemones, represented by the living Zoanthiniaria.

An increasing number of Cambrian calcareous-encrusting organisms have been described in recent years, some of which are assigned to the informal grouping of coral-like forms, the Coralomorpha. Some may be dismissed as cnidarians (e.g. *Archaeotrypa* and *Bija*), others were cnidarian but non-zoantharian (e.g. *Labyrinthus*, *Flindersipora*, and *Hydroconus*), but some genera were probably true zoantharian corals (*Tabulaconus*, *Moorowipora*, *Arrowipora*, and *Cothonion*). These Cambrian corals represent many discrete lineages, but none were probably ancestral to post-Cambrian corals. More material is needed to ascertain the affinity of the tabulate, sac-like Lower Cambrian *Cysticyathus*, and the Late Cambrian coral-like form *Cambrophyllum*. (From Scrutton 1997a.)

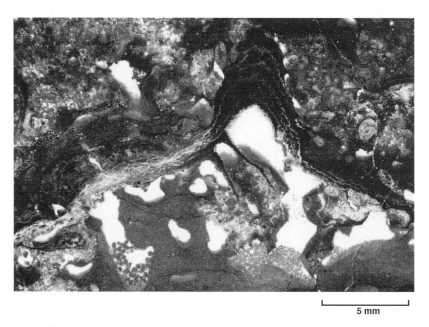

5 mm

Fig. Box 3.5 The coralomorph *Hydroconus*, growing downwards within a Lower Cambrian reef crypt. Note the preservation of pellets within the burrow system that formed beneath of shelter of the coralomorph. Lower Cambrian (Atdabanian), western Mongolia.

Lower Cambrian reefal communities usually developed as small bioherms (Fig. 3.8(a)), sometimes superimposed to form large patch reefs (Fig. 3.8(b)), which grew in intertidal and subtidal areas of shelf edges or seaways. Subtidal reefs growing on ramps could reach vast sizes, for example the Botoman-Toyonian microbial reefs of the Shady Domomite, Virginia, are up to 30 m thick and 3.5 km wide (Barnaby and Reed 1990). Although

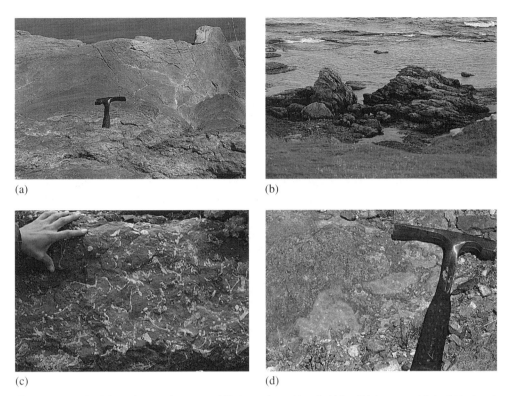

Fig. 3.8 Lower Cambrian archaeocyath sponge–calcified cyanobacterial reefs. (a) Small bioherms overlain by thick, domal stromatolites. Zuune Arts, western Mongolia. Hammer = 32 cm long. (b) Coalesced bioherms forming large patch reefs. Labrador, Canada. (c) Vertical section through a reef dominated by open cup-shaped archaeocyath sponges. Labrador, Canada. (d) Reef showing the development of large framework cavities. Zuune Arts, western Mongolia. Hammer head = 18 cm wide.

archaeocyaths grew between normal and storm-wave base there is no evidence of archaeo-cyath talus which would indicate growth in the turbulent zone, although some forms appear to have settled in stabilized oolitic sediments. The frequently toppled positions of most Early Cambrian reef-building metazoans suggests that they could not withstand highly energetic conditions.

Lower Cambrian reefs are mud-rich, often bearing clotted micritic fabrics. They show no succession apart from the initial stabilization of substrates by the growth of calcified cyanobacteria (Hart 1994). Where archaeocyaths were present, bioherms were often dom-inated by only one or two branching species, implying the rapid colonization and subsequent growth of only a limited number of larval spat falls (Wood *et al.* 1992a, 1993). Archaeocyaths never occupy more than 50% of the total rock volume (Fig. 3.8(c)), and they appear to have flourished in terrigenous-rich as well as pure carbonate sediments.

Rapid synsedimentary cementation is also an important feature of Lower Cambrian reefs (Fig. 3.9), and probably imparted greater rigidity to the community than the binding action of the organisms themselves. Lower Cambrian build-ups thus owe much of their wave resist-ance to geologically imparted, particularly diagenetic, phenomena.

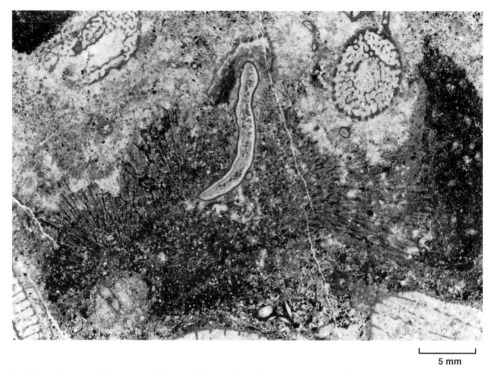

5 mm

Fig. 3.9 Early cements in Lower Cambrian reefs showing fans of needle-like crystals, formerly aragonite, growing around a cribricyath tube. Zuune Arts, western Mongolia.

Stromatactis-bearing reefs, often rich in microbial micrite, and siliceous and calcareous sponge spicules appeared in the middle Tommotian, especially in more tranquil, deep waters up to a depth of 60 m (James and Gravestock 1990). Stromatactis is of controversial origin (see Box 3.6), but may commonly be a constructional growth-framework cavity system modified by simultaneous sediment removal by winnowing (Pratt 1982, 1995).

Box 3.6: **Stromatactis**

Stromatactis is a common feature of many Palaeozoic reefs, especially in micritic or wackestone lithologies and in those parts of reefs inferred to have formed in deep or low-energy settings. Stromatactis consists of a system (or "swarm") of cement-filled laminar voids with flat or undulose lower surfaces and irregular or digitate upper surfaces, and can range from a few centimetres to over 1 m in diameter. The cements are commonly layered, often radiaxial calcite, and with or without geopetal sediment fill.

Stromatactis clearly has multiple origins, but strangely it is not known from reefs younger than the Permian. After much debate, some stromatactis structures have been shown to be shelter cavities, whose ceilings are defined by the growth of the primary reef framework, such as within microbialites and thrombolites, or under metazoans (e.g.

Playford 1984; Lees and Miller 1985; Pratt 1995; Kirkby and Hunt 1996). In many cases, further modification of the growth framework cavity system—by simultaneous sediment removal by winnowing—can be evoked, as stromatactis commonly contains large bioclasts, often attached to the roofs of the stromatactis (Pratt 1982, 1995). In cases where stromatactis structures are parallel, contorted, or fractured, it has been proposed that dilation followed by fracturing of microbially bound and lithified micrite triggered by seismic activity might be responsible (Pratt 1995). Others have suggested that stromatactis represents the cemented voids formed within burrows or a sponge framework, by the decay of such soft-bodied organisms (Bourque and Gignac 1983), or by the dissolution (Playford 1984; Kerans 1985) or recrystallization of skeletal organisms such as stromatoporoids, bryozoans, or cephalopods.

   An inorganic origin has also been proposed whereby stromatactis represents a system of cavities which developed beneath submarine cemented crusts on the surface of reefs by winnowing of uncemented sediment (e.g. Bathurst 1982). This hypothesis is supported by

1 cm

Fig. Box 3.6 Stromatactis (light-coloured) forming within a predominantly microbial reef. A solitary rugose coral can be seen upper left; some cavities have ceilings formed by frondose bryozoans, whose cross-sections appear as rows of black dots (arrowed). Upper Carboniferous, Otto Fiord Formation, Ellesmere Island, Arctic Canada. (Photograph: B.R. Pratt.)

the finding of such a process operating on the lithoherms of the Florida Straits (Neumann *et al.* 1977). Heckel (1972) suggested that stromatactis formed by dewatering of the sediment, while others regard these structures as having formed in response to slumping and compaction, or by dissolution during deep burial (Bridges and Chapman 1988).

The absence of stromatactis in non-reef-associated, fine-grained carbonates does not support an entirely non-biological origin, and the common presence of geopetal sediments and fibrous cement infill mitigates against a replacement or dissolution origin. It is likely that many stromatactis structures will prove to have a reef-framework origin, forming within frameworks dominated by inconspicuous reef-builders such as microbially bound sediment.

Many crypts are present in Lower Cambrian reefs (Fig. 3.8(d)), forming under overhangs, beneath coalesced calcified cyanobacterial bushes, or within storm-generated rubble. These housed rich cryptic communities (Fig. 3.10). Indeed, Lower Cambrian reefs were differentiated into distinct open surface and cryptic communities from their first appearance, and a well-developed and distinctive cryptos continued to be present throughout the Lower Cambrian (see Case study 3.2). These early cryptic communities were surprisingly diverse with archaeocyath sponges, calcified cyanobacteria, and a microburrowing metazoan being the most ubiquitous and abundant elements (Zhuravlev and Wood 1995). Coralomorphs, spiculate sponges, and various problematica were also common crypt dwellers, and scalloped surfaces suggestive of rasping or boring activity are locally present within some crypts.

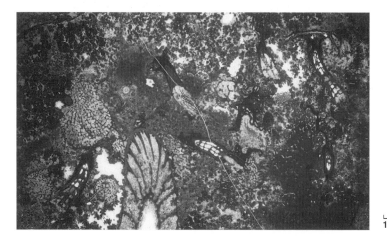

1 mm

Fig. 3.10 Photomicrograph showing a rich Lower Cambrian (Atdabanian) cryptic community from Oymuran, Middle Lena River, Siberia. The crypt is formed by the calcified cyanobacteria *Renalcis jacuticus* Korde (upper left) and *Epiphyton scapulum* Korde (upper right). These cyanobacterial shrubs have been encrusted by abundant heterotrophs: the coralomorph *Khasaktia vesicularis* Sayutina and archaeocyaths (*Neoloculicyathus sibiricus* (Sundukov), *Dictyocyathus bobrovi* Korshunov, and *Erismacoscinus oymuranensis* (A. Zhuravlev). Pockets of micrite within the crypt have been extensively microburrowed.

Case study 3.2: Radiocyath–archaeocyath–cribricyath bioherms, Zuune Arts, western Mongolia (Lower Cambrian (Atdabanian): ~535 Ma)

At Zuune Arts in western Mongolia, a series of shallow, subtidal bioherms developed on a broad, gently inclined shelf in a rapidly subsiding epicontinental sea. Of the four reef communities distinguished, the radiocyath–archaeocyath–cribricyath community—thought to have developed in relatively low-energy waters—is the most diverse.

The community is dominated by the large branching radiocyath *Girphanovella georgensis*, the branching archaeocyath *Cambrocyathellus tuberculatus*, and small mounds of the calcified cyanobacteria *Epiphyton* that coalesced to form crypts. Crypts up to 10 cm³ also formed under the branching radiocyaths, which harboured a rich biota of both juvenile and adult archaeocyaths, cribicyaths, and coralomorphs, together with calcified cyanobacteria. Series of attached individuals, such as the archaeocyaths *Robustocyathus abundans*, *Ajacicyathus mongolicus*, and *Cambrocyathellus tuberculatus*, and cribricyaths, formed chains suspended from the outer walls of radiocyaths. The convoluted mushroom-shaped archaeocyath *Okulitchicyathus* grew in a variety of habits: either free-standing in small aggregations, as the pendent cryptic form, or as a binding agent between adjacent radiocyath cups. The remaining cavity area is filled by fibrous cements, particularly botryoids (hemispheric growths of radiating, needle-like crystals), and red micritic sediment which is often clotted and microburrowed by an unknown deposit-feeder. Pockets of skeletal packstones rich in chancelloriid sclerites together with angular quartz grains are present within the bioherms. (From Wood *et al.* 1993.)

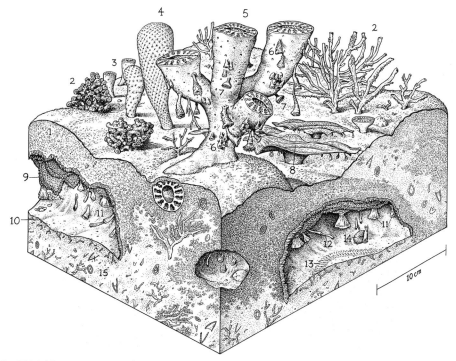

Fig. CS 3.2 (a) Reconstruction of radiocyath–archaeocyath–cribricyath community, Zuune Arts, western Mongolia (Atdabanian; ~535 Ma). 1: *Renalcis* (calcified cyanobacterium); 2: branching archaeocyath sponges; 3: Solitary cup-shaped archaeocyath sponges; 4: chancelloriid; 5: radiocyaths; 6: small archaeocyath sponges; 7: 'coralomorphs'; 8: *Okulitchicyathus* (archaeocyath sponge); 9: fibrous cement; 10: microburrows (traces of a deposit-feeder); 11: cryptic archaeocyaths and coralomorphs: 12: cribricyaths; 13: trilobite trackway; 14: cement botryoid; 15: sediment with skeletal debris. (Copyright, John Sibbick)

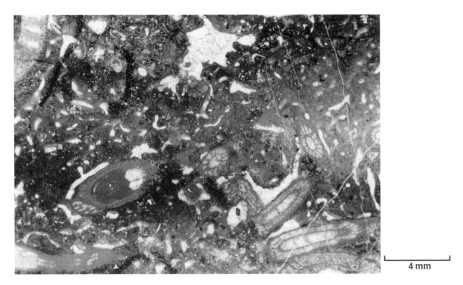

4 mm

Fig. CS 3.2 (b) Microburrows produced by an unknown lower Cambrian deposit-feeder within internal sediment. Note how the burrows follow around the edges of skeletal debris.

Crypts housed a substantial proportion of the overall reef biodiversity of Lower Cambrian reefs (Zhuravlev and Wood 1995), with individual reef communities often containing between 30 and 80 skeletal species.

The Archaeocyatha differentiated both systematically and functionally early in their history into two distinct communities. Whilst open surfaces were dominated by solitary ajacicyathids and Irregulars with branching organizations, crypts preferentially housed solitary Irregulars (archaeocyathids) and solitary chambered forms (coscinocyathids and kazachstanicyathids) (Zhuravlev and Wood 1995). Such archaeocyaths were often abundant crypt-dwellers even when uncommon or absent in the open-surface bioherm community. The archaeocyath species *Zunyicyathus* sp., *Dictyofavus* spp., *Altaicyathus notabilis*, *Polythalamia americana*, and *P. perforata* were obligate crypt-dwellers (found only within crypts), as were all cribricyaths.

In temperate areas, for example in Avalonia, orthothecimorph hyoliths and '*Ladatheca*' (coleolid) thickets produced 'worm' reefs of solitary, aggregating organisms, which were replaced by *Coleoloides* in intertidal–peritidal areas (Landing 1993). Orthothecimorph hyoliths were common in Nemakit-Daldynian and Tommotian shallow marine limestones, and are sometimes preserved by early reef cements as vertically orientated cones (Landing 1993; Riding and Zhuravlev 1995).

In summary, Lower Cambrian reefs probably occupied relatively low-energy shallow and deep subtidal seas. They were essentially mud-rich, cyanobacterial communities that housed diverse metazoan communities dominated by suspension-feeders. These communities were ecologically complex, and differentiated into distinct open-surface and cryptic communities; indeed the light requirements of many calcified cyanobacteria appear to have been sufficiently low for them to colonize near-surface cavities. The growth of sessile reef metazoans (primary consumers) must have been supported by bacterio-, zoo-, or phytoplankton (such as acritarchs, *incertae sedis* organic-walled microfossils that lack the diagnostic features of dinoflagellates) which may have represented the main primary producers of the ecosystem. In addition to calcified cyanobacteria, benthic microbial communities were probably also composed of biofilms of biodegrading and photosynthetic bacteria. The detrivorous fauna which appeared in the Nemakit-Daldynian continued to be present within these reefs, and grazing and predation pressure appears to have remained insignificant. The rigidity of Lower Cambrian reefs was enhanced substantially by the rapid growth of synsedimentary cements.

This reef ecosystem persisted for 10 Myr until the virtual demise of the archaeocyaths at the end of the Lower Cambrian. Archaeocyaths, and other sessile reef-dwelling metazoans, underwent a precipitous decline in diversity from the Botoman to early Toyonian, together with many members of the mobile benthos. This resulted in the extinction of the distinctive Lower Cambrian reef ecosystem by about 520 Ma (Section 5.3.1).

3.3 Middle Cambrian to Middle Ordovician (Late Llanvirn): 520 to ~470 Ma

During the Late Cambrian, several cratons (Laurentia, Kolyma, Siberia, Kazhachstan, and north and south China) and the northern margin of the supercontinent known as Gondwanaland continued to occupy low latitudes. By the Middle Cambrian, subsidence rates due to thermal cooling had diminished (Levy and Christie-Blick 1991) and inferred mid-ocean ridge spreading led to a global marine transgression which reached its peak during the Late Cambrian. Predominantly terrigenous sediment settings of the Middle Cambrian had given way to extensive carbonate ramp systems, and this combined with high sea levels led to the formation of substantial, epicontinental seas.

After the extinction of the archaeocyath sponges, reefs during the Middle and Late Cambrian continued to be built by a variety of bush-like and mat-forming calcified cyanobacteria in shallow subtidal areas, and predominantly by microbes in deeper waters. Indeed, with a few Cambrian exceptions (e.g. *Girvanella–Epiphyton* reefs from the earliest Late Cambrian of the Rockslide Formation, Mackenzie Mountains, Canada), deep-water reefs were entirely microbial from the Early Cambrian to Early Ordovician.

Many shallow reefs of the Middle Cambrian to Late Ordovician were built by complex intergrowths of thrombolites with early cements in subtidal waters shallowing to stromatolites with calcified cyanobacteria in more energetic intertidal to peritidal areas. *Renalcis* and *Epiphyton* formed upright bushes or cryptic growths (Fig. 3.11(a)) which often intergrew with *Girvanella* crusts. Burrowing metazoans, brachiopods, and trilobites dwelt within the reefs, and some show evidence of locally abundant micro- and macroboring. Many were small isolated bioherms with relief of a few decimetres (Figs 3.11(b) and 3.12(a)), that initiated on flat-pebble conglomerates or early lithified horizons, and often bore sharp contacts with the surrounding sediments. These reefs could, however, reach large sizes: some from the Wilberns Formation (Sunwaptan), Llano Uplift, Texas reached tens of metres in diameter (Fig. 3.12(b)), and those within the Cathedral Formation in Laurentia, and in the Changhia Formation, in northern China, were substantial. Many thrived in very wide, interior platform seas and shelf margins, where circulation may have been partly restricted (Kennard and James 1986).

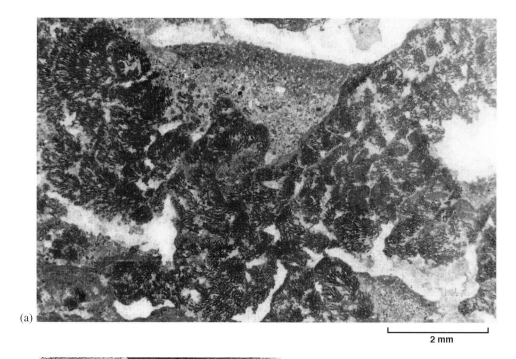

2 mm

Fig. 3.11 Middle Cambrian (Amgan) calcified cyanobacterial (*Ephiphyton*) reefs. Amgar River, south Siberian platform, Russia.
(a) Photomicrograph showing detail of *Ephiphyton* bush growing outwards to form a primary cavity that has subsequently been filled by geopetal internal sediment. (b) Coalescing, small bioherms approximately 0.5–1 m in diameter. (Photographs: A. Yu. Zhuravlev.)

(a)

(b)

A few new calcified cyanobacteria appear during this period, such as the early Middle Cambrian *Amgaella* from the Siberian Platform and the latest Late Cambrian *Seletonella* and *Mejerella* from Khazakhstan, but their role in reef-building was relatively insignificant. Lithistid demosponges, however, became more common in reef communities. Monospecific

(a)

(b)

(c)   1 cm   (d)   2 mm

Fig. 3.12 Late Cambrian reefs and biota from the Wilberns Formation, Llano Uplift, Texas. (a) Small bioherms at Squaw Creek. Bioherms are approximately 1–2 m in diameter. (Photograph: B.R. Spincer.) (b) Large reef at White's Crossing, Mason County. (Photograph: J.A.D. Dickson.) (c) Thin section showing the lithistid sponge *Wilberniscyathus* encrusted by eocrinoid holdfasts (E), and attached to and overgrown by *Girvanella* crusts (G). (Photograph: B.R. Spincer.) (d) Thin section through a small cavity formed and secondarily colonized by *Girvanella* (G). The internal sediment consists of two generations, the first is rich in terrigenous sediment (mainly quartz grains) and skeletal debris; the second has been extensively pelleted and burrowed. (Photograph: B.R. Spincer.)

aggregations of a branching anthaspidellid lithistid *Rankenella* have been reported from the latest Middle Cambrian of northern Iran, growing intimately with bacterial or algal laminated thrombolites (Hamdi *et al.* 1995). These reefs sometimes initiated upon extensive brachiopod shell beds (A. Yu. Zhuravlev and P.D. Kruse, personal observations). Common, cup-shaped lithistids (*Wilberniscyathus*) have also been described from Late Cambrian reefs from the Llano Uplift (Spencer 1996). These are sometimes encrusted by multiple generations of eocrinoid holdfasts (Fig. 3.12(c)). These reefs also contained common, but small growth framework cavities with cryptic cyanobacteria (Fig. 3.12(d)). Their palaeoecology is reconstructed in Fig. 3.13.

Abundant trilobites were present in the inter-reef areas of the Late Cambrian, particularly representatives of the Anomocaridae, Dorypygidae, Plethopeltidae. These were probably deposit feeders and scavengers. Some species, such as *Stigmacephaloides curvabilis*, appear to be restricted to reef settings (B.R. Spencer, unpublished PhD thesis). The first chitons appeared in Late Cambrian (Franconian) and together with molluscs were possibly the first *grazers* (herbivores which crop very close to a substrate, so ingesting substantial portions of living plant tissue, associated small invertebrates, as well as underlying structure) (Runnegar *et al.* 1979).

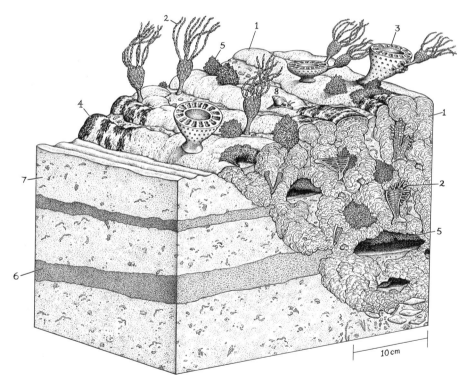

Fig. 3.13 Reconstruction of Upper Cambrian reef community. Llano Uplift, Texas. 1: Thrombolite; 2: eocrinoids; 3: lithistid sponges (*Wilberniscyathus*); 4: calcified cyanobacterial mats (*Girvanella*); 5: calcified cyanobacterial bushes (*Renalcis*); 6: horizons rich in ooids; 7: wackestone/packstone sediments; 8: gastropod. (Modified from Spencer 1998; copyright, John Sibbick.)

Lower Ordovician shallow-water reefs were essentially a continuation of those ecologies established in the Late Cambrian. Reefs were generally small, but extensive examples are known from the early Tremadoc of Newfoundland (Case study 3.3). Many of these reefs were mud-rich (up to 75%) with limited relief of a few decimetres to one metre, and many lack substantial talus. As in the Late Cambrian, subtidal reefs were built by free-standing thrombolites, with stromatolites dominating in shallow intertidal to peritidal areas. The thrombolites also occurred as postmortem encrustations, which served to impart rigidity together with pervasive early cementation. The hard substrates offered by these algal mounds supported a diverse metazoan fauna of stalked and cup-shaped solitary metazoans, together with some new encrusting metazoans (Fig. 3.14). Most of these were probably sponges, such as the lithistid *Archaeoscyphia*, the receptaculitid *Calathium*, and the labechiid (?) stromatoporoid sponge *Pulchrilamina*. Also important were crinoids and the tabulate coral *Lichenaria*, which may be an ancestor for later Ordovician tabulate corals (Scrutton 1984; Box 3.7). *Renalcis* was present, often in cryptic niches, and new groups of algae also appeared during this time, as did trepostome bryozoans, and new dwelling and vagile (mobile) groups, including calcitic brachiopods, cephalopods, rostraconchs, gastropods, and trilobites. Thrombolite mounds often bear burrowed internal sediments, testament to deposit-feeding activities of this fauna which used the reef cavities for shelter (Pratt and James 1982). Abundant and diverse gastropods were probably epifaunal grazers, resulting in no apparent detriment to algal mat growth. Cephalopods were sediment–water interface carnivores.

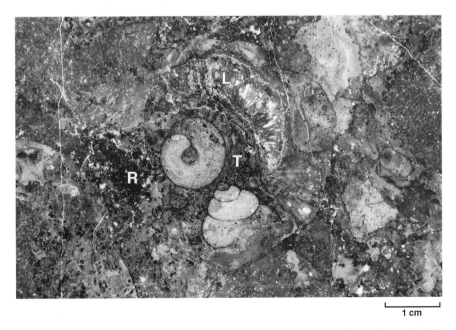

1 cm

Fig. 3.14   Lower Ordovician (Tremadocian) reef; *Renalcis* (R), the tabulate coral *Lichenaria* (L), and thrombolites (T), with gastropods lodged in framework cavities. (Photomicrograph: B.R. Pratt.)

Case study 3.3: St George Group, Newfoundland, Canada (Early Ordovician (Tremadocian): 495 Ma)

The Tremadocian reefs of Newfoundland range from stromatolites in intertidal areas to small thrombolitic heads and extensive banks and complexes many metres thick and hundreds of metres in lateral extent in subtidal waters.

The thrombolites are associated with two different metazoan communities. In shallow subtidal areas, scattered archaeoscythiid (lithistid) sponges grew in a variety of morphologies and reached up to 0.5 m in size. These were associated with receptaculids and the encrusting stromatoporoid *Pulchrilamina* which formed coalesced mounds up to 0.6 m thick. More complex communities grew in deeper waters, composed of a thrombolite–coral–*Renalcis* association together with the tabulate coral *Lichenaria*, and an associated vagile fauna of trilobites, nautiloids, gastropods, and rostraconchs, together with sessile crinoids and brachiopods. *Renalcis* formed cryptic encrustations within framework cavities as did laminated microbialites. The thrombolites and *Renalcis* were able to roof over cavities up to 20 cm wide, this then became infilled with internal sediment which supported an abundant burrowing fauna. (From Pratt and James 1982.)

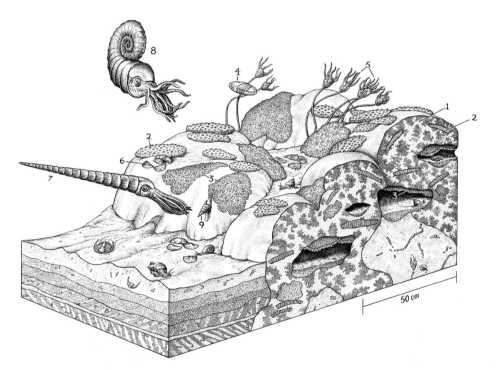

Fig. CS 3.3   Lower Ordovician reef reconstruction: Thrombolite–*Lichenaria*–*Renalcis* community. 1: Living algal mats and thrombolite heads; 2: *Lichenaria* (tabulate coral); 3: *Renalcis* (calcified cyanobacterium); 4: swimming trilobite; 5: crinoids; 6: brachiopods; 7: straight nautiloid; 8: coiled nautiloid; 9: grazing gastropod. (Modified from Pratt and James 1982; copyright, John Sibbick.)

Pulchrilamina was often multilayered, reaching 50 cm thick and 25 cm wide, and sometimes dominated on the top of reefs, presumably in the shallowest waters (Toomey and Nitecki 1979). *Pulchrilamina* was able to encrust soft sediment and would therefore act as a binding agent over unconsolidated sediment. *Archaeoschyphia* reached heights of 25 cm and was up to 13 cm wide, whilst *Calathium* never grew more than 15 cm high. Such dimensions represent similar low tiering (height above the substrate) to Lower Cambrian reef communities, although some crinoids extended well beyond this height.

Similar microbial–calcified cyanobacterial–skeletal metazoan communities persisted from the Tremadoc to the Llanvirn. Lithistid-dominated reefs are known from Argentina (Cañas 1995) as well as North and South America (Toomey and Nitecki 1979), and Late Tremadocian reefs from Hubei, China (Zhu *et al.* 1993). This suggests that this Lower Ordovician reef ecosystem was widespread.

Box 3.7: **Post-Cambrian Palaeozoic corals (Class Anthozoa, subclass Zooantharia)**

Seven orders of Palaeozoic corals are currently recognized, of which the Rugosa (Mid Ordovician–Permian) and Tabulata (Early Ordovician–Permian) are the most important reef-associated groups. These orders did not have a common skeletal ancestor, and probably arose as separate skeletonization events from the Zoanthiniaria. The calcitic skeletons of the Rugosa and Tabulata are predominantly of solid and non-porous construction.

Each individual or polyp resided in a cup-like depression, known as a *calyx*. The polyp mouth was central, surrounded by tentacles. The Rugosa are defined by *septa* (plates which radiate from the base of the calyx), that are serially inserted in four sectors (quadrants). The rugose corals may be solitary or colonial, and were probably not ancestral to the Scleractinian corals (mid-Triassic to Recent). Tabulate corals are exclusively colonial, and normally show no or only weakly developed septa.

Almost all rugose and tabulate corals developed a thin, solid wall, known as an epitheca or holotheca surrounding and enclosing the skeletal surface up to the edges of the corallites. Most were effectively free-living on or partially within soft substrates, with some solitary rugose corals showing adaptations to this mode of life such as outgrowths (sometimes known as fixing structures) from the calyx for stabilization, or flattened morphologies.

2 cm

Fig. Box 3.7 (a)   A tabulate coral (upper surface).

2 cm

Fig. Box 3.7 (b) A colonial rugosan coral.

Deep-water mud-mounds, such as those from the Ibexian Survey Peak Formation, southern Rocky Mountains of western Canada, are also composed of digitate thrombolites up to 1 cm wide, intergrown with anthaspidellid sponges. The thrombolites contain *Girvanella* and clotted micrite (Pratt 1995). Stromatactis-bearing mud-mounds also remain common throughout this interval.

3.4   Middle Ordovician to Upper Devonian (Frasnian): ~470–~360 Ma

During the Middle Ordovician, many continental margins were destroyed by plate collision and mountain-building, while major intracratonic downwarps began to develop. Reef formation became widespread in these new basins, whilst continuing on some passive margins.

The Middle Ordovician records a significant radiation of marine skeletal biota; there is a near-threefold increase in the number of known families (Sepkoski 1993; Droser *et al.* 1996). During the early Llanvirn, bryozoans became important in the tropics (Box 3.8): for example, 12 higher taxa are recognized within the Llanvirnian Chazy Group (which outcrops from Alabama to Quebec). Tabulate and rugose corals, and stromatoporoid sponges radiated during the middle to late Llanvirn, although most diversification occurred in level-bottom, soft-substrate communities. By the Late Llanvirn, at least 13 genera of stromatoporoids (Webby 1994), and six genera of tabulate corals (Scrutton 1984) had evolved. Rugose corals appeared in the Caradoc.

As a result of this major radiation, the middle Palaeozoic is widely cited as a time when algal communities were replaced by encrusting, heavily skeletonized metazoans in shallow waters (Fagerstrom 1987). However, thrombolitic, stromatolitic, microbial, and calcified cyanobacterial communities continue to be associated with reefs throughout this interval.

Box 3.8: **Reef-associated bryozoans**

Bryozoans grow as colonies of minute, clonally produced modular units termed zooids encased within chambers that are generally calcified. Bryozoans show a diverse range of growth forms including erect (branching), free-living, and encrusting types. Two major classes are known, the stenolaemates and gymnolaemates. (A third class—the phylactolaemates—is freshwater and, being uncalcified, lacks a significant fossil record.)

Stenolaemates appeared in the Lower Ordovician, and persisted throughout the Palaeozoic with a few representatives from several orders surviving the end-Permian extinction. Rediversification occurred in the Jurassic, entirely within the possibly polyphyletic group known as the cyclostomes. The major group within the gymnolaemates, the cheilostomes, appeared in the Jurassic and underwent an explosive radiation during the Cretaceous–Eocene. Three orders are recognized, which are almost certainly not monophyletic (McKinney and Jackson 1989).

By the Middle Ordovician, bryozoans had become important in reef communities—indeed all major taxa of bryozoans have been at some time associated with reefs (Cuffey 1974). Three major groups of bryozoans, however, have been particularly important:

(1) massive, encrusting and branching trepostomes and cystoporates during the Ordovician to Devonian;

(2) large branching and frondose fenestrates (including fenestellids) and acanthocladids (Cryptostomata) in both small reefs and deep mud-mounds during the late Carboniferous–Permian; and

(3) encrusting and massive cheilostomes during the late Cenozoic.

1 cm

Fig. Box 3.8 Frondose bryozoans such as this fenestellid became important members of reef communities during the Permo-Carboniferous.

Llanvirnian reefs were dominated by soft-sediment encrusting bryozoans (trepostome and cyclostomes) and the tabulate coral *Lichenaria*, although receptaculitids were locally import-ant. Other persisting taxa from the Late Cambrian remained common in shallow-water reefs, such as *Girvanella*, lithistid sponges (mainly archaeoscyphiids and anthaspidellids), and solenoporacean algae (calcareous red algae). Reef dwellers include common gastropods, trilobites, molluscs, crinoids, and ostracods.

Deep-water, mud-rich reefs have been documented from ramps and deep shelves, but apart from lithistid sponges, crinoids, and encrusting bryozoans, other sessile metazoans appear to be absent (Pratt 1995). These reefs are often constructed by stromatolites or laminar, clotted thrombolites with recognizable *Girvanella* and cryptic communities of *Epiphyton* and *Renalcis*.

By middle–late Llanvirn times, reefs became larger—up to 9 m thick and 150 m diameter. *Solenopora*, stromatoporoids, lithistid sponges, or bryozoans became very common, with individual stromatoporoids reaching up to 3 m diameter. The first bryozoan reefs appeared in deep, cold waters, perhaps related to the presence of microbial mounds, although unusual shallow reefs dominated by trepostome bryozoans, tabulate corals, and abundant eocrinoid holdfasts with microbial cavity encrustations are known from the Laval Formation in Quebec (Pratt 1989*a*). Shallow-water reefs became widespread by the late Middle Ordovician; they are known from Laurentia, Baltia, Siberia, Kazakhstan, and Australia and show relatively high community diversities. There is a general replacement of passive filter- and suspensions-feeders by active suspension-feeders (Bottjer and Ausich 1986) during this time, with an increasing tiering of communities (Ausich and Bottjer 1982). Growth-framework cavities within Middle Ordovician reefs preserve a diverse encrusting fauna (Kobluk 1981*b*).

By the late Ordovician (Caradoc), colonial rugose corals and tabulates had rapidly diversified in shallow waters to develop the first coral patch reefs, together with stromato-poroids and also a suite of characteristic algae known as the *Wetheredella* group. These groups may have displaced the bryozoans. Rich and diverse coral thickets were present, which together with receptaculitids and brachiopods formed extensive communities. In the late Caradocian, a middle Palaeozoic reef ecosystem dominated by tabulate and rugose corals, stromatoporoids, and other sponges, calcareous red algae, *Renalcis* and *Girvanella*, and other non-skeletal algae became established. Many articulate brachiopods lived attached to debris on relatively undisturbed substrates, both associated with reefs and level-bottom communities. Crinoids clustered around the lower parts and peripheries of reefs. A major diversification in macroboring organisms also occurred during the Ordovician—from one taxa described from the Lower Ordovician, to five by the Middle Ordovician, and seven by the Late Ordovician (Kobluk *et al.* 1978). Biota characteristic of the Silurian appeared in the Ashgill, especially during the interglacial interval of the Gamachian/Hirnantian (Copper 1997). Large pinnacle reefs built by microbialites are also known from the Late Ordovician (Webby 1994).

Palaeozoic corals and stromatoporoids were not always contributors to reef frameworks. They grew as branching, laminar, tabular, or massive domal growth-forms which colonized soft, firm, or lithified, substrates. Many formed assemblages composed of isolated individu-als rather than an intergrown or interlocked framework, and it is doubtful whether this fauna was able to grow in the surf zone as do modern scleractinian corals (Wells 1957). Some tab-ulate corals and stromatoporoids could, however, reach substantial sizes, with the largest described being several metres in diameter. Sandy and gravel-rich sediments associated with reefs often contain reworked stromatoporoids several centimetres high, which were perhaps exhumed during storms.

The double latest Ordovician cooling, and subsequent glaciation and sea-level fall, appeared to have little effect on reefs (Section 5.3.2; Copper 1997), although early Silurian reefs are limited in both size and occurrence (P.M. Sheehan 1985; Copper 1988). The beginning of the Silurian is marked by a widespread transgression.

Silurian reefs were abundant and diverse, occupying a variety of settings from nearshore, to mid-, outer-shelf, and to deep-water settings, marking a period of some 75 Myr of globally warm climates when high sea-level stands flooded the continents and faunas were cosmopolitan in distribution. Reefs were widespread by the middle Silurian, with the largest complexes known in the Great Lakes area, the circum-Arctic Cornwallis–Greenland regions, and the southern Urals where a reef belt extended for more than 2200 km (Copper and Brunton 1991).

Few biota were apparently restricted to reef environments during the Silurian (Watkins 1993), although illaeniid trilobites may have been exclusively reef-dwelling (Lowenstam 1957). Microbialite continued to be an important reef-builder and agent of framework preservation (Bourque 1989; Webb 1996). Calcified cyanobacteria, stromatolites/thrombolites, and algae grew in restricted environments, low-diversity coral–bryozoan communities occupied marginal settings, and diverse stromatoporoids, tabulate corals, bryozoans, crinoids, *Solenopora*, and various calcified cyanobacteria, together with thromboids and stromatolites formed reefs in normal marine shallow-water areas (Clough and Blodgett 1989). Deep-water mud-mounds with stromatactis were fairly widespread and sometimes exhibit a vertical zonation in response to shallowing, culminating in stromatoporoid-rich reefs. Many are associated with cyanobacteria and lithistid sponges (Brunton and Copper 1994), such as those known from the Ludlovian Douro Formation, Somerset Island in Arctic Canada (Narbonne and Dixon 1984). Distinctive stromatolitic–sphinctozoan (aphrosalpingid) sponge communities formed both small patch reefs and extensive barrier reefs in continental and island arc complexes during the mid-Ludlovian, now exposed in SE and SW Alaska and the Ural Mountains (Soja 1994; Soja and Antoshkina 1997).

Reef dwellers in shallow marine reefs include rugose corals, brachiopods, bryozoans, crinoids, molluscs, dascycladacean algae, and trilobites, and mottled textures within the reef matrix suggest the presence of bioturbators during reef growth. Such reefs usually had low synoptic relief, up to a maximum of 5 m. Cavities formed beneath laminar and convex massive stromatoporoids, from which cryptic crinoid, bryozoan, and brachiopod biotas have been described (Spjeldnaes 1974; Copper 1996).

As an interesting aside, hydrothermal vent communities with monoplacophoran molluscs, inarticulate brachiopods, vestimentiform and polychaete tube-worms are known from the Silurian (Little *et al.* 1997). Brachiopods and monoplacophorans were common in early Palaeozoic vent communities, suggesting that modern vent communities are not refuges for Silurian shelly vent taxa, and that some temporal shifts in vent-community composition have occurred since the Palaeozoic.

Llandovery–Wenlock reefs developed mainly in intracratonic basins and some settings at the edge of cratons in northern Europe (see Case study 3.4), North America, and Greenland (Brunton and Copper 1994). These reefs often grew within extensive meadows of crinoids in carbonate ramp settings. By the end of the Llandovery, reefs expanded to a global distribution

5 cm

Fig. 3.15 Polished slab cut vertically through a Silurian (Wenlockian) reef, showing labechiid stromatoporoids arching over a crinoidal grainstone substrate.

and the diversity of reef communities increased. Brunton *et al.* (1997), have documented three major reef-building episodes, two in the Wenlockian and one during the Ludlovian. The Wenlockian reefs were tabulate–stromatoporoid sponge-dominated (Fig. 3.15), whereas the Ludlovian communities were dominated by diverse calcareous and siliceous sponges, and algae.

Case study 3.4: Wenlock, Shropshire, England (Silurian (Wenlockian): ~ 430 Ma)

The Wenlockian reefs of Shropshire, England, have been well documented (Scoffin 1971; 1972*b*). In this area, abundant patch reefs averaging 12 m wide and 5 m thick grew on a level shelf on the seaward fringe of a carbonate platform, showing elongation parallel to the platform edge. The reefs are flanked by thin-bedded, fine-grained, crinoidal packstones and micrites.

The reefs grew as a result of the binding activity of low-relief domal and laminar skeletal metazoans and microbialites. Tabulate corals (*Halysites*, *Heliolites*, and *Favosites*)—which account for 10–20% of the rock volume—bryozoans (*Hallopora* and *Rhombopora*), and stromatoporoid sponges (*Stromatopora* and *Actinostroma*) grew aggregated on slightly raised mounds of coarse crinoid debris on an unlithified, but stable sea floor. These metazoans were encrusted by stromatolites, tabulates (*Thecia* and *Alveolites*), stromatoporoids (*Labechia*), and algae (*Girvanella*, *Wetheredella*, and *Rothpletzella*). The stromatolites show banded and peloidal textures, and formed elongate domal structures several centimetres thick on the upper surfaces on metazoan skeletons or directly upon the micrite sea floor. Some encrusted primary framework cavities. Amongst and within this framework dwelled crinoids, ostracods, brachiopods, and gastropods. There is some patchy development of a cryptic biota, especially fistuliporid bryozoans.

Some metazoans grew directly upon soft substrates, and the presence of abundant gastropod burrows penetrating to a depth of 60 cm in inter-reef areas is supporting evidence that the reef grew on a stable, soft substrate. The presence of truncated micrite within branches of broken skeletal organisms, however, suggests early lithification of reef matrix, although there is no evidence for cavity cementation prior to internal sediment accumulation. Micrite sediment continuously infiltrated the framework during reef growth, swamping organisms at the reef margins. The fine grain size and grading of internal sediments is evidence for an open framework with restricted cavities which were progressively sealed as the matrix accumulated.

These reefs are thought to have developed in quiet, stable waters of a maximum 30 m depth. That conditions were tranquil is evidenced by the *in-situ* nature of the vast majority of sessile reef biota (which had no means of secure attachment to the sea floor), the lack of traction current structures, the accumulation of fine-grained internal sediment near the reef surface, and the absence of talus formation except for limited storm-talus development. Reef surfaces were irregular, but generally horizontal to convex, and had low relief of between 0.5 m and 3 m. The reefs intergrew with mechanically deposited sediments at their periphery, and flourished during times when terrigeneous-derived, clay-minerals were not abundant. The reefs were, however, periodically blanketed by clay, usually bentonites, which resulted in death of the community.

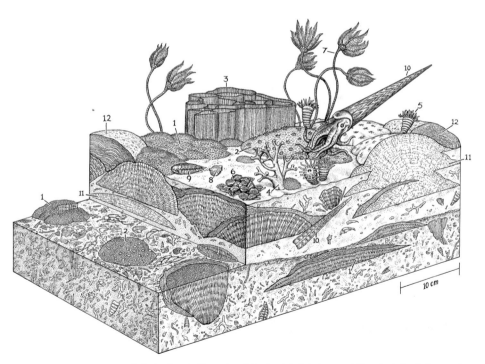

Fig. CS 3.4 Reconstruction of Silurian (Wenlock) patch reef, England. 1: Tabulate coral (*Favosites*); 2: tabulate coral (*Heliolites*); 3: tabulate coral (*Halysites*); 4: bryozoan (*Hallopora*); 5: streptelasmatid rugose coral; 6: spirifid brachiopod (*Atrypa*); 7: crinoid; 8: brachiopod (*Leptaena*); 9: trilobite (*Dalmanites*); 10: orthocone nautiloid; 11: stromatoporoid (*Actinostroma*); 12: thrombolite. (Modified from McKerrow 1978; copyright, John Sibbick.)

The Devonian, in particular the mid–late Devonian (Givetian–Frasnian), is considered to represent possibly the largest global expansion of reefs in the Phanerozoic (Copper 1989). During this interval, the climate was equable and sea level high. Extensive reef tracts, in size exceeding those of the present day, are known from Canada and central Asia, and large reefs are also found throughout Europe (Gischler 1995), western North Africa, South China, south-east Asia, and most famously from the Canning Basin, Australia (see Case study 3.5).

Mid–Late Devonian reefs were diverse and widespread, and dominated volumetrically by microbialites, calcified cyanobacteria (*Renalcis*), and large, heavily calcified metazoans—stromatoporoid sponges, tabulate corals—together with calcified algae such as *Sphaerocodium*, and, to a far lesser extent, rugose corals. Receptaculitids and lithistid sponges are common in foreslope or deeper water reef communities or in low-energy shallow settings (see Case study 3.6). The faunal diversity of shallow-water reefs is high, with metazoan reef-builders showing a tremendous variety of complex morphologies (e.g. Fig. 3.16). Brachiopods were common, sometimes nestling within crypts, but especially attached to the undersurface of laminar/ tabular stromatoporoids or tabulate (often alveolitid) corals (Szulczewski and Racki 1981; Copper 1996). Some were cementers (e.g. *Davidsonia* and *Rugodavidsonia*); others such as atrypids attached by means of spines (Fig. 3.17). Some cryptic tabulate corals are known from the undersurfaces of platy and domal stromatoporoid from Lower Devonian reefs (Pratt 1989*b*). A further radiation of bioeroders occurred in the middle to Late Devonian—11 boring ichnogenera have been identified from the Lower Devonian reefs (Kobluk *et al.* 1978).

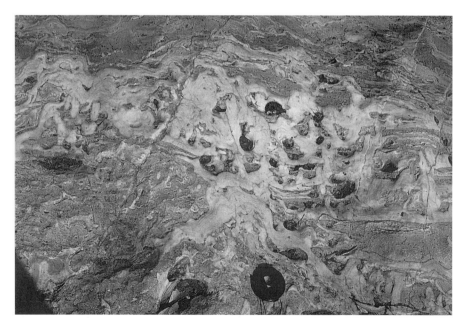

Fig. 3.16 Large stromatoporoid (?*Actinostroma* sp.) growing within back-reef sediments. This individual produced columns in its central part and lateral outgrowths that hovered over the sediment. The lower surfaces of these outgrowths have been encrusted by *Renalcis* algae and infilled by later generations of sediment. Upper Devonian (Frasnian), Windjana Gorge, Canning Basin, western Australia. Lens cap = 52 mm.

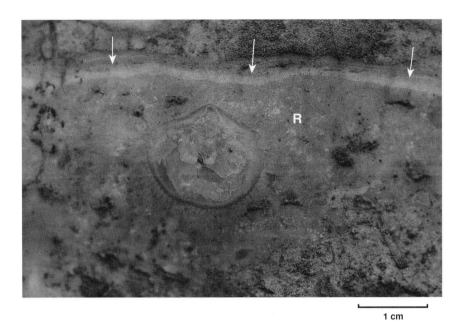

1 cm

Fig. 3.17   Cryptic spiny atrypid brachiopod attached to pendent *Renalcis* (R), which is attached to the undersurface of *Stachyodes australe* (arrowed). Upper Devonian (Frasnian), Geikie Gorge, Canning Basin, western Australia.

Because so many Devonian reef tracts are subsurface and/or dolomitized, exact reef geometry and ecology are poorly known for many examples. However, it is clear that early marine cements are volumetrically important in many Frasnian reefs, in part reflecting the abundance of substantial framework cavities formed by large skeletal metazoans: up to 50% cement by volume is known from the Golden Spike, Canada, and synsedimentary radiaxial cements account for 20–50% by volume for the reefs of the Canning Basin (Kerans *et al.* 1986; Hurley and Lohmann 1989; Figs 3.18(a) and (b)). The importance of early cementation in the Canning Basin is manifest by the numerous, huge reef talus blocks (up to 200 m) incorporated into fore-reef strata and basin debris flows (Fig. 3.18(c)), and the extensive development of neptunian dykes and other fractures subparallel to the reef front. Likewise, substantial growth of microbialites as free-standing mounds, heads, and columns (Figs 3.19(a)–(c)), or as encrusting components (Fig. 3.19(d)), are volumetrically important components of many Frasnian reefs (Mountjoy and Riding 1981; Clough and Blodgett 1989; Tsein 1994*a*, *b*; see Case study 3.5), as are growths of the calcified cyanobacteria *Renalcis* (Figs 3.18(b) and (d)) and *Rothpletzella* [= *Sphaerocodium*], (Mountjoy and Riding 1981; Webb 1996). The presence of spur-and-groove structures is testament to the growth to sea level and wave-resistant capacity of these reefs (Fig. 3.20).

Eifelian and Early Givetian mud-mounds, ridges, and 'atolls' are known, with particularly spectacular exhumed examples described from northern Africa (Brachert *et al.* 1992; Wendt 1993; Wendt *et al.* 1997). In the Ahnet Basin, southern Algeria, these mud-mounds formed in deeper water settings (100–200 m) and achieved heights up to 100 m (Wendt *et al.* 1997) (Fig. 3.21). These mud-mounds formed rapidly in approximately a few 100 kyr, and probably initiated upon crinoid and small tabulate coral thickets. The reefs are composed of

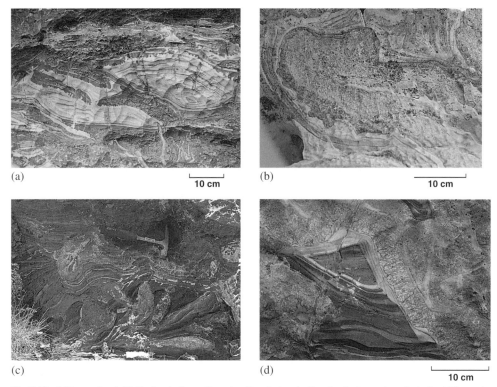

(a)   10 cm   (b)   10 cm

(c)   (d)   10 cm

Fig. 3.18   Evidence of early lithification in Upper Devonian (Frasnian) reefs, Canning Basin, western Australia. (a) Layered fibrous calcites with primary cavities formed beneath the thin stromatoporoid *Stachyodes australe*. Geikie Gorge. (b) Thick crusts of *Renalcis* followed by layered calcite cements growing around the plate of a foliaceous stromatoporoid. Windjana Gorge. (c) Reef-derived talus blocks with the marginal slope. Macintyre Hills. Hammer = 32 cm long. (d) A primary cavity formed by thick crusts of fibrous cements growing around the calcified cyanobacterium *Renalcis*, which is itself encrusting a branching stromatoporoid sponge (*Stachyodes* sp.), right. Windjana Gorge.

massive wackestones, and contain abundant stromatactis but no large skeletal metazoans or green algae, although thickets of slope-hugging crinoids, small *in-situ* corals, trilobites, disarticulated crinoids, brachiopods, and rare bryozoans and sponge spicules are found. The lack of compactional features or bioturbation, the steepness of the mound sides, and the presence of encrusting organisms are all indicative of early lithification. The presence of some peloidal textures and thin, cyanobacterial micritic rims suggest that the mounds probably formed by microbial precipitation together with the growth of calcified cyanobacteria.

In Morocco, asymmetrical mud-mounds up to 250 m high formed on gently sloping ramps at moderate depths of several tens of metres in a transition zone between carbonate shelf and pelagic platform (Wendt 1993); they were contemporaneous with shallow-water stromatoporoid/coral reefs. These mud-mounds also formed rapidly, in one or two condodont zones (a few 100 kyr). The tops of the mounds were maintained at relatively shallow-water depth and kept pace with subsistence. Crinoids and sponge spicules are abundant, with subsidiary, small branching colonies of tabulate corals, solitary rugose corals. Large stromatoporoids grew on the tops of the largest mounds, and stromatactis-like structures are common. Early lithification of the mounds is also apparent, perhaps due to microbial activity, with substantial brachiopod accumulations forming in large cavities.

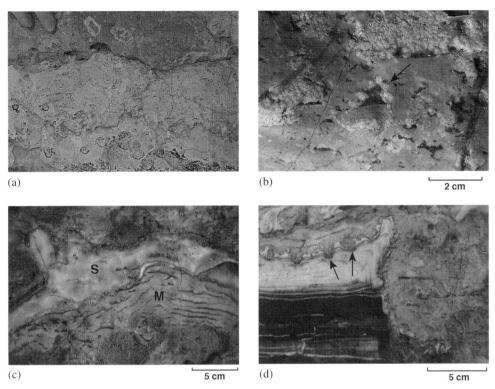

(a)

(b) ⊢ 2 cm ⊣

(c) ⊢ 5 cm ⊣

(d) ⊢ 5 cm ⊣

Fig. 3.19 Microbialite structures from the Upper Devonian (Frasnian) Canning Basin reefs, Windjana Gorge. (a) The domal upper surfaces of microbialite, overlain by a bed rich in oncolites. (b) Detail of microbialite, showing small cavities lined by cryptic, pendent growths of the calcified cyanobacterium *Renalcis* (arrowed), and later filled with geopetal sediment and cements. (c) A small mound of microbialite (M) with well-developed laminar stromatactis that has been encrusted by a stromatoporoid sponge (S). (d) Microbialite forming the ceiling and walls of a framework crypt. The microbialite had been encrusted by bushy growths of the calcified cyanobacterium *Renalcis* (arrowed), before the cavity was infilled by layered internal sediment and fibrous cements.

Fig. 3.20 Growth under wave impact: spur-and-groove structures in an Upper Devonian atoll. The spurs are separated by grooves 20–30 m wide. Teichert Hills, Canning Basin, western Australia.

Fig. 3.21 Spectacular Middle Devonian (Givetian) mud-mounds that reach 30 m in height. Azzel Matti, Algeria. (Photograph: J. Wendt.)

Case study 3.5: Canning Basin, Australia (Upper Devonian (Frasnian): ~365 Ma)

The Canning Basin reef complexes (fringing, barrier, patch-reefs, and atolls) developed on the shallow north-eastern side of the fault-bounded flanks of the north-west–south-east trending Fitzroy Trough (George *et al.* 1994), which formed during significant crustal extension in Middle and Late Devonian times. The reefs are spectacularly exposed in a wide belt which extends for some 350 km and are up to 50 km wide, which preserves reef communities of Givetian to Famennian age. These reefs record a history of some 15 million years of continuous reef-building reflecting on a fault-controlled, reef-rimmed platform fringing the adjacent mountainous Proterozoic landmass of the Kimberley Block (Playford *et al.* 1989).

Many different communities grew within the Frasnian reef complexes of the Canning Basin: small branching stromatoporoids (*Stachyodes* and *Amphipora*) flourished in the relatively sheltered, low-energy areas behind the crest-flat margin, and in lagoonal patch reefs. Domal (e.g. *Actinostroma*), tabular, and laminar stromatoporoids together with microbial and algal fabrics are also characteristic of the back-reef area. Abundant laminar to domal stromatoporoids and lithistid sponges occur in particular beds within the slope–marginal sediments. Due to relative inaccessibility and poor outcrop, the reef-margin is not well described, but appears to be dominated by the algae *Rothpletzella*, *Renalcis*, *Sphaerocodium*, and microbial micrite, together with abundant, large tubular, lithistid sponges.

Back-reef sediments show distinctive shallowing-upwards cycles which are interpreted to reflect the lateral zonation of four communities (Fig. CS 3.5(b)):

1. The onset of carbonate sedimentation, probably induced by deepening, is marked by the colonization of large stromatoporoids on stabilized coarse clastic sediment. These include domal (*Actinostroma* sp.), and inferred whorl-forming *foliaceous* (?*Actinostroma* sp.), and platy-columnar growth forms. Many appear to have initiated upon crinoids, and their considerable elevation above the substrate is reflected by the accumulation of the same geopetal infill within their tiered growth. These stromatoporoids were heavily encrusted by *Renalcis* and microbialite, particularly on sheltered undersurfaces (see Fig. 3.18(b)).

2. The next zone is characterized by thickets of the branching stromatoporoid (*Stachyodes* sp.), and thin, laminar stromatoporoids that arched over the sediment. These show either encrusting collars of *Renalcis*, or cryptic *Renalcis* attached to sheltered undersurfaces.

3. This was followed by the extensive growth of large mounds of microbialite, which were encrusted by the stromatoporoid (?*Clathrocoilona spissa*).

4. Columnar heads of stromatolites develop as the sediments became more shallow, energetic, and dominated by very coarse sands, together with patches of large oncolites (coated grains with irregular and overlapping laminae, which were probably also microbially mediated) and large gastropods.

Marginal slope communities show a further range of communities. Here, the reef rock fabrics are dominated by large-scale (0.2–1.5 m in width) cavities, which are primary growth framework cavities that formed by a variety of domal, tabular, or laminar stromatoporoid sponges. Of particular note are those formed by single, very thin (2–8 mm), laminar forms (mainly *Stachyodes australe*), which formed arching, hollow domes up to 0.3 m in height and 1.5 m in diameter over the sediment surface to enclose flat-based cavities (Figure CS 3.5(c)). The undersurface of these stromatoporoids often supports a cryptic community dominated by abundant, pendent growth of the calcified cyanobacterium *Renalcis*, which imparts an irregular, stromatactis-like texture to the upper surface of the cavity. Such an ecology yields the tabular stromatoporoid–*Renalcis* fabric described ubiquitously from the Canning Basin reefs. Attached to the *Renalcis* are rare, cryptic lithistid sponges and spiny atrypid brachiopods. Remaining cavity space is filled with early marine, finely-banded, fibrous cements (particularly radiaxial calcite) and interbedded with often multiple generations of geopetal sediment containing peloids and ostracod debris.

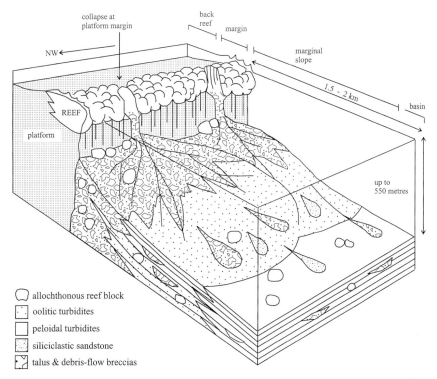

Fig. CS 3.5 (a)   Schematic reconstruction of the Late Devonian reef margin, Canning Basin, western Australia. (Modified from George *et al.* 1997.)

Fig. CS 3.5 (b)   Reconstruction of an Upper Devonian shallowing-upwards back-reef cycle: Windjana Gorge, Canning Basin, western Australia. 1: Stromatolites; 2: domal stromatoporoid (*Actinostroma*); 3: inferred whorl-forming foliaceous stromatoporoids (?*Actinostroma* sp.); 4: calcified cyanobacterium (*Renalcis*); 5: fibrous cement; 6: geopetal sediment infill; 7: platy stromatoporoid; 8: crinoids; 9: branching stromatoporoid (*Stachyodes* sp.); 10: laminar stromatoporoid; 11: Encrusting stromatoporoid (?*Clathrocoilona spissa*); 12: microbialite; 13: coarse clastic sediment; 14: gastropods; 15: oncolites. (Copyright, John Sibbick.)

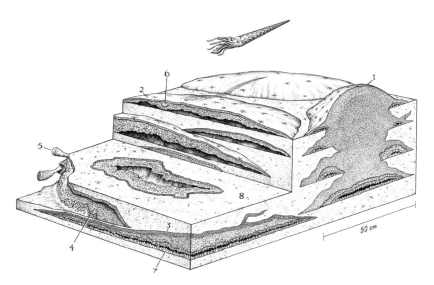

Fig. CS 3.5 (c)   Reconstruction of an Upper Devonian marginal-slope stromatoporoid community: Geikie Gorge, Canning Basin, western Australia. 1: Domal stromatoporoid (e.g. *Actinostroma* sp.); 2: laminar stromatoporoid (e.g. *Stachyodes australe*); 3: tabular stromatoporoid; 4: *Renalcis* (calcified cyanobacterium); 5: stalked lithistid sponge; 6: spiny atrypid brachiopod; 7: radiaxial fibrous calcite cement; 8: sediment. (Copyright, John Sibbick.)

Fig. CS 3.5 (d) Large primary cavities filled with early cements formed below laminar and domal stromatoporoids. Marginal-slope, Geikie Gorge. Late Devonian (Frasnian).

Fig. CS 3.5 (e) A finely branching stromatoporoid, *Stachyodes* sp. with collars of the encrusting calcified cyanobacterium *Renalcis*. Back-reef, Windjana Gorge. Late Devonian (Frasnian).

10 cm

Fig. CS 3.5 (f)   Large lithistid sponges surrounded by microbialite near the reef margin. Geikie Gorge, Western Australia. Late Devonian (Frasnian).

Case study 3.6: Anatomy of a Devonian (Frasnian) patch reef: Glenister Knoll, Laidlaw Range, Bugle Gap, Canning Basin, western Australia (~375 Ma)

Glenister Knolls are a series of patch reefs some 250 × 100 × 50 m that outcrop at the end of the Laidlaw Range. These reefs show a strong, vertical zonation of communities. Abundant orthocone nautiloids and goniatites are found in the basinal sediments near the reef sides. Also near the base of the reef were crinoid thickets, associated with large brachiopods, lithistids, and receptaculitids, bound with early fibrous cements. Exposed at the reef bases are mounds of columnar stromatolites several metres in diameter which grew with attached crinoids and receptaculitids, together with large gastropods. Domal and laminar stromatoporoids form the bulk of the reef, and thickets of branching stromatoporoids and rugosans grew in more shallow waters. The most shallow parts of the reef appear to be entirely composed of laminar tabulate corals, associated with large brachiopods. Possible groove structures are present radiating from the centre of the reef which are rich in brachiopod debris.

Fig. CS 3.6 (a)   Aerial view of Glenister Knolls.

Fig. CS 3.6 (b)   Columnar stromatolites which grew at the base of the reef, to which were attached abundant crinoid holdfasts. Hammer = 32 cm long.

5 cm

Fig. CS 3.6 (c) Receptaculitids (right) and lithistid sponges (left), together with large brachiopods, bound by early fibrous cements.

20 cm

Fig. CS 3.6 (d) Cement- and internal sediment-filled primary reef cavity formed beneath a large, laminar stromatoporoid.

At the end of the Givetian (375–370 Ma), a precipitous drop in atmospheric CO_2 began, which continued until ~320 Ma, marking a dramatic decline in global temperatures. Accompanying this dramatic climatic change, was an extinction of shallow marine taxa in a series of step-down extinctions, particularly at the Frasnian/Famennian boundary (Section 5.3.3).

3.5 Upper Devonian (Famennian) to Permian: ~ 360–251 Ma

Latest Devonian (Famennian) shallow-water reefs are relatively rare (known only from the Canning Basin, Alberta and Omolon, and north-eastern Russia), in striking comparison to the widespread distribution of Frasnian reefs. However, deep-water mud-mounds and shallow-water stromatolites are more widespread, built mainly by calcified cyanobacteria (*Renalcis*) and other microbial communities. Such non-skeletal reefs persisted into the late Carboniferous and Permian, particularly on the northern margin of Pangaea (G.R. Davies *et al*. 1989).

Notwithstanding the near complete loss of large, heavily calcified metazoans, reef development was continuous across the Frasnian/Famennian boundary in the Canning Basin, western Australia. There, reef growth was dominated by microbialites—which had been major framework builders during the Frasnian—where fenestrae and stromatactis voids were encrusted on upper surfaces by cryptic *Renalcis* and large lithistid sponges grew attached to the microbial mounds (Fig. 3.22).

The Carboniferous represents a supposedly unusual phase in the history of reef-building—even though it lasted some 65 Myr. Encrusting organisms remained generally uncommon (P.D. Taylor 1990), and as a result the Lower Carboniferous (Mississippian) is often cited as a period of 'recovery' of shallow marine reef communities. Reef-building is widely assumed

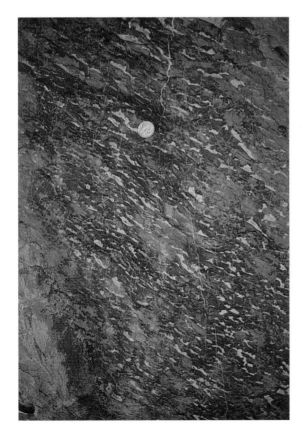

Fig. 3.22 Upper Devonian (Famennian) reefs from Windjana Gorge, Canning Basin, showing abundant stromatactis. These structures probably represent large mounds of microbialite. Coin = 2 cm diameter.

to be represented solely by relatively deep, microbial mud-mound development in Europe and North America. Despite, however, the survival of major Frasnian reef-builders such as microbial communities and *Renalcis*, Lower Tournaisian reefs are very rare, known only from Russia.

The Carboniferous was a time of great climatic and sea-level fluctuations with polar glacial events and depressed global temperatures (S.M. Stanley 1988). Continents were fused into one landmass which straddled the equator, so restricting circulation and excluding the possibility of any equatorial currents (Copper 1989). A considerable range of skeletal and microbial reefs are known from the Viséan (Webb 1994), and mud-mounds were abundant during the middle to late Tournaisian. These are known as Waulsortian mounds, and form a group of reefs that flourished particularly during the Lower Carboniferous, although Waulsortian-like reefs are known from the Late Carboniferous of Ellesmore and Axel Heiberg Islands of the Canadian Arctic (G.R. Davies *et al.* 1989). 'Waulsortian' derives from the Belgium locality where such mud-mounds were first described, but it remains a term used with little consistency or clarity and encompasses reefs of quite variable composition.

Waulsortian mounds are distinctive reefs characterized by a core sediment containing many generations of micrite, micrite-supported cavity systems infilled by marine cements (including stromatactis), and fenestrate bryozoans that are generally considered to lack any skeletal framework. The mounds are proposed to have initiated in deep waters below the wave base (up to 280 m) on the basis of regional sedimentology and the general absence of photosynthetic organisms, but they could grow into fairly shallow waters in higher energy regimes (Somerville *et al.* 1992). (Palaeoberesellids—probable green algae—are known, however, from some late Asbian mud-mounds.) In north-western Europe, most developed in distal ramp settings which were common in the Lower Carboniferous (early Dinantian = Tournaisian and Viséan), although a variety of morphologies are known, reflecting growth on different positions on a ramp and on individual depositional histories (e.g. Lees and Miller 1985). Late Dinantian (Asbian and Brigantian = Serpukhovian) mounds colonized more shallow waters on platform margins and intraplatform ramps as such settings became available after the break-up of ramp systems (Bridges *et al.* 1995). Waulsortian mounds were commonly surrounded by crinoidal, grainstone flanking beds, and the principal skeletal components show marked variation of composition and abundance with depth (Bridges *et al.* 1995). Although all are characterized by sessile suspension-feeders, deeper mounds are dominated by frondose bryozoans and spiculate sponges, crinoids are abundant at intermediate depths, and thrombolites, together with corals and trepostome bryozoans, are prevalent in shallow shelf settings.

The green clay cores of some deep-water mounds are common, and often associated with abundant well-preserved crinoids and frondose bryozoans (Ausich and Meyer 1990). Bryozoans and crinoids were probably capable of baffling and trapping locally fine-grained mud (especially from slightly turbid water) due to unidirectional cilia-generated currents (McKinney *et al.* 1987), but they were not ubiquitous components of Waulsortian reefs and were often equally common in level-bottom sediments, although perhaps of a different species composition. Moreover, the frondose bryozoans commonly occur within Waulsortian reefs surrounded by cement, that is with little associated micrite matrix. It is difficult to understand how bryozoan-baffling alone could create slopes up to 50 degrees and reefs up to 100 m high; such steep slopes suggest that an unrecognized reef framework must be present. Sediments surrounding deep-water mounds are typically thinner and contain significantly higher quantities of fine-grained siliclastic sediment than that of the mound itself; mounds in shallow settings show a massive, muddy appearance at variance with the coarser-grained surrounding sediments, indicative of mobile, winnowed sediments and high or turbulent water energies (Pickard 1996). Such observations indicate that the carbonate mud within the mounds was generated locally and *in situ* under a stabilizing biological control, and with little export. The origin of Waulsortian mounds is controversial, but most authors now agree that diverse and complex microbial processes were important, especially in the shallow-water settings (e.g. Lees and Miller 1995), whilst the contribution of skeletal organisms is variable and probably related to depositional setting and opportunistic colonization (see Case study 3.7). Peloidal and clotted micrites (thought to have been primarily calcite) and micritized cavity walls are common in many Dinantian (Tournaisian and Viséan) build-ups, and these are considered to be primary structures as cavity systems encrusted by early marine cements are commonly developed within them. The micrites are thought to have formed within surficial microbial mats or biofilms that trapped bioclasts and which stabilized the accreting reef surface, allowing the development of steep depositional slopes. Encrusting, often cryptic stromatolites and thrombolites, and occasional *Renalcis* colonies offer direct evidence of microbial activity (Pickard 1996). Stromatolites and other laminated encrustations, formed both by microbial calcification and the trapping of grains within a microbial mat, are known (Fig. 3.23 (a)). These often encrust the microbial micrite, and are testament to the primary origin and early lithification of the micrites (Kirkby 1994). Likewise, encrusting organisms are often associated with the skeletal stromatolites (Figs 3.23(b) and 3.24), such as encrusting foraminiferans (e.g. *Tetrataxis*), serpulid worms, and vermetid gastropods (e.g. Webb 1987), and the micrite may show *Trypanites* borings (Fig. 3.23(c)). Such features suggest that Lower Carboniferous mud-mounds were complex microbial structures probably formed by diverse bacterial and algal communities, with highly undulating upper surfaces and cavity-riven interiors.

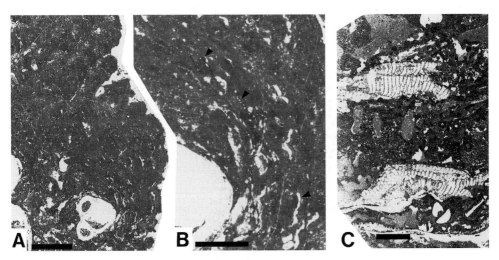

Fig. 3.23 Microbialite (thrombolite) reef framework from Lower Carboniferous Pitkin Formation, northern Arkansas, USA. Scale = 2.5 mm. (A) Thrombolite nucleated upon the tabulate coral *Multithecopora*. (B) Detail showing encrusting foraminiferans (arrow heads). Scale = 1 mm. (C) Encrusting bryozoan and *Trypanites* borings are testament to the early lithification of this structure. Scale = 2.5 mm. (Reproduced with permission from G.E. Webb, *Sedimentology*, and Blackwells Science Ltd.)

Fig. 3.24 Vertical section through microbialite, with attached epibenthos: a bryozoan (B) and serpulid worms (arrowed). Note the geopetal internal sediment to the left, the microbial encrustation on the upper surface of the lowermost serpulid, the cement-filled voids, and trapped bioclastic debris (brachiopods and frondose bryozoans). Lower Carboniferous Pitkin Formation, northern Arkansas, USA. Scale = 2.5 mm. (Reproduced with permission from G.E. Webb, *Sedimentology*, and Blackwells Science Ltd.)

The initiation of deep-water, mud-mound growth remains a mystery, but some have suggested that low sedimentation rates may favour the growth of microbial communities, as mounds seem to form preferentially during transgressions and high sea-level stands when decreased sedimentation rates would be predicted. Cold, nutrient-rich waters may also have aided rapid inorganic cement precipitation and the growth of microbes and suspension-feeding metazoans. Some Early Carboniferous mud-mound development coincides with areas influenced by oceanic upwelling (Wright 1991).

Some workers have suggested that the carbonate realm was dominated by ramp systems during early Dinantian time (Wright and Faulkner 1990), where build-ups are isolated. Shelves world-wide were often argillaceous, with mud-mound development restricted to areas far removed from heavy terrigenous input (e.g. Horbury 1992). As a result, carbonate production was displaced to offshore locations and the presence of extensive crinoid banks may lend support to the influence of high nutrient levels in such settings.

Case study 3.7: Muleshoe Mound (Lake Valley Formation), Sacramento Mountains, New Mexico, USA (Lower Carboniferous (upper Tournaisian): ~345 Ma)

Muleshoe Mound (110 m high and 400–500 m wide) comprises classic Waulsortian mound sediments, and has long been considered a subeuphotic, low-energy reef. However, recent detailed petrographic analyses, mapping of sediment types, and regional correlation now confirm that Muleshoe grew at a shallow depth and under significant depositional energies (Kirkby 1994; Kirkby and Hunt 1996).

Muleshoe Mound is a composite structure of five distinct and unconformable units, which are thought to represent successive growth episodes of mound colonization. These units

Fig. CS 3.7   Muleshoe Mound. Sacramento Mountains, New Mexico, USA.

record a shift from predominantly upwards (aggradational) to lateral (progradational) growth. Reef growth may have been initiated by colonization of the antecendent relief generated by localized lenses of crinoidal packstones, compaction, or localized tectonic processes.

The framework of Muleshoe Mound is composed of rigid micrite masses with rounded, bulbous shapes and thrombolitic fabrics that are lined by early marine cements. The thrombolites are composed of abundant peloids, which are interpreted as microbial precipitations forming within an organic (?algal) precursor. These created primary cavities whose rigidity was enhanced by extensive early cementation. The form of the thrombolites varies according to depositional energy as evidenced by changes in bioclast composition and orientation. In lower growth phases, no such growth orientation is evident, but, in later growth phases (= shallower waters), there is often a pronounced high-angle orientation of the digitate micrite masses and intervening *in-situ* bryozoan fronds that matches the regional orientation of other current indicators such as crinoid segments. Bryozoan colonies over a metre in height been recorded, and fan- to vase-shaped frondose bryozoans mark lateral changes through the mound in response to changes in depositional energy. Flanking beds are common, and consist of grainstones which drape the reef slopes. These were probably deposited as grain flows and resedimented material generated from within the reef. Flanking beds were partially cemented during periods of hiatus. Talus units are common on flanks, as are slumped strata. The presence of graded crinoid grainstones and scour features on the build-up crest is interpreted as evidence that the growth of Muleshoe Mound was modified by storms.

The intermound and basin strata of Muleshoe, as well as other mud-mounds in the Lake Valley area, and in other Lower Carboniferous mound complexes in Alberta and Montana, are dominated by dysaerobic and anaerobic strata alternating with thin, oxygenated horizons. This infers ocean stratification, which indicates a tendency to ocean anoxia (an absence of oxygen) during the Tournaisian, and has been suggested to be related to the ecology or diagenesis of mound growth.

Tournaisian shallow-water reefs are rare, but some are known from the Rangari Limestone, New South Wales, Australia (Webb 1994). These bioherms were constructed by microbialites together with encrusting corals. Similar reef communities are known from the Viséan and

Serpukhovian. Viséan reefs formed extensively on the southern and western margins of Laurussia; in fact there was a tremendous diversity of shallow-water reefs at this time, especially during the Asbian and Brigantian (Late Viséan), which were constructed variously by often high-diversity communities of skeletal metazoans, including bryozoans (frondose and encrusting), lithistid sponges, calcified cyanobacteria (e.g. *Renalcis*), and tabulate (e.g. *Lithostrotion*, *Siphonodendron*, and *Multithecopora*) and rugosan corals (Webb 1987). Algal growth is often pervasive, and considerable rigidity was imparted to these communities by microbial encrustations. This was a time when many localized, endemic reef biotas developed in isolated tectonic areas (see Case study 3.8), in response to the formation of shallow shelf settings owing to accumulation of detrital carbonates (Webb 1994).

Case study 3.8: 'Cracoean' reefs, northern England (Lower Carboniferous (Late Viséan): ~ 335 Ma)

During the middle to late Viséan, shallow-water carbonate platforms and ramps formed over rift-controlled structural highs in the tropics of the southern hemisphere (Leeder 1987). Reefs, known as 'Cracoean' (after the local village of Cracoe in North Yorkshire), commonly formed marginal to rimmed shelves in northern England and have been described in detail by Mundy (1980, 1994). In places, continuous tracts were present constructing substantial frameworks over 30 m thick and covering areas in excess of 3000 m²; in other areas they were represented by large isolated reefs immediately basinward of the margins.

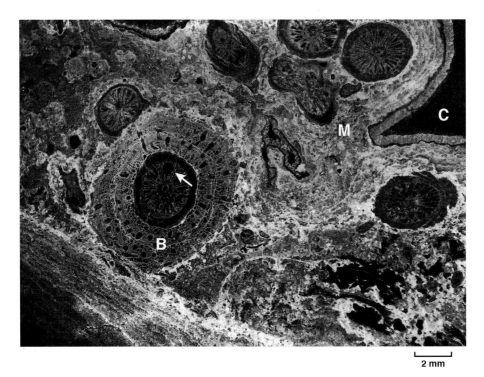

2 mm

Fig. CS 3.8 (a)   Part of a thicket of solitary rugose corals (*Cyathaxonia cornu*) which has been encrusted by microbialite (M), with the formation of small, growth framework cavities (C) lined by marine cement. The central coral shows a *Trypanites* boring (arrowed), and encrustation by a fistuliporan bryozoan (B). Stebden Hill, N. Yorkshire. (Photomicrograph: D.J.C. Mundy.)

The reef biotas were diverse: over 500 species of macrofauna are described, together with common foraminifera, condodonts, dascycladacean algae, and cyanophytes. The framework was dominated by encrusting, laminated microbialite, which was probably constructed by an algal mat community including the cyanobacteria *Ortonella*. This lithified early, and was colonized by a variety of small encrusters including juvenile bryozoans and foraminifera (*Tetrataxis*). Lithistid sponges (*Hapliston* and *Hindia*), frondose bryozoans (fenestellids), and favositid corals (*Michelinia*, *Emmonsia*) attached to the microbialite surfaces. Encrusting bryozoans (*Tabulipora* and *Fistulipora*) formed multiple encrustations on the corals, and aggregating groups of solitary rugose corals were common. The framework supported a unique shelly fauna of specialized, attached (often spiny, but also cementing) productid brachiopods and cementing bivalves (*Pachypteria*). Also common were free-living productids and small vagile organisms including gastropods and trilobites.

Cavities were common within the microbialite and internal sediments (typically peloidal packstones and grainstones), which contain concentrations of ostracodes and the trilobite *Griffithides acanthiceps*. The ostracodes were free-living forms, and both these and the trilobite may have sheltered in cavities during moulting. The presence of only one species of trilobite in crypts from the 17 known from Cracoean sediments suggests that this form was a reef crypt dweller.

Highly localized bioerosion is present, consisting of *Trypanites* up to 3 mm in length, and microborings attributable to microbial endoliths.

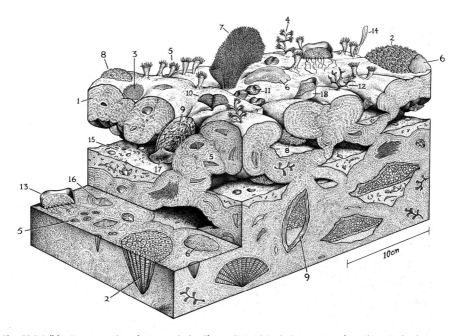

Fig. CS 3.8 (b) Reconstruction of a Lower Carboniferous (Late Viséan) 'Cracoean' reef, northern England. 1: Stromatolite/microbialite; 2. tabulate coral (*Michelinia*); 3: tabulate coral (*Emmonsia parasitica*); 4: tabulate coral (*Cladochonus*); 5: rugose coral (*Cyathaxonia*); 6: bryozoan (*Fistulipora*); 7: frondose bryozoan (*Fenestella*); 8: lithistid sponge; 9: cementing bivalve (*Pachypteria*); 10: brachiopod ('*Reticularia*'); 11: rhynchonellid brachiopod (*Stenoscisma*); 12: robust fenestrate bryozoan (*Thamniscus*); 13: attached productid brachiopod (*Limbifera*); 14: attached productid brachiopod (*Proboscidella*); 15: myodocopid ostracod concentration (*Cypridinella*, *Entomonchus*); 16: trilobite (*Griffithides*); 17: strophomenid brachiopod (*Streptorhynchus*); 18: strophomenid brachiopod (*Leptagonia*). (Modified from McKerrow 1978 and Mundy 1993; copyright, John Sibbick.)

A unique type of Lower Carboniferous shallow reef is known from south-eastern Canada, where restricted biotas, including problematic worm tubes, grew in association with evaporites and metallic sulphide mineralization. The absence of typical marine biotas may be due either to hypersaline conditions in a restricted setting, or to proximity to hydrothermal vents.

During the early Namurian, reefs declined dramatically, caused by the loss of carbonate-platform habitats as a result of terrigenous sediment influx onto shelves, tectonic uplift, and regression. Early Namurian reefs are rare, and few reefs (except the Akiyoshi Atoll, Japan) survived the Lower–Upper Carboniferous boundary. Mid-Carboniferous reefs have been described from a seamount terrain in Japan (Sano and Kanmera 1996), which are constructed by bryozoans, rugose corals, and chaetetids.

The mid-Carboniferous saw the appearance of new reef-associated organisms: particularly the phylloid ('leaf-like') algae (Box 3.9), the alga *Palaeoaplysina*, and problematic tubular algae (palaeosiphonoclads and aoujgalids; Mamet 1991) and the problematic organism *Tubiphytes* (Box 3.10). These additional elements became major constituents of Late Carboniferous and Early Permian reefs, but peloidal micrites continued to be an important component of the reef matrix. Small dascycladacean algal reefs are also known from the Early Permian (Flügel 1994). Phylloid algae grew up to 15 cm in height and diameter, and formed a variety of growth forms, including cup-shaped, upright, and reclining (see Fig. 3.25). Such algae were closely aggregating, and both densely carpeted level bottoms

Fig. 3.25 Reconstruction of a composite phylloid algal reef community. 1: *Archaeolithophyllum* (phylloid algae); 2: cup-shaped *Eugonophyllum* (phylloid algae); 3: leafy *Eugonophyllum* (phylloid algae—the root-like holdfasts are conjectural); 4: frondose bryozoan; 5: crinoids; 6: echinoderms; 7: brachiopods; 8: brachiopods; 9: encrusting foraminifera; 10: encrusting bryozoans; 11: fusilinid foraminifera; 12: shark; 13: ostracodes; 14: micrite-infill; 15: *Tubiphytes*; 16: gastropod. (Copyright, John Sibbick.)

and formed patch reefs. They probably grew rapidly and so were able, successfully, to colonize areas of newly available stabilized substrate. Multilayered, peloidal, geopetal sediment infills within primary cavities are common, whose early lithification together with abundant submarine cement imparted considerable rigidity to the reefs. Phylloid algae are associated with a minor encrusting biota of foraminiferans (e.g. the agglutinating form *Minammodytes*) and bryozoans, grazing gastropods and echnoids, and brachiopods (e.g. spirifid brachiopod *Composita subtilita*). In an analysis of Lower Permian phylloid algal reefs from Texas, Toomey (1976) suggested that their dense growth excluded other sessile organisms, limiting many to encrusting epiphytic or dwelling habits. Crinoids and frondose bryozoans remained common within some phylloid algal reefs.

Box 3.9: **Phylloid algae**

'Phylloid' algae refers to late Palaeozoic calcified (aragonitic) algae with a variety of platy, cup-shaped, and encrusting leaf-like shapes of mixed taxonomic affinity. From a few well-preserved examples, some representatives are known to have affinities with udoteacean (Chlorophyta), ancestral coralline algae (Rhodophyta) (Wray 1977; Kirkland *et al.* 1993), and possibly Sqamaraceae (Wray *et al.* 1975).

Phylloid algae are characteristic of many Late Carboniferous–Early Permian reefs, which are often overlain by fusiline foraminiferan-rich, cross-bedded packstones and wackestones. The reefs had characteristically low depositional relief, and consisted of either *in-situ* phylloids, or fragmentary accumulations of algal plates orientated subparallel to bedding. Platy, encrusting morphologies were free-lying, and probably bound unconsolidated sediment. Erect platy or cup-shaped forms were often closely packed and orientated, perhaps into the oncoming current.

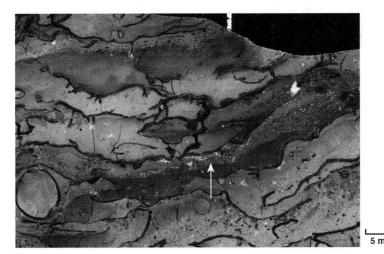

Fig. Box 3.9 Free-lying phylloid algae, with cryptic encrusting bryozoans (arrowed). These algae may have been partially rooted to the soft substrate by spines that projected from the lower surfaces. Lower Permian, Sacramento Mountains, Texas. (Photograph: G. Forsythe.)

Box 3.10: *Tubiphytes* (Carboniferous to Cretaceous)

Tubiphytes, or more correctly *Shamovella* Rauser–Chernousova (Riding 1993), is a common, but small (typically a few millimetres), problematic encrusting and branching organism associated with both shallow-and deep-water tropical reefs as well as more temperate habitats from the Early Carboniferous to mid-Cretaceous. *Tubiphytes* is commonly irregularly cylindrical in form, with a smooth external surface, and bears a densely flocculent, micritic microstructure, often secreted in a series of layers or packages around an axial hollow core or tubular structure. It is likely that many different taxa have been artificially united under the name *Tubiphytes* (Senowbari-Daryan *et al*. 1993).

The disputed affinity of this form stems from the lack of any distinctive, diagnostic features. Cyanobacterial (Mazlov 1956), hydrozoan, poriferan (Riding and Guo 1992), and foraminiferan–cyanobacterial affinities (Flugel 1981; Pratt 1995) have all been proposed. The lack of a convincing *sponge filtration system* does not support a poriferan origin, but the observation that the central cores of *Tubiphytes* can consist of foraminifera belonging to the Nubeculariidae and Fischerinidae, together with dense tangles of microbial filaments and clotted micrite lends some support to a foraminiferal– cyanobacterial origin (Pratt 1995). However, recent, scanning electron-microscopic observations of the skeletal structure of *Tubiphytes* from Lower Permian reefs suggests that biomineralization was under closer biological control than would be predicted for a microbial precipitate (G. Forsythe, personal communication). Of central importance, however, to this debate is whether the axial canals are an integral part of the organism, or chance nucleation sites (e.g. foraminifera) for the *Tubiphytes* organism. The jury remains out.

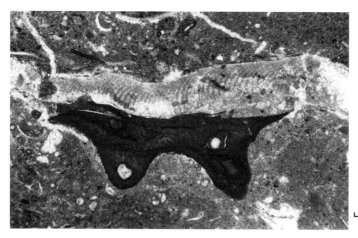

1 mm

Fig. Box 3.10 *Tubiphytes* encrusting the undersurface of a phylloid algae (*Archaeolithophyllum*). (Photomicrograph: G. Forsythe.)

In the Lower Permian reefs at La Colmena, in the Sacramento Mountains, New Mexico, USA, large heads of *Archaeolithoporella*, up to 2.5 m in height, formed reef knolls, whose sides and underhangs were colonized by tubular sphinctozoan sponges, encrusting bryozoans, and *Tubiphytes*. These were surrounded by thickets of upright phylloid algae (G. Forsythe, personal communication).

Reefs possessing features almost identical to the classic Waulsortian mud-mounds persist into the late Carboniferous and Permian (Lees 1988), especially along the northern margin of Pangaea, which are rich in fenestellid bryozoans and *Tubiphytes*, and marine cements (G.R. Davies and Nassichuk 1990). The late Lower Permian (Artinskian) climatic cooling-event led to the disappearance of shallow-water, algally dominated reefs, but microbial activity remained important in temperate deep-water reefs.

Large chaetetid calcified sponges (see Box 3.4) occupied muddy, shallow-water settings during the late Carboniferous, and some aggregated to form small bioherms of low relief (Fig. 3.26). Many attached to small ephemeral debris, such as spiny productid brachiopods, on otherwise muddy substrates, but show no mutual encrustation and are associated with little other biota or early cementation.

Like the Carboniferous, the Permian was a time of climatic cooling and regression, the latter eliminating many shallow shelf seas. Most late Palaeozoic reefs occur in and around the margins of intracratonic basins. During the Early Permian, the continents merged to form a single supercontinent known as Pangaea. S.M. Stanley (1988) suggests that remaining shelves must have been strongly affected by currents on the west side of Pangaea and by mid-ocean gyres carrying cool waters to the equator.

Late Permian reefs are characterized by frondose bryozoans (including fenestellids), and calcified sponges, particularly inozoans and sphinctozoans (see Box 3.4), as well as by *Tubiphytes* and various encrusting algae including *Archaeolithoporella*. Volumetrically, Late Permian reefs were dominated by bioclastic sediments and early cements, but still formed substantial rimmed margins with well-developed zonation from fore-reef talus to

(a)

Fig. 3.26 Upper Carboniferous chaetetid sponges. Gerard Quarry, Kansas. Hammer = 32 cm long. (a) Aggregating domal individuals forming a small bioherm; (b) isolated chaetetid growing within muddy sediments. The ragged margins suggest that this individual grew during episodic sedimentation. (Fig. 3.26(b) overleaf).

(b)

Fig. 3.26

reef slope, crest, reef flat, and back-reef lagoon. In many examples, including the famous Capitan Reef of Texas and New Mexico (Fig. 3.27; Case study 3.9), the stabilization of sediments and aggregating reef communities by rapid cementation was crucial to the formation

Fig. 3.27 View looking north-west over the Permian Capitan Reef, Texas and New Mexico. The reef marks a prominent topographic boundary between the basinal sediments in the foreground, and the lagoonal sediments behind. (Photograph: J.A.D. Dickson.)

of the reef rock (e.g. Tucker and Hollingworth 1986; Wood *et al*. 1996). Cementation was apparently concentrated along the shelf edge and slope.

Case study 3.9: Capitan Reef, Texas and New Mexico, USA (Late Permian (Upper Guadalupian): 280 Ma)

The Permian Capitan reef, West Texas and New Mexico, forms one of the finest examples of an ancient, rimmed carbonate shelf. The reef, as expressed in the massive Capitan Limestone and associated carbonate platform, defines the margin of the Delaware Basin, and marks a prominent topographic boundary between deep-water basinal deposits and shallow shelf sediments to the north-west. The Capitan reef is the youngest of a series of shelf-margin complexes which developed around the Delaware Basin over a total period of some 12 Myr (Garber *et al*. 1989). The reef forms prograding beds of generally 20–40°, which may be locally vertical (Bebout and Kerans 1993).

The Capitan Limestone ranges from 100 to 200 m thick and has been subdivided into Lower, Middle, and Upper Members. The Capitan Limestone contains a diverse and distinctive biota estimated at some 350 taxa (Fagerstrom 1987), which includes abundant calcified sponges (sphinctozoans and inozoans), putative algae, bryozoans, brachiopods, *Tubiphytes*, and *Archaeolithoporella*.

Until recently, established ecological reconstructions emphasized the role of branching or solitary organisms (sphinctozoan sponges, bryozoans, and *Tubiphytes*) as upright bafflers of sediment, and massive putative algae (*Collenella*, *Parachaetetes*, and *Solenopora*) in the construction of the Capitan reef, together with the binding and encrusting contribution of *Archaeolithoporella* and extensive early marine cementation. However, at least five reef-building communities are known from the Middle and Upper Capitan Limestone (Wood *et al*. 1996):

(1) phylloid algae (Upper Capitan);

(2) *Tubiphytes*–sponge (Upper Capitan);

(3) *Tubiphytes*–*Acanthocladia* (Middle Capitan);

(4) frondose bryozoan–sponge (Lower, Middle, and Upper Capitan);

(5) platy sponge communities (Middle and Upper Capitan).

Much of the Permian Capitan reef was strongly differentiated into open surface and cryptic communities. Indeed, most of the preservable epibenthos was housed within the cryptos and zonation developed only in the most shallow parts of the reef. Most sphinctozoan sponges did not grow upright but rather were pendent crypt-dwellers, as were nodular bryozoans, and rare solitary rugose corals and crinoids. Indeed, most members of the preservable cryptos appear to have been obligate crypt-dwellers. Much of the Middle Capitan reef framework, and those parts of the Upper Capitan inferred to have occupied waters deeper than about 30 m, was constructed by a scaffolding of large frondose bryozoans, together with the subsidiary platy sphinctozoan *Guadalupia zitteliana*. Bathymetrically shallow areas of both the Middle and Upper Capitan reef were, however, characterized by large platy calcified sponges. In parts of the Upper Capitan, some of these sponges (*Gigantospongia discoforma*) reached up to 2 m in diameter and formed the ceilings of huge cavities which supported an extensive cryptos.

The relatively fragile Capitan reef remained intact after the death of the constructing organisms, as rigidity was imparted to this community by a postmortem encrustation of *Tubiphytes* and *Archaeolithoporella*, together with substantial amounts of encrusting micrite inferred to be of microbial origin. The resultant cavernous framework was partially infilled with sediment and preserved by synsedimentary intergrowth of cement botryoids and

Archaeolithoporella. Extensive cement precipitation within the reef framework was probably favoured by a number of factors, including deep anoxia which generated upwelling waters with elevated alkalinity. (From Wood *et al*. 1996.)

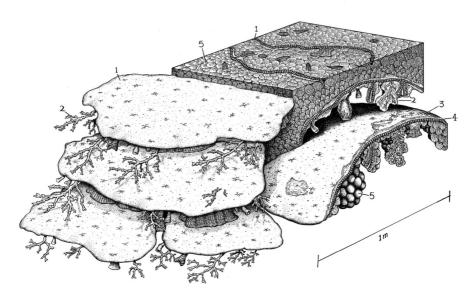

Fig. CS 3.9 (a)   Reconstruction of the Upper Permian Capitan Reef: Platy sponge community 1. *Gigantospongia discoforma* (platy sponge); 2: solitary and branching sphinctozoan sponges; 3: *Archaeolithoporella* (encrusting ?algae); 4: microbial micrite; 5: cement botryoids. (Copyright, John Sibbick.)

Fig. CS 3.9 (b)   Reconstruction of the Upper Permian Capitan Reef: Frondose bryozoan–sponge community 1. Frondose bryozoans (*Polypora* sp.; *Goniopora* sp.) 2: solitary sphinctozoan sponges; 3: *Archaeolithoporella* (encrusting ?algae); 4: microbial micrite; 5: cement botryoids; 6: sediment (grainstone–packstone). (Copyright, John Sibbick.)

(i)

2 cm

(ii)

1 cm

Fig. CS 3.9 (c)   Weathered surfaces parallel to reef growth from the Permian Capitan reef. (i, ii) Middle Capitan Limestone, McKittrick Canyon. (iii, iv) Upper Capitan Limestone, Mouth of Walnut Canyon. (i) Pendent sphinctozoan sponges attached to a frondose bryozoan (arrowed). These metazoans have been encrusted by pale grey micrite. Remaining space is filled by intergrown *Archaeolithoporella* and cement botryoids. (ii) A frondose bryozoan (arrowed) to which is attached a pendent sphinctozoan sponge. The whole has been encrusted by layered micrite of a probable microbial origin (M). Remaining space is filled by intergrown *Archaeolithoporella* and cement botryoids.

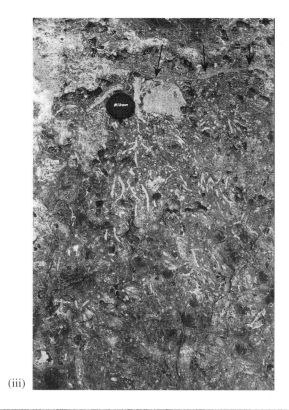

(iii)

(iv)

2 cm

Fig. CS 3.9 (c)

(iii) Huge cavity formed by *Gigantospongia discoforma* (arrowed) supporting an extensive cryptos, including a branching individual of *Lemonea* sp. (left) and the compound *Guadalupia explanata* (right). Lens cap = 52 mm. (iv) Large individual of branching *Lemonea* sp. attached to the undersurface of *Gigantospongia discoforma*.

The causes of the end-Permian extinction remain elusive, but they were certainly complex and related to a rapid drop of sea level which eliminated many shallow marine carbonate habitats, followed by extensive volcanic eruptions which may have increased climatic instability, and finally by a rise in sea level which may have caused anoxic waters to flood newly available, shallow marine habitats (see Section 5.3.4). This profound extinction event resulted in the global loss of between 80 and 95% of all species, and removed sessile suspension- and filter-feeding reef benthos: lost were the tabulate and rugose corals, bryozoans, and phylloid algae, as well as many calcified sponges, crinoids, and brachiopods.

In summary, Palaeozoic reefs were dominated by gregarious, suspension-feeding skeletal metazoans and sometimes calcified algae, but microbial frameworks remained very important. By the mid-Palaeozoic, calcified reef metazoans with often laminar or domal morphologies had reached considerable sizes. Bioeroders were only abundant locally, and the common preservation of intact frameworks does not suggest growth into turbulent zones except in some cases where microbial frameworks were extensive. In some Palaeozoic reefs, calcified invertebrates were subsidiary, and pervasive, early cementation was of considerable volumetric importance, imparting rigidity to otherwise fragile communities. Palaeozoic reefs were capable of forming highly cavernous structures with strongly differentiated open surface and cryptic biotas.

3.6 Triassic: 251–206 Ma

During the Triassic, the break-up of Pangaea initiated the formation of a large equatorial seaway, the Tethys sea, which continued to develop throughout the Mesozoic. Many reefs grew on its margins.

Reefs are virtually unknown for a 7–10 Myr interval in the Early Triassic (Flügel 1994), which has been attributed to a period of 'reorganization' of the reef community following the end-Permian extinction (P.M. Sheehan 1985). Schubert and Bottjer (1992) suggest that stromatolites became abundant in normal marine shallow seas following the extinction, but Pratt (1995) suggests that some of these stromatolites grew in deep-shelf settings. They are associated with encrusting oysters, siliceous sponges, and crinoids.

Non-stromatolite reefs first reappear in the Middle Triassic (Anisian), although some Late Scythian reefs are found in one of the many displaced terranes of volcanic origin thought to have been exotic to North America during the early Mesozoic. This suggests that during the Triassic, the ancestral Pacific was strewn with volcanic islands such as those now found within the circum-Pacific region (Hallam 1986). Low-relief Anisian bioherms of small, aggregating metazoans are known from the Tyrolean Alps, Italy and Hungary, Iran, and southern China. Middle Triassic reefs have a similar ecological caste to Permian reefs, consisting of calcified sponges (particularly sphinctozoans), *Tubiphytes*, and algae such as *Archaeolithoporella* together with abundant early marine cement. Phylloid algae have been reported from the Stikinia Terrane of the Yukon (Reid 1987). This late Palaeozoic style of reef-building continued into the succeeding Ladinian and Carnian (Fois and Gaetani 1984), although the Anisian–Ladinian Latemar reef in northern Italy is dominated by microbialite, with intergrown *Tubiphytes* and *Terebella* (sessile foraminifera), together with minor, encrusting sphinctozoan sponges, solenoporacean algae, bryozoans, and serpulids (Harris

1993). The microbial framework, which is clotted or peloidal with abundant *Girvanella* filaments, forms irregular, tabular crusts with digitate projections, with primary framework, and stromatactis, cavities (Pratt 1995). Some stromatactis cavities are encrusted by *Renalcis*-like micrite, and early cements such as botryoids are common.

Scleractinian corals appeared in the Anisian (Box 3.11). Most of these early representatives, at least until the Carnian, were small, solitary or phaceloid (simple branching) forms which inhabited soft substrates in deep or otherwise protected settings (Fois and Gaetani 1984). Higher energy reefs continued to be built by calcareous sponges, bryozoans, and solenoporacean algae. Anisian to Ladinian reefs range from serpulid, algal, and microbial communities with large primary cavities bearing abundant early cements (Senowbari-Daryan *et al.* 1993), to sponge, algal, and coral-dominated communities. For example, the Wetterstein Limestone platform north of Innsbruck, Austria, contains patch reefs in the lower part and a shallow-water reef complex in the upper part, known as the Hafelekar reef complex. The Hafelekar complex consists of a reef, with the reef front sloping at 5°, together with a wide reef-flat with skeletal sand-shoals separating the reef from a lagoon. Areas of the complex exposed to high energy were dominated by massive corals, solenoporacean algae, and some calcified sponges: delicate corals, branching sphinctozoan sponges, branching *Tubiphytes*, and some bushy codiacean algae grew in more sheltered areas (Brandner and Resch 1981). These communities were bound by microbial micrite and *Tubiphytes* (Pratt 1995).

Upper Triassic (Carnian) patch reefs are well-known from the Cassian Formation, in northern Italy, and are interpreted to have formed on the seaward slopes on the lower parts of carbonate platforms (Wendt 1982). These reefs are dominated by thrombolitic and stromatolitic frameworks that contain foraminifera and calcified algae such as *Ortonella*, with a rich biota of cryptic calcified sponges. Unusual deep-water (<100 m) hexactinellid–*Terebella* (worm tube)-microbialite reefs up to 60 m high have been described from the Carnian of China (Wendt *et al.* 1989). These are associated with cylindrical, branching, conical, and tabular lithistid sponges up to 70 cm in length, as well as rare serpulids, crinoids, and bryozoans.

A marked faunal change-over occurred from the Carnian to the Norian with a rapid extinction and coincident radiation producing a dramatic increase in diversity. This is noted not only in the scleractinian corals (when 90% of all species going extinct), but also in the calcified sponges, pectinid bivalves, conodonts, and brachiopods. This Carnian extinction resulted in the loss of the Permian 'holdovers' and the rise in dominance of scleractinian corals, and coincided with world-wide regression following a Ladinian–Carnian maximum, high sea-level stand which flooded an estimated 16–19% of continental margins (Hallam 1981). This reorganization represents the start of the assembly of the modern coral-reef community. The late Triassic also marks the start of radiation of vagile carnivores, which resulted in a prolonged radiation of the boring and encrusting lifestyles common in the shallow marine communities today.

The development of Norian reefs coincided with a renewed expansion of shallow-shelf seas recognized throughout Tethys (Hallam 1981). In particular the southern margin of Tethys was bordered by extensive carbonate platforms. In the Austrian area, the slow but continuously subsiding Dachstein platforms accumulated thick sequences of cyclically deposited sediments, known as the Dachstein Limestone, and the fringing reefs that developed on these platform margins facing deeper channel or oceanic waters were often correspondingly extensive and very thick, up to 1200 m (Zankl 1969). These reefs developed on ramps that faced the open Tethys, with patch reefs developing seaward and vast tidal flats behind. Also, extensional tectonic activity towards the close of the Triassic transformed the wide tidal flats into interior platform basins situated behind large reef tracts. These elongate lagoons, such as the Kössen basin, deepened to form ocean circulation and tidal systems allowing well-developed reef systems to become established.

Box 3.11: **Scleractinian corals (Class Anthozoa, subclass Zooantharia): Middle Triassic (Anisian) to Recent**

Scleractinian corals (skeletal hexacorals) appeared in the Anisian, and represent a further independently calcified clade of zoanthiniarian sea anemones. The vast majority of scleractinian corals bear porous skeletons of crystalline aragonite. The earliest faunas were systematically diverse: 8 to 12 species are known from the Anisian, belonging to at least 7, but possibly 9, families (Veron 1995), and some show complex skeletal features and colonial morphologies. The typical Mesozoic genera *Isastraea* and *'Thecosmillia'* were already present. Such diversity suggests that the scleractinian corals must have had a not-inconsiderable, soft-bodied history prior to the first appearance of skeletal forms in the Middle Triassic. This is confirmed by molecular phylogenies that show that the scleractinian corals probably arose as an unskeletonized group, within which two subordinal groups later diverged and independently acquired calcareous skeletons (Romano and Palumbi 1996). Molecular phylogenies also suggest that living non-skeletal hexacorals such as sea anemones (actinians and corallimorpharians) are more closely related to living scleractinians than some scleractinians are to each other (Buddemeier and Fautin 1996a).

The scleractinian calyx bears septa, which are inserted cyclically, but the vast majority of scleractinians lack an epitheca. Polyps may be connected to neighbours in the colony by a thin sheet of tissue, known as the coenosarc: polyps and coenosarc thus constitute a veneer of living tissue over the skeleton that they have secreted. The coenosarc is secreted by the edge zone which confers the ability to encrust extensive hard substrates. Corals may

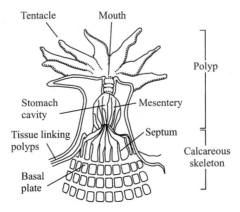

Fig. Box 3.11 The scleractinian coral animal.

be solitary, or may form branching, encrusting, or massive colonies of clonally generated individuals.

Living scleractian corals are found on the abyssal plains, on shelves at cool, high latitudes, in deep-water coral banks, as well as in tropical–subtropical carbonate banks, platforms, and reefs. Multitudes of symbiotic, unicellular algae are present within the tissues of shallow-water, subtropical- to tropical-living, reef-building corals; it is transferral of algal photosynthetic products to the coral host that provides the key to the prodigious biological productivity and carbonate production of modern reef corals.

During the Norian, corals, as other reef benthos, increase in both diversity and in ecological roles. It has been thought that, at this time, corals such as the large, branching genus 'Thecosmillia' (probably referable to Retiophyllia) became reef frame-builders and major contributors to sediment production. The branches of 'Thecosmillia' are often encrusted by foraminifera (e.g. Alpinophragmium perforatum) and the probable red alga Lithocodium. 'Thecosmillia' shows a gradation from delicate branching to robust branching forms with decreasing depth and wave energy.

Norian reefs are globally very widespread, being known from Europe, Asia as far as Japan, and within the Cordilleran region of western North America. A variety of reef types are known: from isolated mounds with probably 10–50 m vertical relief which grew in shallow, basinal environments; to those which formed on platform margins; to poorly known, deeper water, stromatactis-rich mud-mounds. The cavities in these reefs are full of peloids and problematica, such as the possible red algae Baccanella floriformis, Microtubus communis, and microbial coatings.

The palaeoecology of late Triassic, shallow coral reefs is hotly debated. Part of the large carbonate platform formed in central and north Austria platform, known as the Steinplatte, an Upper Norian reef near Tyrol, Austria, has become a classic model for the wave-resistant rim reef (e.g. Wilson 1975). However, recent analyses have challenged these ideas and have suggested that a rigid, frame-resistant framework was absent even in the latest Triassic reefs, with most representing bioclastic mounds or small patch reefs (up to 10 m in diameter and 5 m in height) that grew under limited storm-influence but rapid rates of sea-floor lithification (Stanton and Flügel 1989; Fig. 3.28). Stanton and Flügel (1989) argue that this is the result of reef growth below sea level, at about 10 m depth, such that the uniform environment over a wider area resulted in a more random distribution of reef communities. This idea is supported by the absence of either pronounced zonation or an abrupt platform-edge break comparable to that found in modern reefs, and also by the lack of any orientated coral growth into incoming waves. These conclusions have also been largely confirmed by study of the Höchkonig reef, Northern Calcareous Alps, Austria, by Satterley (1994), discussed in Case Study 3.10.

By the end of the Triassic, coral reefs suddenly disappeared. Of the 50 scleractinian genera known globally, only 11 are recognized to have survived into the Liassic (Beauvais 1984). There is good evidence for an extensive and significant regression in many localities at this time which lead to the widespread emergence and karstification of the reef complexes (Satterly et al. 1994; Section 5.4.5).

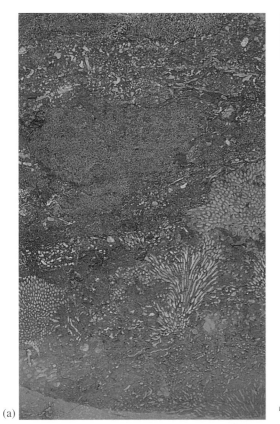

Fig. 3.28 Branching scleractinian corals forming patch reefs, Tropfbruch, near Halleim, Salzburg, Austria. Upper Triassic (Norian). (Photographs: E. Flügel.)

(a)

50 cm

15 cm

(b)

Case study 3.10: Höchkonig reef , Northern Calcareous Alps, Austria (Upper Triassic (Norian): ~ 210 Ma)

The Upper Triassic Dachsteinkalk of the Höchkonig Massif, 50 km south of Salzburg in Austria, represents an exceptionally thick (700 m) platform-margin reef complex. Various environmentally zoned reef communities developed across the shelf until a reef margin of 11° to 5° separated the platform from the reef slope. Coadiacean algae were very abundant through the reef platform and their probable degradation to micrite provided the main source of carbonate sediment.

Wave-resistant stromatolites formed a barrier between a proximal, oncolite-rich lagoon and a back-reef area. Megalodontid bivalves formed extensive accumulations in the lagoons. Here, coral patch reefs and calcified sponges appear. Central platform areas are dominated by reef-derived, wave-generated, coarse bioclastic debris with isolated small patch reefs (accounting for only some 10% of the central reef area) of *in-situ* sphinctozoans, inozoans, and corals, which probably developed in waters less than 10 m deep. The patch reefs are formed by thickets of the branching corals '*Thecosmillia*' and '*Montlivaltia*', together with massive corals, some of which reached several metres in diameter. Some branching corals probably grew rapidly, and reached over 10 m high, but bore thin branches only 5–10 mm in diameter. It is probable that they were not free-standing, but encased and supported by sediment. Some of these communities were monospecific stands, others were with other high-growing corals together with small solitary corals, sponges, and bryozoans. Although bivalves were present, they were not important occupants of these reef environments. The reef margin itself probably grew at a depth of 30 m, and was subject to episodic collapse which introduced debris flows and large blocks of lithified patch-reef material into the upper slope sediments. In contrast to the abundance of branching forms in more shallow waters, the corals that dominate the reef slope are solitary, or platy, domal or massive. Sponge

Fig. CS 3.10 Reconstruction of the Höchkonig reef and adjacent platform area. (Modified from Satterley 1996, by kind permission of SEPM Society for Sedimentary Geology).

diversity increases on the lower reef slope, with large tubular inozoan sponges (e.g. *Peronidella fischeri*) up to 10 cm in diameter and up to 1 m long. This species may also have formed isolated stands in the back-reef areas. Thin-branching *Thecosmillia* grew in water depths of 150–200 m, and cryptic sponges grew at the base of the reef slope at depths of 200 m or more. Sphinctozoans may also have colonized crypts within stabilized, coarse reef-slope breccias (A. K. Satterley, written communication).

The generation and supply of carbonate sediment from the central reef area to surrounding areas modified the composition of the fore slope. The reef margin processes included the mass movement of unconsolidated sediment down the low-angle reef slope (controlled by the rate of sediment supply), and erosion of the margin by slumping caused the initiation of debris slows and the formation of slump scars. Dislodged and rotated blocks of reef rock are common at the boundary between the central reef area and the fore-reef in some localities. (After Satterley 1994.)

3.7 Jurassic to Cretaceous: 207–65 Ma

Reefs were globally rare for 4–10 Myr following the end of the Triassic (G.D. Stanley 1988). During the Lower Jurassic (Hettangian–Sinemurian), there was minimal shelf-carbonate development and few reefs. Sinemurian reefs are known from volcanic islands in the ancestral Pacific, and Late-Triassic coral species such as *Phacelostylophyllum rugosum* appear to have survived at these sites (G.D. Stanley and Beauvais 1994). Scleractinian corals occupied the abyssal plains, cool, high-latitude shelves, deep-water coral banks, as well as tropical–subtropical carbonate banks, platforms, and reefs by the Toarcian (Roniewicz and Morycowa 1993), and coralline algae reappeared at this time. By the Middle Jurassic (Pliensbachian), many new reef coral species had evolved. In Middle Jurassic reefs, endemic taxa predominate, together with a few remaining Norian taxa.

Beginning in the Toarcian, a major faunal change-over occurred. According to Beauvais (1986), all Triassic genera abruptly disappeared. This marked a substantial phase of scleractian coral radiation, which has been suggested by some to mark the differentiation into photo-symbiotic and non-photosymbiotic forms (G.D. Stanley 1981). Whilst only 21 genera are known from the Liassic, 100 are recorded from the Middle Jurassic, and in many regions of Europe, spectacular shelf-edge coral reefs are known (Heckel 1974).

Shallow-water, Jurassic reef communities were highly variable (for example see Insalaco *et al.* 1997), dominated by scleractinian corals (Case study 3.11), stromatoporoid sponges, thrombolites, and microbialites, as well as some unusual examples formed by oysters and *Solenopora* (see Case study 3.12). Most Jurassic reefs can be classified into three broad community types which grew in increasing water depths (Leinfelder 1994):

(1) coral–calcified sponge–solenoporacean algae–microbialite;

(2) deep-water siliceous sponge–thrombolites;

(3) thrombolites.

Case study 3.11: Coral patch-reef community, England (Jurassic (Oxfordian): ~150 Ma)

Coral patch reefs are known from a variety of localities in central and southern England. They are dominated by the colonial massive corals *Thecosmillia*, *Isastrea*, and *Thamnasteria* which could reach up to 1.5 m diameter. Locally, bioerosion was intense, with corals heavily bored

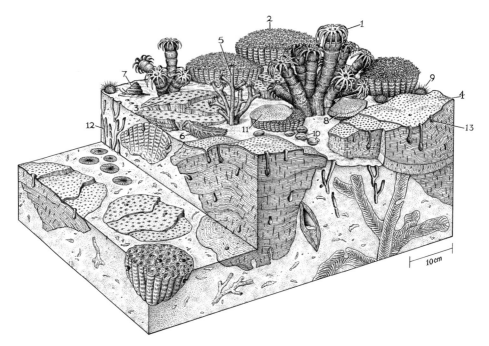

Fig. CS 3.11 Reconstruction of Jurassic (Oxfordian) coral patch-reef community, England. 1: Scleractinian coral (*Thecosmillia*); 2: scleractinian coral (*Isastrea*); 3: scleractinian coral (*Thamnasteria arachnoides*); 4: scleractinian coral (*Thamnasteria concinna*); 5: scleractinian coral (*Rhabdophyllia*); 6: bivalve (*Lopha*); 7: trochid gastropod; 8: pectinid bivalve (*Chlamys*); 9: sea urchin (*Cidaris*); 10: terebratulid brachiopod; 11: bryozoan; 12: scleractinian coral (*Cladophyllia conybeari*); 13: boring bivalve (*Lithophaga*). (Modified from McKerrow 1978; copyright, John Sibbick.)

by the bivalve *Lithophaga* and sponges (clionids). Pectinid bivalves lived on the soft substrates, and brachiopods were common inhabitants of cryptic niches, along with occasional nests of terebratulid brachiopods. Reef debris is often encrusted by serpulid worms and *Exogyra* oysters. The thick-shelled grazing gastropod, *Bourgetia*, and sea urchins, were common.

Case study 3.12: Oyster–algal patch reefs, Portland, England (Upper Jurassic (Portlandian): ~145 Ma)

Patch reefs, up to 3.5 m high and 7 m across grew on stabilized oolite shoals at the top of the Portland Limestone Formation, on the Isle of Portland, southern England. They appear to have initiated during periods of reduced sedimentation rate. The principal reef-builders (providing 55–70% of reef volume) were a variety of intergrown encrusters: cementing bivalves (*Liostrea*, and the highly versatile *Plicatula*), solenoporacean red algae (with conspicuous growth banding), and exceptionally large, multilaminar encrusting bryozoans (*Hyporosopora*). The reefs were highly porous, and the remaining reef volume was filled by sediment, mostly in the form of laminated synsedimentary peloidal cement of probable microbial origin. A diverse subsidiary fauna of small cementers, nestlers (byssate bivalves and terebratulid brachiopods), lithistid and other demosponges, vagile groups (such as burrowing worms and gastropods), and 11 taxa of macro- and microborers have been identified. The borers locally removed up to 50% of the total reef volume, and vacated borings also provided sites for the precipitation of further peloidal cement. Boring was particularly intense around the edges of the reefs, so forming distinct edge zones with multiple genera-

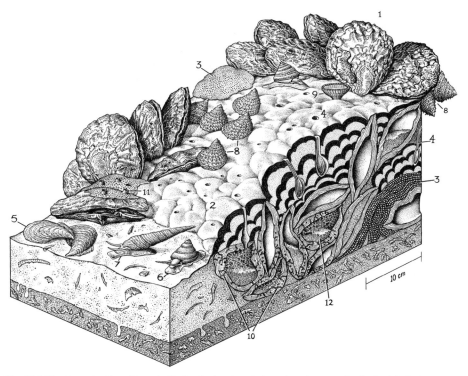

Fig. CS 3.12   Reconstruction of Jurassic (Portlandian) oyster–algal patch-reef community, England. 1: Oyster (*Liostrea*); 2: calcareous red algae (*Solenopora*); 3: thick bryozoan encruster (*Hyporosopora*); 4: boring bivalve (*Lithophaga*); 5: bivalve (*Isognomon*); 6: gastropod (*Pleurotomaria*); 7: gastropod (*Aptyxiella*); 8: spiny bivalve (*Plicatula*); 9: lithistid sponge; 10: sponge borings; 11: boring sponge (*Cliona*); 12: pellets and fungal hyphae. (Modified from McKerrow 1978; copyright, John Sibbick.)

tions of boring, peloidal cement precipitation, and reboring. Detrital aprons of reef-derived material extend away from the reefs, probably as a result of periodic erosion.

These large reefs are noteworthy in being both high-diversity and bivalve-dominated, and for the excellent preservation by natural casts of an abundant and diverse boring biota. This assemblage records the radiation of boring organisms, especially those associated with reefs, which took place in the Upper Jurassic in response to the rise of predators (the Mesozoic Marine Revolution). Other Late Jurassic reefs, however (e.g. Kimmeridgian *Praexogyra* reefs from Portugal) do not show such a well-developed boring biota even after accounting for differential preservation. This suggests that the Portlandian reefs might have grown under specific environmental conditions (such as very low sedimentation rates or nutrient enrichment). Environmental peculiarities might also be indicated by the rarity or complete absence of typical reef-builders (corals) and other major marine groups (brachiopods and echinoderms). (From Fürsich *et al.* 1994.)

The distribution of these reef types appears to have been determined by relative energy and light levels, together with sedimentation rates. Shallow reefs in the form of patch reefs, and shelf-margin ramp or barrier complexes were dominated by corals and could be highly diverse, with up to 30 genera present. Some patch reefs were dominated by branching forms (Fig. 3.29), others by massive corals and microbialites (Fig. 3.30). Middle Jurassic reefs

Fig. 3.29 Thickets of the phaceloid scleractinian coral *Thamnasteria dedroidea*, Haudainville, France. The hammer = 32 cm high. (Photograph: E. Insalaco.)

Fig. 3.30 Jurassic microbialites which developed within the primary cavities of the reef framework. The 'pillow' surfaces have a clotted fabric with a scant encrusting biota. L'Epine, Novion-Porcien, France. Hammer head = 19 cm diameter. (Photograph: E. Insalaco.)

show abundant microbialite in both photic and aphotic settings (Leinfelder *et al.* 1993), especially in low-energy settings or where sedimentation rates were low, and reefs with limited microbiate and thrombolite did not develop distinct relief (Leinfelder *et al.* 1993). Stromatoporoid sponge reefs were particularly common around the southern margin of Tethys during the Middle Jurassic to Late Cretaceous (Case study 3.13).

Case study 3.13: Stromatoporoid sponge reefs, Makhtesh Gadol, southern Israel (Jurassic (Callovian): ~165 Ma)

Reefs were well developed throughout Tethys during the Middle Jurassic to Late Cretaceous, and many were constructed by stromatoporoids (calcified demosponges). A variety of these reefs occur in the middle to upper Callovian Matmor Formation, which is spectacularly exposed at Makhtesh Gadol (Hathira Anticline), southern Israel. These include monospecific aggregations of branching forms (*Shuqraia*, *Promillepora*, and *Parastromatopora*) and domal forms (*Actostroma*) that colonized soft substrates, forming low-relief, low-diversity patch reefs or thickets, which sometimes covered substantial areas. There is no evidence of any ecological succession or wave-resistant framework. These reefs grew in relatively shallow,

Fig. CS 3.13 (a)   Large individuals of *Promillepora* sp. which have been bored by the bivalve *Lithophaga*. Hammer = 32 cm long.

Fig. CS 3.13 (b)   An aggregation of closely branching *Shuqraia* sp. Hammer = 32 cm long.

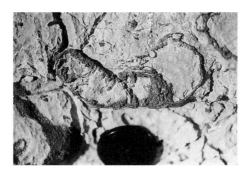

Fig. CS 3.13 (c)   A massive individual of *Dehornella crustans* showing reoriented growth (right) after toppling. Lens cap = 52 mm.

fairly open conditions, as evidenced by fine-grained carbonate lithologies consistent with soft-bottom, low-energy conditions, predominance of red (gymnocodiacean) algae over the relatively rare dascyclads, abundant echinoderms and diverse molluscs, as well as miliolid, lituolid, and epistominid foraminiferal faunas (C. Benjamini, unpubl. data).

In contrast, isolated, large (up to 0.35 m diameter) domal *Dehornella* and *Promillepora* heads, together with smaller domal corals, grew under more agitated conditions. These stromatoporoids were established on intermittently disturbed oolitic shoals, and show repeated growth reorientation due to wave or current disturbance. Up to 50 prominent growth bands (latilaminae) are present, giving an indication of considerable longevity for many individuals. However, only limited mutual encrustation is noted, with no resultant framework formation.

Upper Jurassic shelf-margin reefs in Slovenia were built by corals and stromatoporoids (Turnšek *et al.* 1981). The fore-reef slope contains substantial talus blocks, while the reef margin shows distinct outer and inner zones. The outer zone is characterized by diverse massive and phaceloid corals, and stromatoporoids, while the large bivalve *Diceras* (an ancestor of the rudists) is found in the most shallow part of the reef together with gastropods, *Tubiphytes*, and algae. The inner zone is dominated by diverse branching and encrusting corals together with small, nodular stromatoporoids.

Late Jurassic, deep-water mud-mounds formed by columnar, digitate, or crust-like thrombolitic and stromatolitic intergrowths were common in many areas of the western Neo-Tethys, in both deep intrashelf and ramp settings (Scott 1988). These were often associated with erect *Tubiphytes*, the worm *Terebella*, and siliceous sponges (hexactinellids and lithistids). The nodular limestones of the 'Ammonitico Rosso' known from Italy and France, which developed intermittently between the Toarcian and Tithonian, may also represent columnar microbial deposits up to 10 m thick that formed on drowned and rifted carbonate platforms in hundreds to thousands of metres of water (Pratt 1995). These deposits contain pelagic bioclastic debris such as ammonoids and radiolaria, and the pelagic bivalve *Bositra buchii* (= *Posidonia alpina*).

While siliceous sponge sediments with hexactinellids grew in deeper waters, lithistids occupied shallower areas. Some dictyid hexactinellids grew as large vases up to 2 m in diameter and 1 m high (Leinfelder 1994). Many of these sponges are covered with thick thrombolitic and microbial crusts. The sponges are also frequently encrusted by serpulids, foraminifera, bryozoans, thecidean brachiopods, and rare bivalves (Wagenplast 1972), some of which are bored (Reitner and Keupp 1991). Thrombolitic build-ups developed as either columns or irregularly surfaced hemispheres nucleating around metazoan skeletal debris.

Thin, well-bedded, low-diversity accumulations of platy microsolenid corals were common and widespread during the Upper Jurassic (Fig. 3.31(a)). They grew relatively slowly in low-energy, deeper waters with low rates of sedimentation, and appear to have

(a)

(b)

Fig. 3.31 Late Jurassic coral growth forms. (a) Thin, platy, microsolenid scleractinian corals. These are inferred to have grown at depth, but still within the photic zone. Liesberg, Swiss Jura. Coin = 2.5 cm diameter. (b) Foliaceous coral colonies. Haudainville, France. Possibly *Actinaraea*, which is often covered with a rich encrusting, cryptic biota. Pen = 14 cm long. (Photographs: E. Insalaco.)

been adapted to low light levels (Insalaco 1996; see Section 8.4.2(d)). These corals were associated with a rich encrusting and boring biota, and associated bivalves and echinoids. Attached to the coral undersurfaces were serpulids, thecidean brachiopods, calcified sponges, and bryozoans. Also known from the Upper Jurassic are unusual foliaceous coral colonies (Fig. 3.31(b)).

Late Jurassic reef communities persisted into the Early Cretaceous, but slope communities became less common, while the large, specialized bivalves known as rudists (Box 3.12) became abundant in shallow waters (Scott 1988). In the Early Cretaceous, coral–stromatoporoid–rudist communities were the most common shelf-margin reefs, dominated by diverse corals, branching and encrusting stromatoporoids, siliceous sponges, and green algae. Thin, low-relief sponge bioherms are known from the Scotia shelf, which developed at some 60–100 m depth under gentle current conditions. These reefs were dominated by lithistids, calcified sponges, and some hexactinellids (Jansa *et al.* 1982). *Tubiphytes* is present, and boring by bivalves as well as encrustation by algae, serpulids, and foraminifers is common. The first shelf-margin reefs to be dominated by rudists appeared in the Hauterivian (Masse and Philip 1981).

Box 3.12: **Rudist bivalves (Jurassic to Cretaceous (Maastrichtian))**

The Jurassic–Cretaceous rudists were heavily calcified heterodont bivalves which show a modification of the hinge and ligament system resulting in complete uncoiling of both valves (Skelton 1991). Rudists differ from most other bivalves in being inequivalve. The large (right) valve is conical, cylindrical, or coiled, and the upper (left) free valve is flattened. Paired internal muscle scars show that rudists could open and close these lid-like upper valves. The large valve is often divided by tabulae which served to support the organism as it grew and migrated upwards.

Most rudists were semi-infaunal, soft-sediment dwellers, although some were wholly epifaunal and included cementers. Sediment type, sedimentation rate, and water energy appear to have been the main determinants upon morphology and ecology, and it has been proposed that high rates of sediment flux (as well as the lack of available hard substrata) on platform margins is responsible for the ecology of rudists (Gili *et al.* 1995).

Rudists often colonized storm-generated bioclastic debris, probably during periods of reduced net sedimentation. Many formed aggregations of loosely to tightly packed groups, sometimes with a preferentially growth direction inclined downstream into food-bearing currents. Of these, the most gregarious were forms that bore conical or cylindrical morphologies where the entire commissure exhibited upward growth, especially hippuritids and radiolitids (the 'elevators' of Skelton and Gili 1991). Such forms grew under conditions of low- to moderate-energy levels in muddy environments, but perhaps with positive net sedimentation rates, with their valves largely embedded within and supported by soft substrate (Gili *et al.* 1995; Steuber 1997). Hippuritid rudists grew rapidly, with some forms achieving extension rates of 20–40 mm per year. As a result, calculated mean carbonate production rates are comparable to those recorded from modern coral reefs (Steuber 1996, 1997).

Rudists were often extensively bored postmortem by fungi, sponges, and algae. Of note, however, is the lack of encrusters and cements upon rudists. This may be due to: (1) most of the encrusters being soft-bodied, or (2) that rudists bore chemical defences to prevent encrustation, or (3) some extrinsic environmental controls. Rudist evolution is marked by several crisis (during the mid-Aptian, end-Cenomanian, and throughout the Maastrichtian) to final extinction towards the end of the Maastrichtian. After each crisis, remaining rudists diversified in adaptive radiations into the same array of morphological types.

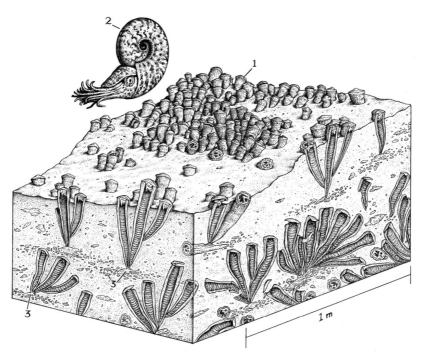

Fig. Box 3.12 Reconstruction of a rudist aggregation (Late Cretaceous). 1: Rudists (*Vaccinites* sp.); 2: ammonite; 3: shell lags. (Copyright, John Sibbick.)

The Middle–Late Cretaceous climate was equable, with sea levels 100–200 m higher than today, ice-free poles, and seasonally warmer climates at higher latitudes (Barron *et al.* 1981). Estimates suggest that 20–40% of the continents were flooded, and there is no evidence for rapid, glacioeustatic, sea-level oscillations of the type that provided the karstic substrates for reef growth during the Pleistocene. Eustatic flooding of the continents as well as high rates of tectonic extension of passive margins around Tethys was conducive to the formation of broad, shallow carbonate platforms, often with low-angle slope margins (although some reached up to 40°). Episodes of progradation, were common in such settings, and progradational or aggregational complexes developed where margins faced onto deep marine and ocean basins. Some Tethyan basins became restricted such that anoxic waters developed, and in addition, widespread oceanic anoxic events were fairly common during the mid-Cretaceous (Jenkyns 1980). Vogt (1989) has shown that volcanogenic upwelling of nutrified, anoxic water may also have been common at this time. The warmer middle Cretaceous climate and higher sea level coincide with an increase in the accumulation rates of organic carbon; as a result approximately 60% of the world's oil reserves are known from the Cretaceous (Irving *et al.* 1976).

Reefs with bound skeletal components are not characteristic of the Middle–Late Cretaceous. Although species diversity is higher than during the Jurassic, corals appear not to have formed shallow framework reefs, and usually inhabited soft, often muddy, substrates in deeper outer shelf settings, sometimes together with coralline algae in the Late Cretaceous. Stromatoporoids and chaetetids continued to form small, low-diversity patch reefs throughout the southern margin of Tethys. Microbialite-dominated reefs occurred in deeper water settings, including the subeuphotic, forming on ramps during the Turonian and Coniacian, and in the Aptian–Albian on shelf foreslopes (Pratt 1995). These were dominated by clotted and peloidal micrite and the alga *Bacinella*, which commonly encrusted corals, chaetetids, and lithistid sponges in shallow, upper parts (García-Mondéjar and Fernández Mendiola 1995). Calcified sponges referable to living genera (e.g. *Acanthochaetetes* and *Vaceletia*), lithistids and lychniscose hexactinellids, as well as small solitary scleractinians, thecidean brachiopods, bryozoans, and serpulids were common cryptic faunas within some coral and microbialite reefs. These forms may have shown the very slow growth rates and long longevities of similar modern communities within coral reef crypts.

The extensive platforms of this time were dominated by rudists, some of which were highly gregarious. They formed dense, usually monospecific aggregations, known from the Albian to the Maastrichtian (Fig. 3.32), that yielded huge piles of shells over 10 m thick and

Fig. 3.32 A substantial monospecific aggregation of the large Cretaceous rudist *Vaccinites vesicularis*. Each rudist is about 20 cm long. The view probably shows the superimposition of several communities. Upper Cretaceous (Campanian), Central Oman (Photograph: D. Schumann.)

kilometres in extent. These aggregations formed tabular or lenticular stratified bodies with little or no synoptic relief and no obvious signs of rigidity. But as few rudists were capable of forming elevated frameworks, their aggregations did not form protective rimming reefs such that most of this carbonate was redistributed as extensive bioclastic beds over large areas within platforms. Such rudists were thus both mutually supported and in part mechanically supported by infilling sediment (Gili *et al.* 1995), and their communities are associated with a high percentage of unbound sediment, often containing little associated shelly biota except for abundant rudist fragments, perhaps derived from bioerosion. Gregarious rudists appeared in the Turonian, but formed the most prolific aggregations during the Santonian–Campanian. Many hippuritid rudist associations may have grown in mixed carbonate–terrigenous settings with relatively low rates of sedimentation (Steuber 1997).

There has been much debate as to whether rudists displaced corals from shallow marine settings, or whether they radiated to invade newly available habitats of the very warm, shallow epicontinental seas of the Cretaceous, which may have had restricted circulation (compare Scott 1988 and Gili *et al.* 1995, with Kaufmann and Johnson 1988). Although large corals are known in these settings, they rarely flourished, and most aggregations of laminar or platy forms appear to have been restricted to the margins of platforms in deeper settings, but still within the photic zone. By the Albian, most microbialite reefs occurred in deeper, including subphotic, waters.

Turonian and Coniacian mud-mounds in ramp settings were composed of stromatolites of dense, clotted and peloidal micrite, microbial fabrics (such as calcified filaments and microbial colonies) and *Bacinella*, and assorted bioclastic debris at their shallow tops. These reefs probably grew on open-shelf settings at a few tens of metres depth (e.g. Pascal 1985; Camoin 1995), in contrast to Palaeozoic counterparts (Devonian and Carboniferous) which grew in deep waters (often 50–200 m, but up to 1000 m; Pratt 1995). Cretaceous mud-mounds tend to be lens-shaped, and most are only a few metres thick with low depositional relief and shallow dipping slopes.

Rudist and other mollusc communities declined some 1.5 to 3 Myr before the mass extinction event at the end of the Cretaceous (Kauffman 1984), but the event at the Cretaceous/Tertiary boundary marks a coincident catastrophe on land and in the sea (Section 5.3.6). One-third of families (about 57% genera) of corals went extinct.

3.8 Cenozoic: 65–0 Ma

After the end-Cretaceous extinction, global carbonate production plummeted, at least for 1 Myr, but probably for much longer (Zachos and Arthur 1986). The Cenozoic began with low sea levels, emergent continents, and global temperatures probably slightly warmer than today. The development of circumpolar circulation in the southern hemisphere, however, may have lead to cooler climatic regimes through the Palaeogene. Sea level also fell successively, primarily related to the accumulation of polar ice.

Although over half the genera of scleractinian corals are estimated to have become extinct, nine genera of living photosymbiotic corals survived the extinction event into the Palaeocene and all either became, or were closely related to, important reef-builders in the Cenozoic (Rosen 1998). Coralline algae appear to have survived virtually unscathed and indeed are responsible for the construction of most Palaeocene reefs, together with sponges.

The ecology of the various Tertiary reefs is not well known, as many reefs occur either in tectonically stable areas where they remain buried or in active areas where their relationships are complicated by faulting. Also they occur in tropical areas where they have been affected by deep weathering (James 1983).

Carbonate platforms and reefs were rare in the Palaeocene, and the oldest known are from the Danian (Bryan 1991). Many may not have occurred in the tropics, as they include non-photosymbiotic scleractinian coral thickets from Greenland, and small, low-diversity bryozoan and coral bioherms from Sweden and Denmark. Danian coralgal reefs are also known from the Paris Basin. More extensive reefs became established by the mid–late Palaeocene (Bryan 1991); these were composed of low-diversity assemblages of scleractinian corals (particularly *Stylophora*, together with *Actinacis*) and thin, platy coralline algae on the inferred, deep fore-reef, together with bryozoans—sponges being very important in muddy sediments. Back-reef deposits consist of delicately branching corals and abundant regular echinoderms. A well-described reef found in Libya shows distributional patterns similar to Recent reef complexes (Wray 1969): stromatolites occur in intertidal areas, lagoons are characterized by abundant codiacean and dascycladacean algae. Solenoporacean algae (which disappear at the end of the Palaeocene) occur at the lagoon to reef transition, and the reef itself is dominated by coralline algae. Siliceous sponge and coral-bearing coralline algal (thinly encrusting ?*Archaeolithothamnium*) reefs have been documented from the Late Palaeocene of Alabama (Bryan 1991). These reefs possess abundant micritic cements and bryozoans.

Many Palaeocene and Eocene shelf margins were dominated by accumulations of coralline algae and photosymbiotic benthic foraminifera (Frost 1986), but patch reefs of coral and bryozoans also developed. Fish similar to modern, coral reef faunas are known from the early Eocene (50 Ma).

High-latitude cooling began in the early to middle Eocene, when permanent ice appeared at the poles. However, at this time there was a new radiation of scleractinian corals that spread throughout Tethys. Groups appearing at this time such as the poritids and their relatives the actinids, together with the favids (which had survived the Cretaceous extinction) dominated most coral reef communities throughout much of the Cenozoic (McCall *et al.* 1994). By the end of the Eocene, all modern coral families had appeared. The ecology of Eocene reefs, however, is poorly documented, but distinct back- and fore-reef zones were certainly present in some Eocene reef complexes, and large foraminifera were abundant. Corals reaching over 1 m in diameter have been reported from the late Eocene of Israel (Benjamini 1981). However, extensive coral frameworks were rare, and true rimmed margins are uncommon at this time. The oldest coralline algal ridges have been reported from the middle Eocene of Catalonia (C. Tabernar, in Bosence 1983).

Coral communities declined near the Eocene–Oligocene boundary, but extensive rimming reefs comprising scleractinian corals and coralline algae were present by the Oligocene, with the late Oligocene being a time of widespread and massive reef development (James 1983). Changes in ocean circulation around the Oligocene/Miocene boundary may have resulted in a more vigorous deep circulation of cold waters and upwelling in the Caribbean region, causing some extinction. During the Miocene, Tethys was finally obliterated and the Mediterranean formed as a separate sea. Also, two barriers developed (although their timing is not well established) marking the start of the differentiation of Indo-Pacific and Caribbean coral faunas, which today share no species in common (some authors have, however, suggested that this differentiation was already apparent as early as the Eocene: e.g. Coudray and Montaggioni 1982). First, a marine barrier in the Eastern Pacific is believed to have existed in the Miocene (and possibly earlier) which was later replaced in the Pliocene by the uplift of the Isthmus of Panama, and the second was the emergence of an African–Asian land barrier that became the Middle East in the late Oligocene to Miocene. Reefs were abundant, although possibly with a restricted latitudinal range. Some early Miocene reefs are characterized by the relative abundance and diversity of large, cylindrical solitary corals which are quite absent in reefs today (McCall *et al.* 1994). Roughly half of the Caribbean reef coral genera became regionally extinct during the early Miocene. Low-diversity reefs of latest Miocene age from the western Mediterranean were constructed by branching *Porites* and encrusting coralline algae, and crusts of probably microbial origin (Fig. 3.33).

Halimeda bioherms similar to those found in the Java Sea and on the Nicaraguan Rise, have been described from a Late Miocene (~6 Ma) succession of shelf to basin slope sediments in SW Spain (Martín *et al.* 1997). These aggregations are proposed to have been stabilized by rapid lithification of micritic and peloidal microbial crusts (Fig. 3.34).

During the Pliocene, the compression of climatic belts and the rise of the Isthmus of Panama fully restricted reef growth to the Caribbean and Indo-Pacific regions. As a result,

Fig. 3.33 Branching *Porites* corals. Upper Miocene (Messinian), Majorca. Lens cap = 52 mm.

Fig. 3.34 Detail of upper Miocene *Halimeda* bioherms, in SW Spain. These aggregations of disarticulated plates were stabilized by rapid lithification of micritic and peloidal microbial crusts.

perhaps related to climatic cooling or habitat loss, a major episode of coral faunal turnover took place between 4 and 1 Ma during the Plio-Pleistocene in the Caribbean (Budd *et al.* 1994). The most strongly affected reef corals were genera in the Pocilloporidae and Agariciidae, many of which persist in the Indo-Pacific.

Due to this faunal turnover, late Miocene and early Pliocene communities are entirely distinct from early Pleistocene–recent Caribbean reef communities. Although acroporoids also appeared in the Eocene, they did not become dominant reefs until the early Pleistocene. Today over 150 species of *Acropora* have been described, although only five are dominant on Caribbean reefs (K.G. Johnson *et al.* 1995). Pocilloporoids appear to have dominated Caribbean reefs for 5–6 Myr, but following a 1-Myr transition period of mixed acroporoid–pocilloporoid dominance, acroporoids became dominant in reef communities, although may not have achieved the levels of their extreme Recent abundance until the late Pleistocene (Budd and Kievman 1994). *Pocillopora* became extinct in the Caribbean about 60 ka. With this unexplained rise to dominance of branching acroporoids in the early Pleistocene, and a corresponding decline in massive, domal corals, coral reef communities with a completely modern aspect appeared (Fig. 3.35).

Fig. 3.35 Reconstruction of an Indo-Pacific coral reef. 1: Brain coral (*Leptoria phrygia*); 2: feather star (*Comanthus bennetti*); 3: Parrotfish (*Scarus* sp.); 4: Staghorn coral (*Acropora* sp.); 5: Emperor Angelfish (*Pomacanthus imperator*); 6: Gorgonian; 7: vase sponge (*Callyspongia* sp.); 8: anemone with clown fish; 9: giant clam (*Tridacna gigas*); 10: encrusting corals (*Montipora* and *Hydnophora*); 11: brittle star (*Ophiarachella gorgonia*); 12 and 13: sea urchins; 14: cowrie; 15: sea cucumber (*Thelenota ananus*); 16: sea star; 17: boring bivalve (*Lithophaga*); 18: cement botryoids; 19: internal sediment; 20: cone shell (*Conus textile*); 21: wrasse (*Coris gaimard*.) (Copyright, John Sibbick.)

Part II

Environmental controls

4 The role of physicochemical change

The ambient physicochemical environment and its fluctuations influences reef growth on many scales: from local hourly, daily, or seasonal effects which control the functioning of individual reef organisms; to regional factors that determine the distribution of different reef types; to those slow, but sustained and widespread changes which operate over millennia to millions of years that are responsible for Earth's changing climate, continental and ocean distribution, and ultimately for the geological record of reef-building.

On a geological timescale, reef deposition is often a relatively transient affair. Even seemingly long-lived reef complexes are in fact composites of many superimposed communities formed by successive collapse and recolonization in response to changing environmental conditions. Carbonate platforms build up to—and just above—sea level, and because reef communities also grow at particular depths, all must respond to changes in sea level. Reef growth and carbonate formation will cease when rates of sea-level rise outpace production rates, or when environmental change leads to the cessation of carbonate production or the widespread mortality of reef organisms.

Reefs throughout Earth's history are thought to have responded to changing environments in much the same way, and so the Phanerozoic history of reef-building potentially provides a good record of global change, just as the skeletons of individual reef-builders record the local temperature and chemical characteristics of the sea water in which they grew. However, controversies over the sensitivity of reef biotas and reefs to environmental change (e.g. Done 1991; Richmond 1993), and therefore the utility of reefs as records of such change, demonstrates that our understanding of reefs in relationship to their environment is still in its infancy.

Until recently, reef biotas were thought to respond to environmental change in a consistent and predictable fashion. But it is now clear that modern reef corals show great flexibility of response to their environments allowing species to acclimatize and adapt to considerable physicochemical fluctuations. This raises doubts as to the reliability of reef biotas as records of environmental change. This flexibility is, however, limited by distinct thresholds. Indeed, some corals may well live near their limits of tolerance, for example of sea-water temperature, making them vulnerable to extreme environmental disturbance (i.e. stress) in the short term. But as coral biotas have clearly been relatively stable throughout the many climatic upheavals and extinction events of the past 240 Myr, this suggests that the multifarious ways in which corals respond to short-term environmental change may also be responsible for maintaining the long-term success of the coral reef ecosystem. The difficulty, then, is to determine how environmental change interacts with reef biota on ecological and evolutionary timescales. For example, while moderate rates of sea-level rise will have little effect on the growth of individual reef-builders, this will be one of the primary determinants of the morphology of the resultant reef as it develops over millennia. And while changes of sea level will markedly affect the global abundance and distribution of reefs, this may have little evolutionary significance at the scale of the ecosystem as reef communities will simply re-establish themselves at suitable new locations (Blanchon and Shaw 1995). Here, the relationship between the magnitude and rate of environmental change, the generation time of the reef-building organism—and its ability to adapt to environmental change—becomes important. Longer term changes, however, can exert permanent, evolutionary change on reef ecologies, and there have also been profound, global physicochemical perturbations in the history of life which although currently poorly understood, are clearly responsible for the modification, and sometimes collapse, of entire reef ecosystems.

Anthropogenic causes of global change are now well established, and of accelerating importance locally, regionally, and globally. Of great concern is how natural ecosystems will respond to these changes. On a world-wide scale, there is an alarming deterioration in the health of living coral reefs: some estimates have suggested that up to 70% of modern reefs are under serious threat, mainly from local population pressures that are responsible for increased sedimentation, pollution as a result of nutrient loading, and overfishing (Wilkinson 1993; Wilkinson and Buddemeier 1994). Approximately 350 million people live on islands with coral reefs, and 450 million live within 60 km of coral reefs (SBS 1995). The prospect that global climate change as a result of increases in greenhouse gases might operate synergistically—that is their effects are additive—with local anthropogenic stresses may pose even more, but as yet poorly understood, threats (Wilkinson 1996). So whilst reef corals are well adapted to surviving natural environmental change, uncertainties as to the nature and rate of anthropogenic global change have made predictions about the future response of reef communities more difficult and imprecise.

Analyses of how environmental factors control modern reef growth are often used in an attempt to explain the temporal distribution of reefs in the geological record. Notwithstanding the problems of scale, such an uniformitarian approach can be highly problematic. This is because:

- Most modern carbonate sedimentation began only some 4–8 ka, since the sea-level rise at the end of the last Pleistocene glaciation. Moreover, the extensive carbonate-platform reefs and atolls of today are the product of an unusually prolonged period of stable sea level (some 6 kyr). This, together with relict topography, has exerted a strong influence over modern reef form and the style of sedimentation. Before the current stillstand, sea level fluctuated drastically over the previous few million years in response to the 'Great Ice Age'.

- Global reef distributions have varied considerably though geological time, and present-day sea level is relatively low compared to much of the geological record. The area of shallow-water tropical seas is thus small, resulting not only in a reduced volume of shallow-water carbonates being formed today, but also in an absence of analogues for the very extensive, shallow epeiric seas that were common when sea levels were high. When sea level was even lower than today (e.g. during the last glacial maximum, 21–18 ka), reef growth was further restricted to narrow zones along only the steepest continental slopes.

- As reef-building organisms have changed throughout the Phanerozoic, so have their environmental requirements, mode of growth, and rate of accretion compared to modern coral reefs. Moreover, while bioerosion is an important sediment-generating process on modern reefs, reef communities older than late Jurassic age lacked abundant bioeroding organisms.

- Present-day climate is relatively cool, with well-developed polar ice-caps. During significant periods of geological time, ice-caps were not present, and at times (such as the mid-Cretaceous), the northern hemisphere mean annual surface temperatures may have been 15 to 25 °C warmer than today.

- Sea-water chemistry has changed over geological time. For instance, 60% of modern $CaCO_3$ production is accounted for by calcareous plankton which produces pelagic carbonate-ooze deposited on the deep ocean floor (100–4500 m depth), leaving only around 40% of carbonate production in shallow marine waters. Calcareous plankton evolved during the late Mesozoic and their appearance caused a shift in the sites of carbonate accumulation from shallow cratonic to oceanic areas: before this time, shallow marine carbonate deposits accounted for up to 90% of global production (e.g. Davies and Worsley 1981). Consequently, for much of geological time, marine carbonate distribution—and possibly carbonate saturation levels—were very different to today.

- Shallow-water carbonate accumulation is an integral part of the carbon cycle, and the pattern of global carbonate production is an important indicator of weathering rate and levels of atmospheric CO_2 during the Phanerozoic, as well as providing a record of changing climatic gradients. Moreover, the dominant mineralogical composition of carbonate skeletons and inorganic precipitates has changed through the Phanerozoic, probably in response to long-term fluctuations in sea-water–atmosphere interactions.

This chapter addresses the following questions:

1. What are the environmental factors that govern the distribution of carbonate production and reefs?

2. What is the response of individual reef-builders, reef communities, and reefs as geological constructions to environmental change at different timescales?

3. To what extent can the evolution of reef communities over the Phanerozoic be explained by environmental change?

4. How important is reef deposition for understanding the ancient carbon cycle?

First, let us consider the environmental controls that govern the formation of calcium carbonate in modern seas.

4.1 Physicochemical controls on carbonate production

It is well established that, in modern seas, areas of carbonate production and accumulation generally occur in regions of shallow water and warm temperatures removed from sources of terrigenous input—which introduces clastic sediment, freshwater, and nutrients that inhibit carbonate formation. Indeed, significant local carbonate accumulation can also occur in deep, cold temperate waters and even Arctic latitudes where terrigenous sediments are excluded. However, most organic productivity and carbonate production takes place in less than 10–15 m of water, so that the location of suitable shallow environments is determined largely by tectonic context (Section 2.4). In modern seas, more than 30% of tropical continental-shelf area is less than 30 m deep (Kleypas 1997).

The saturation of sea water with respect to calcium carbonate decreases with depth, with undersaturation with respect to aragonite occurring at 300 m in the Pacific, and 2000 m in the Atlantic. Below 200–300 m, carbonate suffers dissolution, but only at depths of several thousands of metres does the rate of dissolution increase markedly (at the lysocline), and the *carbonate compensation depth* (CCD) is reached where the rate of sedimentation equals that of dissolution. Shallow marine carbonates therefore have a good preservation potential in the geological record, although those buried to a few hundred metres through continuous sedimentation may be vulnerable to dissolution by contact with large volumes of sea water undersaturated with respect to aragonite.

Much debate concerns the relative importance of biologically and chemically controlled factors in predicting suitable sites for carbonate production and, in particular, reefs (e.g. compare Lees 1975 with Opdyke and Wilkinson 1993). Specifically, environmental conditions control the development of reefs in two ways, by:

(1) indirectly controlling the production of calcium carbonate; and

(2) directly limiting the growth of reef-builders.

It is now clear that elevated ambient sea-water supersaturation with respect to calcium carbonate broadly corresponds to sites of carbonate sediment generation, and this is related to a number of factors, most importantly surface sea-water temperature. However, many carbonate-forming marine communities are also sensitive to changes in salinity, light intensity, nutrients, and sediment concentrations or fluxes, so that such local environmental factors will influence the style and rate of carbonate deposition independent of the broad, global distribution of temperature and saturation state. Non-biological processes that remove CO_2 from sea water, such as evaporation and turbulence, can lead to increased $CaCO_3$ production, and skeletal carbonate formation is also substantially enhanced locally by small-scale supersaturation due to photosynthesis, which consumes CO_2 so removing a potentially limiting waste product (Kinzie and Buddemeier 1996).

4.1.1 Temperature and sea-water saturation state

On the basis of measurements of aragonite saturation of surface sea water, surface ocean values of temperature and saturation state have been shown to be positively correlated, and each exhibits a systematic variation with latitude (Fig. 4.1; Opdyke and Wilkinson 1993). This is because equilibrium with the partial pressure of CO_2 (pCO_2) is nearly achieved at all latitudes. In equatorial regions (0–15°), saturation values are depressed relative to this correlation due to lower rates of evaporation, higher atmospheric precipitation, and local upwelling of deep marine water which raises the concentration of CO_2. At higher latitudes, temperature and saturation state decrease approximately linearly. The latitudinal saturation maximum occurs near 20°.

When duration-normalized Holocene accumulation rates are considered as a function of latitude, rates of carbonate deposition are greatest between 15° and 20°, and decrease both polewards and toward equatorial regions (Fig. 4.1). This demonstrates that the latitudinal distribution of carbonate deposits records climatic gradients in the temperature and saturation state of sea water. Of most note, is that the broad patterns of change in accumulation rate with respect to latitude are very similar to those defined by saturation values.

To confirm this relationship, it is informative to consider the rate of inorganic carbonate precipitation in laboratory experiments. Here, the relationship between carbonate accumulation rate (R), latitude, and sea-surface temperature can be described as:

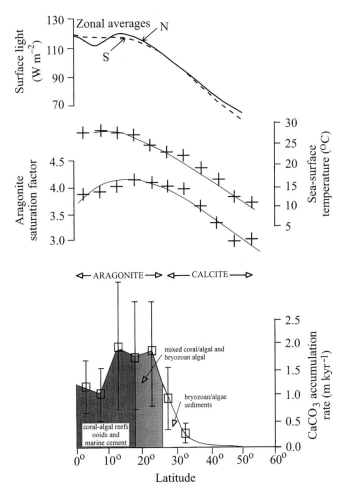

Fig. 4.1 Latitudinal distribution of photosynthetically active surface radiation (Kleypas 1997), temperature and surface-water aragonite supersaturation ratio, compared with Holocene calcium-carbonate accumulation rates and types (Opdyke and Wilkinson 1993). Note the close correlation between aragonite supersaturation and the distribution of carbonate. (Modified from Buddemeier 1997, with kind permission from Macmillan Magazines Limited.)

$$R = k \, (\Omega - 1)^n \, ;$$

where n = order of the 'process', k = a rate constant, and Ω = calcite or aragonite saturation (Opdyke and Wilkinson 1993). This relationship can be used to calculate anticipated changes in the carbonate accumulation rate from different latitudes in relation to temperature and saturation state. What is striking is that rates of carbonate accumulation from many modern environments and those calculated from laboratory growth of carbonate minerals are in very close agreement. These data support the contention that the latitudinal variation of sea-water composition is therefore indistinguishable from biological factors related to the ecological demands of carbonate-producing biota in controlling the broad global distribution, rates, and composition of shallow-water carbonate accumulation (Opdyke and Wilkinson 1993).

However, both biologically and non-biologically controlled (inorganic) carbonates exhibit significant changes in the composition of constituent grains and mineralogy with increasing latitude (Fig. 4.1). The following sections consider the processes responsible for the distribution of three types of carbonate which contribute to reef fabrics: inorganic carbonate, microbialite carbonate, and skeletal carbonate.

Box 4.1: **Reef cements**

Reefs are sites of substantial cement precipitation. Cements are precipitated within voids, both between grains and within cavities, and these will have distinctive crystal fabrics. As different cement types are generally characteristic of different environments, so a sequence of cements (and other diagenetic events—see Section 4.4.2) may form which reflect the evolution of marine pore waters. Different cement types can be predicted when a carbonate sediment passes from one environment to another, such as might occur with sea-level change, burial, or vertical tectonic movements.

Although a wide variety of cement mineralogies are present in modern coral reefs, most are the polymorphs of either aragonite or high-Mg calcite (12–20 mole-% $MgCO_3$). Most aragonite cements form as needle-like (*acicular*) crystals typically 10 µm in diameter and 100—500 µm long which grow elongate parallel to the crystal's c-axis. Aragonite cements grow as fringes, needle meshworks, fans, and hemispherical *botryoids* (which may reach 100 mm in diameter), or as micrite; high Mg-calcite cements form acicular—or bladed— fringes, equant (those with approximately equal dimensions) crystals, micrite (within internal reef sediments), or peloids. Apart from those of faecal or bioclastic origin (see Section 4.4.2), most peloids are probably *in-situ* precipitates, but the possible involvement of bacteria, algae, or the decay of organic matter in their formation remains controversial (see Chafetz 1986; Macintyre 1984, 1985).

Although factors which control cement type and mineralogy are not well understood, they may be influenced by the mineralogy and ultrastructure of the substrate, and the chemistry and temperature of the pore water together with rates of fluid-flow which determines the supply of carbonate ions. Indeed, the availability of carbonate ions appears to determine the rates of nucleation and crystal growth, and accounts for the acicular form of most cements formed in shallow marine environments (where nucleation sites are limited and growth rate rapid); equant crystals form where the carbonate supply rate is much lower (Given and Wilkinson 1985); and micrite cements form where nucleation sites are high but crystal growth rate is limited. Variation in the degree of cementation appears to be strongly controlled by rates of fluid flow through a sediment, which is determined by permeability: cementation is most pervasive on windward reef margins, where there is more active circulation of water through the highly porous and permeable reef.

Many ancient reef cements are *isopachous fibrous fringes*—those where the c-axis of the crystals are significantly elongated and are of the same length; these often show episodic and layered growth. Acicular fibrous cements are common, but most crystals are columnar, with many showing a *radiaxial* fabric which show convergent undulose extinction. Whether radiaxial calcite is a primary or replacement fabric is controversial, but support for a primary origin comes from the discovery of radiaxial calcite in Miocene limestones from Enewetak Atoll (Saller 1986) and the Pleistocene of Japan (P.A. Sandberg 1985*a*). However, the almost complete absence of radiaxial calcite from Quaternary limestones, but abundance in Palaeozoic reefs, suggests that sea-water chemistry may have changed during the course of the Phanerozoic (Section 4.4.4).

Other ancient marine cements do not have exact modern analogues, including cements with low-Mg, columnar morphologies. This lack of modern analogues makes identification of the original mineralogy and fabric problematic, particularly as most ancient cements are now low-Mg calcite. Aragonite is metastable and so is not normally preserved in the geological record, such that petrographic and geochemical evidence must be used to indicate an original aragonitic mineralogy (see texts such as Tucker and Wright (1990) for further details). There appears to have been a secular variation in the dominant mineralogy of marine cements precipitated through the Phanerozoic (see Section 4.4.4).

If there has been little diagenetic change of marine cements, particularly fibrous calcite, then carbon and oxygen isotope signatures will reflect the composition of the sea water from which they formed. From such data, we know that $\delta^{18}O$ has changed throughout the Phanerozoic, with for example Devonian sea water being 2 to 3‰, lighter than modern sea water.

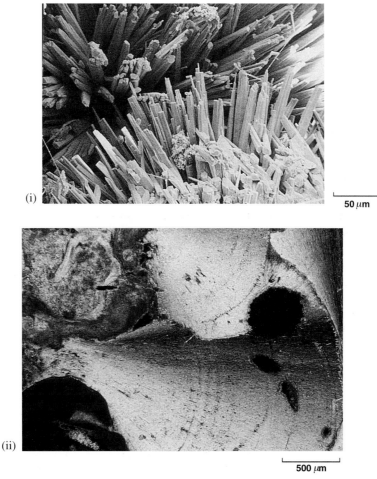

(i)   50 μm

(ii)   500 μm

Fig. Box 4.1(a) Modern cements. (i) Two-year old fibrous aragonite. (ii) Botryoidal aragonite from a coral fore-reef. Exuma Cays, Bahamas. (Photomicrographs: G.M. Grammer.)

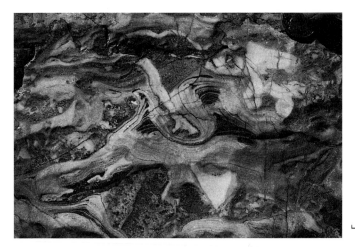

5 cm

Fig. Box 4.1(b) Ancient early marine cement: radiaxial fibrous layers from back-reef sediments of the Upper Devonian Canning Basin, western Australia.

4.1.2 Inorganic carbonate

While temperate shelves are dominated by skeletal carbonates (see below), equatorial carbonates are typified by both aragonitic or magnesium calcitic skeletal carbonates and inorganic components such as ooids, peloids, and marine cement (Schlanger and Konishi 1975). Ooids and peloids form in areas where mean temperatures exceed 18 °C, such that they are restricted generally to latitudes of 15–25° where salinities are slightly higher than normal. However, for reef formation it is the processes of cementation that are of paramount importance.

The precipitation of carbonate cements occurs when pore-fluids are supersaturated with respect to the cement phase and no inhibiting factors are present (such as dissolved phosphate, Mg^{2+}, or sulphate ions). Cementation is driven by CO_2 degassing from agitated sea water, which generates CO_3^{2-} ions from the dissociation of HCO_3^-. Rates of cement growth are very rapid on modern reef environments, not only in the intertidal and very shallow subtidal zones, but also in deeper environments up to 60 to 85 m (Grammer *et al.* in press). Average growth rates of 8–10 mm 100 year^{-1}, but rates of up to >25 mm 100 year^{-1} have been recorded from aragonitic botryoid cements forming in marginal slope deposits in the Bahamas and Belize (Grammer *et al.* 1993). A wide variety of cement mineralogies and morphologies are found in modern and ancient reefs (Box 4.1).

Cementation is pervasive in reef-front and reef-crest zones where high-water flux occurs as a result of the pumping action of waves (James *et al.* 1976). Sea water may be pumped through shelf margin reefs at rates up to 2000 m day^{-1}, in contrast to subsurface pore-fluid rates which are often less than 1–10 m year^{-1} (Buddemeier and Oberdorder 1986). This partially explains the steep, wave-resistant profiles of many modern and ancient reefs: walled reef complexes present a prominent surface to wave and current action so that the force of sea-water flux is high. Reefs with low-angle profiles present less of a barrier and so undergo far less cementation (Kendall and Schlager 1981). The same is true of ancient reefs—the steep-fronted Devonian Canning Basin reefs, western Australia (CS 3.5) and the Permian Capitan Reef, Texas and New Mexico (CS 3.9) are well-cemented, but the low-relief Devonian Swan Hills reef of Alberta shows only minor cementation in the reef crest and front (Wong 1979). Moreover, in the Capitan Reef, the degree of cementation increases as the profile of the reef steepens. (However, it should be noted that the striking steep, lower reef fronts of at least some present-day reefs may be poor analogues for ancient reefs since they have developed on relict topography.)

4.1.3 Microbialite

Carbonate-inducing microbial growth or decay can be a significant component of reef frameworks, and such carbonates are known variously as microbialites, stromatolites, and thrombolites (see Box 3.1). Prokaryotes (bacteria), photosynthetic bacteria (cyanobacteria), eukaryotic algae, diatoms, macroalgae, and soft-bodied metazoans such as sponges that contain populations of bacteria, can all be involved.

While the processes are poorly understood, micrite cement-formation clearly requires particular physicochemical conditions, principally high levels of carbonate supersaturation. The mechanisms are multiple, but three broad groups of factors appear to be important:

(1) The alteration of Ca concentrations, Ca/Mg ratios or alkalinity, or raising pH by either the removal of CO_2 (by photosynthesis or degassing during wave agitation or elevated temperatures (Merz 1992); or the production of ammonia from decaying organic matter (Macintyre 1984); or the weathering of silicates (Reitner 1993);

(2) the presence of specific Ca-binding organic molecules or matrices that serve as templates for crystal nucleation;

(3) mechanical pumping of sea water through pore-spaces by wave or current action.

The direct calcification of microbial, soft-bodied algal, or sponge tissue, and sediment-trapping therein, appear to be minor processes in the Recent compared with biologically induced cement precipitation and the calcification of biofilms. But the relative abundance of inferred microbialites in the ancient reef record, especially before the Cretaceous, suggests that the physicochemical processes conducive to the calcification of microbial communities in normal marine sea waters were far more prevalent than is known today (Section 4.4.4).

4.1.4 Skeletal carbonate

The latitudinal pattern of global carbonate distribution is also related to the taxonomic composition of the dominant sources of skeletal carbonate, and this is controlled by the individual tolerances of the skeletal organisms. Reefs develop wherever both the environmental conditions for carbonate production and those of the constructing organisms are met, with different reef-building communities having different demands. The relative ability of reef-constructing organisms to withstand physical (such as wave energy, turbulence) and biological disturbance (predation and competition) as well as their demand for light will determine their relative position upon the sea floor. The main physical factors which influence this distribution are temperature, salinity, light, and nutrient levels.

Modern temperate shelves are typified by calcitic bryozoans, benthic forams, molluscs, barnacles, and calcareous algae, whereas equatorial carbonates are dominated by aragonitic corals, calcareous green algae, and coralline algae (Lees 1975; Schlanger and Konishi 1975; Fig. 4.1). Many of these organisms are tolerant of extreme salinities, and reefs built by serpulids, oysters, sabellariids, and bryozoans can form in brackish or hypersaline conditions which exclude stenohaline competitors. Reefs built by monospecific stands of vermetid gastropods and sabellariid polychaetes can also grow under conditions of high nutrient flux. Light is clearly not a limiting factor for reefs built by these non-photosynthetic organisms.

But while the skeletal organisms which dominate temperate shelves (and so construct temperate reefs) are eurythermal, most corals and coralline algae cannot thrive below about 14–16 °C, and can usually tolerate only a narrow range of salinities (approximately 25–40 ppt). (In the geological record, rising salinities due to increasing restriction are recorded by a shift from coral-dominated to algal-dominated communities, and eventually to evaporite deposits (Hubbard 1989).) The presence of algal symbionts in reef corals also demand high light intensities, and most corals cannot tolerate sustained, high nutrient levels, which favour the growth of macroalgae (see Section 7.1.11). The linear *Halimeda* mounds found in several off-reef subtropical settings can form where ocean current upwelling of nutrients occurs along shelf margins.

Only those communities capable of some degree of wave resistance such as corals will construct relief at high-energy platform margins and so affect sedimentation across the top of the whole platform. In contrast, *Halimeda* mounds form only at depths greater than 40 m with active currents, and many temperate aggregating organisms prefer quiet water settings of the inner shelf. The setting will determine the ambient energy, such that, for example, low-energy reefs may be found in deep water, shallow-water protected areas, and downslope on gently dipping platform margins.

A limited number of basic environmental parameters can be used to model successfully the distribution of modern coral reef habitat and calcification rates (e.g. Bosscher and Schlager 1992; Kleypas 1997). Temperature, salinity, and, to a lesser extent, nutrients determine the distribution of coral reef communities, where hydrographic exposure enhances coral growth, and light controls the rate of calcification. It should be noted, however, that many environmental variables are inextricably correlated with changes in water depth. These include physicochemical variables such as changes in light penetration, water motion, sedimentation, substrate type, temperature, and salinity, as well as biological factors such as changes in competition and predation pressure. These factors also probably interact, so making individual constraints on reef growth difficult to isolate.

4.2 Factors that govern the distribution of modern coral reefs

4.2.1 Temperature

Few photosymbiotic corals grow below 14 °C, and the limiting high temperatures for corals are sustained maxima of 30–34 °C (e.g. Glynn 1984), although these maxima vary according to geographic region. At least 50% of photosymbiotic corals are estimated to occupy non-reef habitats with an average minimum temperature of 14 °C (Veron 1995). Heat stress does not limit the dispersal of corals, but can cause a breakdown in the coral-photosymbiosis leading to the loss of algal symbionts. Heat stress tends to occur in small areas within equatorial regions, often correlating with abnormal events such as very low tides or reduced tidal flushing.

The species diversity of modern reef corals also broadly correlates with latitude, and the decrease in diversity of scleractinians with distance from the equator is well established (e.g. Rosen 1981; Veron 1995). Coral growth rate and species diversity typically increase with temperature in continuous island arc or other reef settings: for example, the southward decrease in coral diversity along the Great Barrier Reef has been explained by decreasing temperatures (Hopley 1982), and coral species richness in Japan can be predicted almost solely from the sea-surface temperature (Veron and Minchin 1992). In areas with reduced availability of suitable reef habitats, however, the dispersion capability of planulae may be more limiting.

Most coral reef growth is markedly restricted to waters which do not fall below 18 °C for significant periods of time, and so coral reefs are confined to low latitudes (28 °N to 28 °S) except where the buffering effects of warm currents such as the Gulf Stream allows growth in higher latitudes. The transition from mostly aragonitic to low-Mg calcite sediment also occurs broadly at the temperature minimum for reef-building coral (Fig. 4.1). Reef formation by corals is therefore not limited by the low temperature tolerance of corals alone, as confirmed by the fact that the regional concentration of reefs and species diversity of corals are only weakly correlated (Veron 1995). It is not clear, however, as to the relative effect of temperature in suppressing coral reef growth: it may operate either by indirectly reducing the calcification rate (perhaps via daylight values and supersaturation state), or by directly involving interactive ecological processes where the energetic demands of corals are progressively less competitive in comparison with macroalgal communities.

4.2.2 Saturation state

Living, non-skeletal hexacorals such as sea anemones (actinians and corallimorpharians) are probably descended from scleractinian corals (skeletal hexacorals), and molecular phylogenies suggest that they are more closely related to living scleractinians than some scleractinians are to each other (Buddemeier and Fautin 1996a). A comparison of the environmental and ecological preferences of photosymbiotic and non-photosymbiotic corals and photosymbiotic sea anemones can thus provide interesting observations on the constraints of calcification within the hexacoral lineage (Buddemeier and Fautin 1996a, b).

Living, non-photosymbiotic sea anemones and scleractinians have broadly coincident latitudinal distributions that extend from the littoral zone to depths of several thousand metres, and they occupy similar or identical habitats (Buddemeier and Fautin 1996a). Sea anemones can live even deeper as they are not constrained by the solubility of calcium carbonate at depth. While photosymbiotic scleractinians are not limited to the tropics, only in the tropics do they dominate benthic communities; all high-latitude scleractinians are non-photosymbiotic. However, within the tropics, at least 50% of photosymbiotic corals occupy non-reef habitats with an average minimum temperature some 4 °C below the accepted limit for coral reef growth (Section 4.2.1). In coral reefs, photosymbiotic corals dominate the open surfaces, while non-photosymbiotic scleractinians and sea anemones are generally cryptic. By contrast, photosymbiotic actinians extend to high latitudes including polar seas, even though they are most common in the tropics and subtropics. They are also not wholly restricted to low-nutrient waters. This suggests that it is not the requirements of the algal symbiosis that restricts the distribution of photosymbiotic scleractinians to shallow seas with high temperatures and irradiance, and often low-nutrient levels.

When the distribution of photosymbiotic corals and photosymbiotic sea anemones is compared to aragonite saturation levels in the Pacific, it is clear that while neither calcification nor photosymbiosis is latitudinally limited, the distribution of photosymbiotic calcification in hexacorals corresponds closely to the distribution of at least 2–3 times supersaturation with respect to aragonite (Buddemeier and Fautin 1996*b*; Fig. 4.2). These observations support the conjecture that aragonite supersaturation may be an important environmental control on the distribution of symbiotically induced calcification. Photosymbiosis may enable corals to optimise skeletal production from warm, supersaturated waters (Buddemeier and Fautin 1996*b*; Buddemeier 1997).

We have, however, already noted that temperature is a near-perfect proxy for carbonate saturation state (see Section 4.1.1), making the relative effects of these two factors difficult to separate. Moreover, light as well as differing biological interactions are also covariant with these factors, such that interpretations of trends and correlations in terms of any single factor cannot be justified. It is necessary then to confirm inferences from distributional data with direct experimental evidence on the relationship between algal symbiosis, calcification rate, and saturation state.

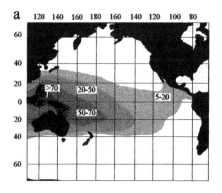

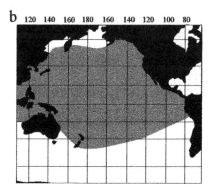

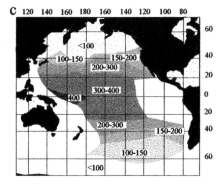

Fig. 4.2 Pacific Basin distributions of: (a) the number of genera of photosymbiotic reef coral (data from Veron 1986); (b) photosymbiotic sea anemones; (c) the average per cent aragonite supersaturation in surface sea water (data from Lyakhin 1968). (Redrawn from Buddemeier and Fautin 1996*a*.)

4.2.3 Nutrients

Most coral reef growth occurs when annual average concentrations of nitrate are less than 2.0 µmol L^{-1}, and phosphate less than 0.20 µmol L^{-1} (Kleypas 1995). High nutrient levels enhance the growth of benthic macroalgae, which compete with corals for space and light because such algae are highly nitrogen-limited (see Section 7.1.11). The seasonal influx of nutrients on inshore reefs which favour blooms of macroalgae can be an important limiting factor on coral reef growth/carbonate production.

Although coral reefs appear to grow in waters low in nutrients, by remaining sessile in a highly active hydrodynamic regime, reef organisms are able to remove nutrients from a source that is constantly being renewed (Atkinson 1992).

4.2.4 Light

Reef growth by corals that derive most of their metabolic requirements from their photosymbionts is clearly light-limited, and appears to be a function of photosynthetically available radiation (wavelengths 400–700 nm) and its attentuation with depth. Although coral-species growth rates are highly variable, all measured reef-coral extension rates decline logarithmically with depth (see Fig. 8.5(a)). Calcification rate is thus a linear function of the photosynthesis rate. Models which estimate the total area occupied by modern coral reefs based on a minimum light-level requirement of 250–300 µE m^{-2} s^{-1}, where E = Einstein, unit of energy (which restricts reef growth to 30 m or less) are close to those estimates based on charted reef locations (Kleypas 1997)—although all such estimates must decide arbitrarily on the definition of a coral reef.

The depth to which light can penetrate water is dependent upon water clarity and the incident angle of the Sun's rays. This varies with turbidity, which is influenced by the amount of suspended plankton (controlled by nutrient flux), dissolved substances, and particulate organic matter. Latitude and turbidity thus determine the depth of the *photic zone*, so governing both the maximum depths at which photosymbiotic corals are able to grow and the relative distribution of coral zones (Hallock and Schlager 1986). Sediment input can cause abrasion of delicate coral tissues and smothering, and will inhibit larval recruitment. Sedimentation can therefore be an important determinant of across-shelf gradients.

4.2.5 Hydrographic exposure

Reefs grow fastest where hydrographic exposure is high, so that reef growth is particularly favoured on topographic breaks and highs. Reef communities attain their greatest biodiversity on the edges of continental shelves and on headlands, rather than on topographically uniform, inner continental shelves.

Numerous aspects of reef growth and form, from reef organism morphology to zonation and global reef distribution, are closely related to the hydrodynamic regime. All reefs will be variously affected by storms, currents, tides, and wave energy. In particular, there is a marked difference between reefs that grow under conditions of wave-action and those that do not. Areas of low wave energy and frequent storm disruption are dominated by open pavements with only scattered coral colonies (Hubbard 1989). As well as favouring carbonate production, high wave agitation provides oxygen and nutrient flux, removes sediment by off-shelf transport, and may inhibit intense predation and herbivory. The high-energy, swell-dominated setting of the open Pacific favours the formation of massive, algal ridge veneers, which are also found in high-energy areas of the Caribbean reef complexes. Reef zonation can also been explained in terms of lateral gradients in hydrodynamic regime (but see Section 7.1.8), and the frequent development of spur-and-groove structures in shallow parts of reefs is a response of corals to swell-dominated or trade wind wave-dominated regimes.

4.2.6 Topography

Substrate availability is one of the most biogeographically limiting physical parameters for reef growth, and pre-existing (antecedent) relief upon the sea floor is often a major determinant of both the location of modern reefs and their geometry. Reefs may form on topographic highs produced by older reefs which were exposed as platforms or shelves (e.g. the patch reefs of Belize and Bermuda). Karst surfaces, which developed globally during prolonged subariel exposure during low sea level some 5–6 ka, also form the foundation for some modern coral reefs. Drowned, Pleistocene erosional terraces have been colonized by Holocene reefs in Jamaica, as have silicaclastic or volcanoclastic depositional topographies such as the Pliocene sands found in Florida and Belize. However, obvious topographic highs have not been detected under many modern reefs.

The near-vertical fore-reefs developed on some modern reefs are a result of the underlying Pleistocene topography, whereas many ancient reefs have fore-slope angles of between 20 and 40°. With no antecedent topography to mould the reef profile, slope repose angles are controlled by sediment grain size (Kenter 1990). Muddy slope sediments build an incline on less than 40°, while breccia- and packstone-dominated slope sediments build up to 35°. The near-vertical relief provided by antecedent topography also facilitates offshore sediment removal away from the reef.

Elevation of topography determines the rate at which a reef will reach sea level. The gross morphology of individual reefs in many areas varies greatly, from elongate, shore-parallel structures to large, deltaic features where the tidal influence is clear, inherited either from antecedent tidally controlled substrate or current tidal regimes.

4.3 Rate of carbonate accumulation

The accumulation rate of reefs is determined by:

(1) the calcification rate of the constructional organisms;

(2) the rate of inorganic carbonate precipitation;

(3) the local erosive regime;

(4) the accommodation space available for reef growth, as influenced by topography, rate of subsistence, and eustatic sea-level change.

4.3.1 Calcification rates

The total calcification of a reef is the sum of calcification by all its constituent species together with inorganic production, and each reef community responds uniquely to the local regime. Generally, reefs growing in agitated waters will show higher calcification rates than those growing under either lower temperatures (although this effect is difficult to isolate as distinct from light or saturation state) or energy conditions, and marked seasonal effects—especially at higher latitudes—will also strongly influence both the maximum depth of reef growth and the total calcification rate. Moreover, different communities within a reef complex grow at different rates: the windward margins of reefs show rapid growth rates compared to leeward margins and algal ridges. Maximum growth rates on modern reefs are, however, often not recorded due to suboptimal conditions (such as temperature stress, nutrification, excess of predators or loss of herbivore population).

Reefs dominated by different species will also show differing rates of calcification. Growth rates of reef corals are highly variable, ranging from a few millimetres per year to 150 mm year^{-1} for *Acropora cervicornis*. For example, in the western Caribbean where growth was dominated by *Acropora cervicornis*, the calcification rate was different to that in the eastern Caribbean where *Acropora palmata* was the main framebuilder (Macintyre 1988). This may be explained by the higher saturating light requirements of *A. palmata*—which produces a reef framework at depths less than 5 m—than *A. cervicornis* which can form a framework up to a depth of 20 m. Such differences in the calcification rate also explain the different responses of reefs to changes in sea level.

4.3.2 Inorganic precipitation rates

We have already noted that cementation is pervasive in areas where high water flux occurs as a result of tidal action or the pumping effects of waves. Water flux through a sediment is determined by permeability, so that cementation is most pervasive on steep, windward reef margins, but reefs with low-angle profiles present less of a barrier and so undergo far less cementation. Here the ability of organisms to withstand the hydrodynamic regime becomes important.

4.3.3 Erosion

Coral growth rate cannot be extrapolated to reef accretion rate due to the pervasive action of physical and biological erosion. Loss of carbonate by highly variable rates of bioerosion and export of sediment off a reef makes carbonate production very difficult to quantify. Mass mortality events may also be followed by very rapid rates of erosion (e.g. Glynn 1997*b*).

4.3.4 Accommodation

As many sediment sequences are deposited close to sea level, the rate of sedimentation is actually constrained by *accommodation*, that is the space available to be filled by sediment. Significant carbonate accumulation is thus ultimately dependent upon topography and either the rate of sea-level rise or the rate of *subsidence*. The accommodation space available on islands or isolated oceanic platforms is very restricted compared to continental margins, and the thickness, size, and development of a carbonate platform and a reef will depend upon these factors, with the thickest reefs forming during periods of protracted relative sea-level rise as long as the reef constructors are able to maintain growth. In contrast, because sea level has been relatively stable over the past several thousand years, many modern reefs have grown to an elevation where further upward growth has not been possible. The accumulation rates of Phanerozoic carbonate platforms are therefore, in part, controlled by the growth rate of platform-edge reefs, although these are probably significantly higher than the growth potential of the platforms (Bosscher and Schlager 1993).

Modern rates of subsidence vary from 0.01–2.5 m 1000 year^{-1}, depending on the type of subsidence. Typical subsidence rates on passive continental margins are low, in the order of 0.01–0.1 m 1000 year^{-1}. Sea-level change due to plate tectonic processes yield a rate of sea-level change of about 0.1 m 1000 year^{-1}, but changes due to fluctuations in ocean-water volumes as a result of changes in the mass of polar glacial ice give far higher rates, up to 10 m 1000 year^{-1} (Donovan and Jones 1979). So while carbonate sedimentation—and reef growth—can keep up with moderate sea-level changes, it cannot maintain production with rapid subsidence (which is often associated with local faulting) or with major rises in sea level through rapid glacial melting (Schlager 1981).

On geological timescales, coral reef distribution can be explained, in part, by the interaction of temperature/saturation gradients with subsidence. For example, as the Pacific crust beneath the Hawaiian islands moves to the north-west, it cools and subsides. South of approximately 28 °N, reefs produce carbonate sufficiently fast to offset the rate of subsidence. But north of this point (known as the 'Darwin Point'), cooler water (less than 18 °C) inhibits coral growth such that reefs can no longer keep up. As a consequence, they slowly drown.

4.3.4 Measuring accumulation rates

As with most sedimentary sequences, the most important factor influencing the carbonate accumulation rate is the duration of time over which the accumulation is measured, rather than method of analysis, latitude, or sediment type (Opdyke and Wilkinson 1993). This is because only a fraction of time is actually represented by the rock record itself, as many breaks in sedimentation occur (hiatuses), mostly due to emergence, which can represent lengthy periods of time. Therefore, the importance of hiatuses will decrease with the increasing time interval over which the rate of accumulation is determined: thinner sequences incorporate either more or less amounts of depositional or hiatal time, but thick sequences will approximate the mean erosion/deposition rate to any particular location. As a result, time duration/accumulation rate data often exhibit an apparently logarithmic distribution. Where optimum conditions exist for carbonate formation—and regardless of the biota involved—the rate of production is fairly constant. A typical, mean accumulation rate of about 1.1 m 1000 year^{-1} is achieved for a 5-kyr duration in late Holocene sections, regardless of whether the carbonate sediment is sand, mud, or reef (Opdyke and Wilkinson 1993).

However, depositional styles and rates of carbonate production are locally highly variable: carbonate accumulation can occur over very broad, shallow shelves or be concentrated around reefs that represent only a small proportion of the shelf area. Globally, reefs account for only 2% of the shelf area between 30 °S and 30 °N (Kleypas 1997), but large, long-lived, frame-building organisms such as corals trap sediment which might otherwise be more widely dispersed by waves and currents, as well as being sites of intense cementation. Accumulation rates are highly variable within a reef complex. Maximum, vertical accumulation rates from coral-dominated reef fronts derived from reef cores can reach an average 10 mm year^{-1} (10 m 1000 year^{-1}), but rates of 9–15 m 1000 year^{-1} have been recorded in the Caribbean (Adey 1978). For the benthic foraminifera, coral, and coralline algal communities on seaward flats they are 0.5–1.5 m 1000 year^{-1}, while those in back-reef lagoons are lower, in the order of some 0.1–0.5 m 1000 year^{-1}. Average carbonate production across all coral reef environments has been estimated at 1.2–1.5 kg CaCO$_3$ m^{-2} year^{-1} (Milliman 1993), but on rapidly growing reef crests production rates can reach an impressive 10 kg CaCO$_3$ m^{-2} year^{-1} (Hatcher 1990).

4.4   Environmental change

Environmental change is highly complex, with numerous feedback mechanisms that involve many environmental variables, such as the close relationship between tectonic activity, climate, and sea level. Rather than treat each physical or chemical factor separately, these phenomena will be considered in the following section in terms of the timescale upon which they operate. This allows some understanding of how environmental change occurring at different rates has been responsible for the various characteristics of reef-building organisms, the development of individual reef complexes, the changing global distribution of reefs, and the geological record of reef-building (see Table 1.2).

From the outset it should be stated that there are no easy generalizations to explain the interaction of reefs with the environment. Each reef will uniquely respond to local environmental variables, making modelling and prediction very problematic. Moreover, some ecological and geological processes are slow and cannot always be detected on the scale of human observation. However, as we shall see, it may be this diversity of response that actually accounts for the robust nature of coral reef ecosystems over geological timescales.

The role of environmental change on reef biotas is considered here on four timescales:

(1) environmental fluctuations operating on ecological timescales (daily to interannual);

(2) short-term geological change operating in terms of thousands of years (kyr);

(3) medium-term geological change operating on tens to a hundred thousand years (10–100 kyr);

(4) long-term geological change, operating in tens of millions of years (10 Myr).

4.4.1   Environmental fluctuations on ecological timescales

Reef environments vary greatly in disturbance. Marginal temperate environments are subject to frequent, but sometimes predictable, disturbance. In contrast, *Halimeda* reefs growing at 30–100 m depth probably live under undisturbed conditions: δ^{13}C isotope data taken from boreholes that record 1000 years of continuous growth show remarkably uniform signals, probably indicating conditions of highly stable light and nutrient levels (Aharon 1991).

Contrary to the previous consensus that emphasized the stability of physical variables at tropical latitudes, the modern, shallow, tropical marine environment is in fact characterized by considerable fluctuations in physical conditions on a daily, seasonal, and annual and interannual basis. These fluctuations are compounded by the topographic heterogeneity of the shallow marine tropical environments, especially that of coral reefs, resulting in a tremendous variation in the physicochemical characteristics received by reef biota living in different habitats within single reef complexes.

These data have radically altered the view that coral reef communities are stable ecosystems that have existed in a state of near-equilibrium for millions of years, arising largely as a result of long-term and diverse competitive evolutionary adjustments under benign environmental conditions (Odum and Odum 1955; Grassle 1973). Reefs are clearly not in equilibrium, but present as highly dynamic communities that are constantly responding to both constant disturbance, and also to wholly unpredictable catastrophic events.

Daily

Although fluctuations on a daily basis are relatively poorly documented, it is clear that tidal influences (diurnal and semi-diurnal) are important in governing daily fluctuations in sea-surface temperature (SST), salinity, water movement and circulation, and irradiance. Although diurnal temperature variations of 1.5 to 4.5 °C are commonplace on shallow reefs, substantial fluctuations in daily temperature ranges from 6 to 14 °C have been recorded over shallow reef flats (e.g. Potts and Swart 1984), with reduced ranges of 0.1 to 0.8 °C found in deeper lagoon waters (e.g. Pickard 1986). Over a 24-h period, Orr (1933) recorded a temperature range from 25.3 to 34.9 °C for a tidal pool on the Great Barrier Reef; the lowest temperature ever recorded in these pools was 18 °C. Maximum and minimum SSTs for the moat at Heron Island are 35 and 16 °C, respectively, with the lowest ever temperature encountered at this locality being 13.5 °C. Continuous temperature records show that shallow-water corals can even be subject to rapid and unpredictable fluctuations of 3 to 4 °C over a 12-h period (Fig. 4.3).

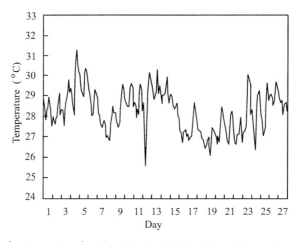

Fig. 4.3 Hourly, sea surface temperatures from thermistors located 0.5 m below the lowest low water in 2-m maximum water depth on a shallow coral reef in Pulau Pari, Indonesia, for January 1995 (after Dunne 1994). This site is exposed to continuous oceanic flushing in addition to periodic extreme low tides.

Seasonal

That seasonality is an important affect in the shallow tropics is evidenced by prominent, seasonal banding patterns in the skeletons modern and ancient reef biotas. While solar radiation is more constant in the tropics than in temperate zones, the tropics show a marked seasonal variability in salinity, tidal influences, and, in particular, precipitation (Fig. 4.4). Ozone concentration also varies markedly with latitude, which can result in higher levels of UV radiation in the tropics compared with temperate latitudes. Seasonality is particularly important in monsoonal areas, where 70% of all modern coral reefs are found (Potts 1983). There is now little evidence to support former conjectures that seasonality increases with increasing latitude, as even shallow waters (3–5 m) close to the equator show seasonal SST ranges of 4 to 5°C (Brown 1997a). Shallow marine environments also suffer seasonal changes in nutrient concentrations, sedimentation, and wave energy, as well as in current regimes and tidal patterns. Many reef flats experience seasonal variations of mean sea level of 0.2 m, with interannual variations reaching 0.3 m between some years.

Annual and interannual

The major intermittent disturbances which occur on an annual and interannual basis in shallow marine tropical environments include ENSO events (El Niño southern oscillation), fluctuations in monsoonal strength, and hurricane activity.

The ENSO phenomenon is an irregular oscillation of the ocean–atmosphere system in the tropical Pacific, which occurs approximately every 3–5 years. This complex phenomenon alters large-scale atmospheric pressure systems, wind systems, rainfall patterns, and ocean currents and sea levels in the tropical and subtropical oceans. It is probably linked with the Asian monsoon, where, during the El Niño phase of the oscillation, the southwest monsoon tends to be weak (Quinn 1992). During the peak of the El Niño, SSTs in the

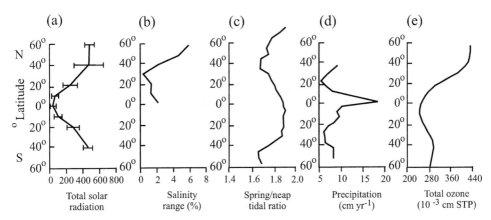

Fig. 4.4 Changes in physical parameters with latitude: (a) total solar radiation (Budyko 1974); (b) salinity range (Moore 1972); (c) spring/neap tidal ratio (Moore 1972); (d) average precipitation for all oceans (Sverdrup *et al.* 1942); (e) average total ozone (Cutchis 1982). (Modified after Brown 1997a, by kind permission from Academic Press.)

eastern tropical Pacific can reach 2 to 3 °C above normal for several months (Glynn and d'Croz 1990). Mean sea level may be also significantly depressed during ENSO-years at sites both within and outside the Pacific, leading to an increase in irradiance during calm conditions. This can lead to considerable coral mortality, where recovery may be slow and take at least several decades (e.g. Glynn 1984). Coral mortality can lead to a total cessation in reef growth due to a shift in macroalgal dominance and an increase in bioerosion (e.g. Glynn 1993). It appears that over the past 40 years, the southern oscillation has become more common throughout the global tropics, with increased frequency and strength (Wolter and Hastenrath 1989).

The monsoon shows significant seasonal variation and is associated with abrupt changes in atmospheric general circulation. Alterations in the timing and intensity of monsoons marks rapid changes in coastal upwelling, water circulation patterns, solar irradiance, nutrient levels, and turbidity due to changes in freshwater run-off.

Hurricanes occur in belts between 2 to 25° north and south of the equator, and can subject reefs to periodic catastrophic damage, particularly from 1 to 20 m depth. Hurricane activity is extremely variable, but on average any particular reef may only be struck by a severe hurricane four to eight times a century (Harmelin-Vivien 1985). Reef recovery is also variable, ranging from 5 to 40 years depending both on the degree of initial damage and the presence of any intervening disturbances (Scoffin 1993).

Environmental variation across a reef

A horizontal traverse across a shallow coral reef will mark great changes in topography, water depth, water energy, irradiance, current speed, and sedimentation style. Even individual coral colonies will experience very different degrees of illumination, water movement, and sedimentation (Chevalier 1971). While many of these changes may be complex and patchy, they are reflected in the zonation of reefs, most prominently in the growth form and species distribution of corals (see Fig. 6.3).

Wave and current strength show great variation across a reef profile, with a significant proportion of incoming wave energy being extracted at the reef crest (Roberts and Suhayda 1983). Currents measured across the reef crest can vary more than twofold depending on the tidal state, from 71 mm s^{-1} at high tide to 173 mm s^{-1} at low tide. Current forces may be more important than waves at some points on a reef, and the boundary between wave-dominated and current-dominated zones corresponds to distinct changes in the coral community. At Grand Cayman, the wave-dominated zone is characterized by robust branching (e.g. *Acropora palmata*), bladed, and encrusting growth forms, whereas delicately branched, massive, and plate-like growth forms (e.g. *Acropora cervicornis* and *Montastraea*) prevail in the current-dominated zone (Roberts *et al.* 1977).

The attenuation of irradiance with water depth may show great local variation according to substrate topography, slope, and exposure, as well as at the microhabitat scale. For example, both the irradiance levels and spectral quality of the light differs on differently inclined faces of a single colony (Brown *et al.* 1994).

Variation in SST has been recorded across some reefs. At Heron Island, Great Barrier Reef, the inner-reef flat showed the greatest temperature range (12.7 °C), compared to the outer reef flat (8.5 °C), the lagoon (6.8 °C), and reef slope (3 °C) (Potts and Swart 1984).

Biotic response

Shallow tropical coral reefs occupy changing and often extreme environments where great local variations in the distribution and abundance of marine species exist, both spatially between local 'patches', and temporally. For example, records show that community composition of corals growing at the same depth on Jamaican reefs was highly variable for several decades before the catastrophic events of the 1980s (Jackson *et al.* 1996).

Modern reef corals are highly dynamic organisms that are able to respond to the considerable, and rapid, physicochemical fluctuation in their environments via both genetic adaptations (such as life history traits—that is the schedule of events that occurs between birth and death) and phenotypic responses (*acclimatization*—those which operate during the lifetime of the individual). It is probable that the physicochemical fluctuations described above also operate synergistically, but the details remain poorly known.

Coral species appear to show extensive intraspecific variation, but considerable phenotypic variability also appears to be widespread, although it is likely to differ markedly between species. The high variability and plasticity of some coral species may reflect tremendous genetic variability of both the coral host and its photosymbionts. This high level of genetic variation has profound consequences for our interpretation of the fossil record, as such variation may leave no trace in the fossilizable skeletons which do not record the all-important behaviour, reproductive, or soft-part morphological differences between different genotypes. As a result of these biological insights, serious doubts are raised as to our understanding of fossil species definitions, and their stability and longevity.

Our current knowledge of these responses have been summarized recently by Brown (1997*a*), who described a host of molecular, physiological, and morphological adaptations in scleractinian corals. Physiological adaptations to a fluctuating environment include changes in the fluxes of photosynthetic pigments and protective UV sunscreens (primarily carotenoids) in the coral algal symbionts, and the production of heat-stress proteins. Indeed, the constant disturbance, and heterogeneity of the coral reef habitat, appear to be necessary for the functioning of a healthy coral reef community and for reef growth, and these factors are probably responsible for the many adaptive processes shown by corals.

The extensive intraspecific variation shown by corals probably permits rapid local adaptation to environmental conditions encountered at particular sites during the lifetime of any one generation. How each reef responds depends upon the physiological tolerance limits of the local coral population, which results in a very wide variety of responses to both environmental perturbation and in the style of subsequent recovery. For example, corals from Anewetak (fomerly Enewetok) have an upper lethal limit of 34 °C, whereas those from Hawaii can survive only up to 32 °C. These thresholds correspond to the 2 °C difference in the normal maximum seasonal temperature between the two areas. In the Arabian Gulf, reef corals are exposed to one of the most rigorous temperature and salinity regimes—the hardiest species can survive exposure to maximum temperatures of 36–38 °C and minimum temperatures of 11.4 °C, and to salinities that can soar to as high as 46 ppt (Brown 1997*b*). At least 24 species of coral have adapted—probably genetically—to these extremes which are way beyond the tolerances of most corals found elsewhere.

It has been postulated that genetic adaptation occurs in response to consistent environmental change, whereas phenotypic adaptation is a response to fluctuating change at a speed that matches that of environmental change (Potts and Garthwaite 1991). If this is true, we need to differentiate between those changes that are long-term and perhaps responsible for directional evolutionary change, and those that are temporary, but which select for plasticity and variability.

Record of environmental change in coral and sponge skeletons

The recognition of annual and interannual density bands within long-lived massive corals promises a proxy record of high-resolution climate and other environmental change. On the basis of assumed continuous, and relatively rapid growth rates, coral skeletal density patterns have been reported to reflect annual cycles in temperature and light with reproductive cycles, nutrient availability, sedimentation, and turbidity (see reviews in Buddemeier and Kinsey 1976; D.J. Barnes and Devereux 1988). Coral skeletons also contain trace elements (such as strontium, barium, and cadmium) and isotopic compositions (carbon and oxygen) which have been inferred to record environmental change. Annual cycles in $\delta^{18}O$ are considered to record sea-water temperature, and $\delta^{13}C$ primary production and photosymbiont activity. Unusual environmental occurrences such as ENSO events have been suggested to introduce anomalies into these annual records of temperature and primary production. The processes of coral growth, however, introduce biases that distort the recording of this environmental data, such that interpretation of the signal is not always straightforward, and is currently far from well understood (see D.J. Barnes and Lough 1996 for a discussion) and isotopic analyses of age and growth rates for fossil material are few.

Although very slow growing (some 0.25 mm year^{-1}), the isotopic compositions of the skeletons of very long-lived calcified sponges, e.g. *Astrosclera*, have also been demonstrated to record changing atmospheric CO_2 levels since the industrial revolution (Wörheide 1998).

4.4.2 Short-term geological change (kyr)

The two major environmental changes of importance to reefs that occur on a scale of a few thousand years are changes in sea level—which primarily concerns the shaping of individual reef morphology and growth—and diagenesis.

Reef morphology and growth

Reefs can initiate during sea-level rise (*transgression*), following exposure during sea-level fall, or when inhibiting factors to reef growth are removed. Most living coral reefs initiated growth during *high stands* (when sea level was at its highest point in any cycle) during the Pleistocene, and, in many reefs, subsequent fluctuations in sea level are accurately recorded by the rate of reef growth, carbonate accumulation, and resultant reef morphology. In the Ancient, many reefs also appear to have formed during the later phases of transgressions, e.g. during the late Llandovery–early Wenlock when sea-level changes coincided with the waxing and waning of early Silurian glaciations (Brunton and Copper 1994).

Modern reef morphologies are primarily the product of their inherited foundations and Holocene sea-level change (see Case study 4.1). In the Caribbean, the rate of sea-level rise decreased 3–5 ka and this has continued slowly to the present day (Hubbard 1989). Many shelf-edge reefs are thus still in the process of 'catching' up, resulting in profiles where sloping fore-reef environments predominate. Along the Great Barrier Reef (GBR), relative sea level stabilized over 6 ka (Thom and Chappell 1975). Here, stable sea level has curtailed reef growth so that reefs have grown to sea level over vast areas: in such conditions of stable sea level, marked lateral zonation has arisen with the development of extensive reef flats (Davies and Hopley 1983). The different morphological form of these reefs effectively creates two different environments, which in turn may control ecological factors such as the differences in styles of zonation, herbivory, and coral recruitment. This has also led to profound differences in sedimentary regime: on the GBR, most carbonate sediment is retained in actively prograding reef flats, whereas in the Caribbean, up to half is exported to the fore-reef over the steep shelf margin. So differences in rates of carbonate production are a reflection of local tectonic and sea-level history.

With stable, or only slowly rising sea level, reefs which grow on the edge of shelf margins often prograde over their own talus on the fore-reef slope. Progradation occurs because the rate of reef growth has matched sea-level change, such that accommodation space for the substantial production of reef carbonate becomes limited. Aggradation takes place when a reef continually grows vertically as a response to sea-level fluctuations, where the amplitude is large and the period short or when reef-builder growth rates barely match the rate of sea-level rise. When sea level is slowly rising, onlapping (retreating) reefs form which migrate continuously in a landward direction by progradation of the reef core over older leeward deposits. Such reefs are known in the Modern during stillstand after reefs have caught up with sea-level rise, but they are rare in the geological record. When sea-level rise is rapid, as during a transgression, backstepping may occur where reefs nucleate progressively upslope (landward) in stages, and pinnacle reefs can develop from former, shallower water patch reefs.

At high stand, upward growth may cease, and lateral growth seaward, especially in a windward direction, may occur such that shallow-water reef zones prograde over and bury deep-water communities. Long-term estimates of progradation for modern reefs have revealed rates of 0.75 m 1000 year^{-1} for volcanic islands of the Lesser Antilles (Adey and Burke 1977). More rapid rates of 0.84 to 2.5 m 1000 year^{-1} have been recorded from the Holocene of St Croix.

Case study 4.1: Galeta Reef, Panama

Data derived from 13 boreholes drilled through a small Caribbean fringing reef off Galeta Point, Panama, have enabled the reconstruction of the growth history of a reef in relation to underlying topography and rising sea level (Macintyre and Glynn 1976; Macintyre 1983).

Galeta Reef initiated upon the erosional surface of a Miocene siltstone about 7000 years ago. A typical Caribbean fringing reef formed, initially reflecting the relief of underlying topography, dominated by *Acropora palmata* in shallow areas and a mixed coral-head community on its seaward slope. By 4000 years ago, reef-flat rubble had begun to prograde over *A. palmata*, which itself started to prograde over the fore-reef. The reef kept pace with rising sea level until about 3000 years ago when its development was restricted by the lack of accommodation space such that it became emergent. A well-cemented, shallow fore-reef pavement formed, behind which developed an extensive reef flat dominated by coral rubble. By 2000 years ago, active framework formation had ceased, and today mangroves have encroached over the reef flat and a thick deposit of talus has built up against the fore-reef.

CARBONATE DEPOSITIONAL ENVIRONMENTS

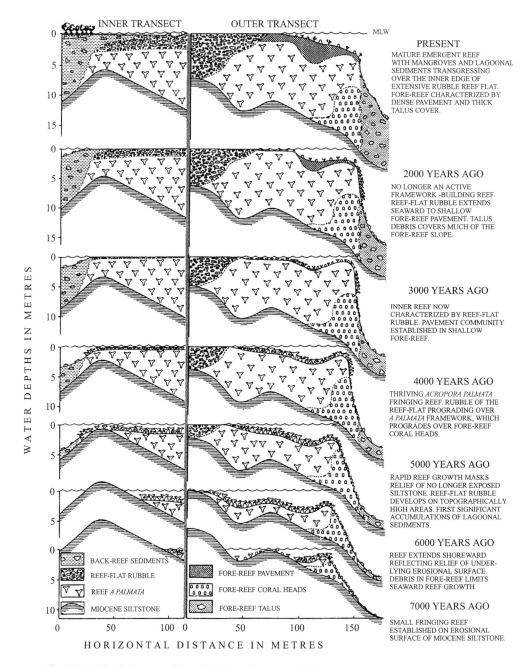

INNER TRANSECT   **OUTER TRANSECT**   MLW

PRESENT

MATURE EMERGENT REEF
WITH MANGROVES AND LAGOONAL
SEDIMENTS TRANSGRESSING
OVER THE INNER EDGE OF
EXTENSIVE RUBBLE REEF FLAT.
FORE-REEF CHARACTERIZED BY
DENSE PAVEMENT AND THICK
TALUS COVER.

2000 YEARS AGO

NO LONGER AN ACTIVE
FRAMEWORK -BUILDING REEF.
REEF-FLAT RUBBLE EXTENDS
SEAWARD TO SHALLOW
FORE-REEF PAVEMENT. TALUS
DEBRIS COVERS MUCH OF THE
FORE-REEF SLOPE.

3000 YEARS AGO

INNER REEF NOW
CHARACTERIZED BY REEF-FLAT
RUBBLE. PAVEMENT COMMUNITY
ESTABLISHED IN SHALLOW
FORE-REEF.

4000 YEARS AGO

THRIVING *ACROPORA PALMATA*
FRINGING REEF. RUBBLE OF THE
REEF-FLAT PROGRADING OVER
A.PALMATA FRAMEWORK, WHICH
PROGRADES OVER FORE-REEF
CORAL HEADS.

5000 YEARS AGO

RAPID REEF GROWTH MASKS
RELIEF OF NO LONGER EXPOSED
SILTSTONE. REEF-FLAT RUBBLE
DEVELOPS ON TOPOGRAPHICALLY
HIGH AREAS. FIRST SIGNIFICANT
ACCUMULATIONS OF LAGOONAL
SEDIMENTS.

6000 YEARS AGO

REEF EXTENDS SHOREWARD
REFLECTING RELIEF OF UNDER-
LYING EROSIONAL SURFACE.
DEBRIS IN FORE-REEF LIMITS
SEAWARD REEF GROWTH.

7000 YEARS AGO

SMALL FRINGING REEF
ESTABLISHED ON EROSIONAL
SURFACE OF MIOCENE SILTSTONE.

WATER DEPTHS IN METRES

BACK-REEF SEDIMENTS

REEF-FLAT RUBBLE

REEF *A.PALMATA*

MIOCENE SILTSTONE

FORE-REEF PAVEMENT

FORE-REEF CORAL HEADS

FORE-REEF TALUS

HORIZONTAL DISTANCE IN METRES

Fig. CS 4.1 The development of a small Caribbean fringing reef off Galeta Point, Panama, in relation to underlying topography and rising sea level (Modified from James and Macintyre 1985.)

Rising sea level favours vertical growth as long as the rate does not outpace coral growth rates (Neumann and Macintyre 1985). The growth rate potential of corals is very high— greater than most tectonic and subsistence rates—but when sea-level rise is fairly rapid, shallow-water reef communities, such as acroporoid corals, can be buried by deeper water communities ('catch-up' reefs). If sea-level fall is rapid, the reef will be isolated, exposed and die, followed by subaerial diagenesis. Some reefs which appear to have ceased growth abruptly ('give-up' reefs) have probably not been drowned by excessive sea-level rise, but rather have succumbed to adverse environmental conditions, such as stressful hot or cold temperatures or eutrophication (although there may in some cases induced by rapid sea-level rise), and may show a deepened succession terminated by an extensively bored and submarine lithified cap (Schlager 1981). For example, the Cretaceous guyots (flat-topped seamounts) of the Pacific drowned sequentially over a 60-Myr interval, while being trans-ported northwards by the Pacific plate through a narrow equatorial zone (1–10 °S), suggest-ing that they succumbed to currently undetermined adverse environmental conditions (Wilson *et al.* 1998). Other reefs may have been killed by the turbid, heated, saline, or nutrified bank waters formed by restricted circulation behind the reef. Where reefs have kept up with sea level throughout their growth ('keep-up' reefs), their internal structure reveals a continuous sequence of shallow-water reef communities that have successfully tracked sea level. Response to sea-level change varies greatly, however, even within the same reef. For example, at Papeete, keep-up reefs typify the windward margin, catch-up reefs the leeward margin, and give-up reefs the muddy patch-reef margins (Montaggioni 1988).

These concepts can be directly applied to ancient reefs. The geometry of individual ancient reefs is controlled by the nature of sea-level fluctuations during their growth and the ability of the reef community to respond to this change. Alternating reef and detrital reef-flat facies will be typical of reefs which tracked sea level but did not exceed it. Dominantly boundstone fabrics will characterize reefs that grew in intermediate water depths below the surf zone. Catch-up reefs will show repeated shallowing-upwards sequences.

Most coral reefs were able to continue growth despite the 0.2 m per decade rises in sea level that occurred between 14 and 6 kyr (Digerfeldt and Hendry 1987). Current predictions of a sea-level rise of 45 mm per decade (Callander 1995) are low compared to these Holocene rates, and indeed should have a beneficial outcome by increasing accommodation space over previously limited continental reef flats. However, such a sea-level rise will have serious impacts on low-lying coral reef islands (Wilkinson and Buddemeier 1994) where no lateral migration is possible. Under such circumstances, the upward growth of reef flats will not match these rates of sea-level rise, and will probably also be insufficient to prevent the destructive intrusion of sea water from storms so exacerbating the effect. Indeed, as sea-water temperature rises, tropical storms may become more frequent, and their range may expand to higher latitudes (Ryan *et al.* 1992), affecting reef growth and structure. Climate change will affect alterations to oceanic currents (so changing larval distribution patterns), rainfall distribution, sedimentation run-off, and salinity, but climate change models are not yet sufficiently precise to make detailed predictions for any individual reefs.

High temperatures and increased UV radiation are widely believed to operate synergistic-ally in the collapse of the algal symbiosis in corals and other invertebrates, causing the mass expulsion or *in-situ* degradation of the algae and resulting in *bleaching*. Extreme bleaching events can lead to widespread coral mortality, sometimes with a resultant community shift from a benthos dominated by corals to one dominated by macroalgae (T.P. Hughes 1994) and bioeroders (Glynn 1996). These ecological observations suggest that coral bleaching is a climate change impact that may directly damage reefs on a global scale, reducing reef accretion rates and carbonate production (Glynn 1993, 1996).

It is probable, however, that some reef-building taxa will survive, as numerous reef-building coral species have survived three historical periods of global warming (the Pliocene optimum 4.3–3.3 Ma; the Eemian interglacial 125 ka; and the mid-Holocene 6–5 ka) when atmospheric CO_2 and sea temperatures exceeded those of today (MacCracken *et al.* 1990). Although some living coral species and their photosymbionts may be operating at their physiological limits, no species' extinctions have yet been documented, and a warming of sea-surface temperatures might even result in a greatly increased diversity of corals at some higher latitude locations (Veron 1995).

Diagenesis

Diagenesis involves all the changes that occur between the formation of a sediment to the time when changes can be considered to be metamorphic—usually due to the high temperatures and pressures associated with deep burial. In carbonate sediments, diagenesis is the transformation into a limestone or dolomite. This includes many complex processes such as replacement (the transformation of one mineral into a more stable one), dissolution, the cementation of voids, compaction, micritization (the transformation of crystalline carbonate to micrite which often occurs under the action of boring microbes), and neomorphism (replacement by a different crystalline form). Major controls on diagenesis are the composition and mineralogy of the sediment, pore-fluid chemistry and rates of flow, the geological history of the sediment, and prevailing climate. Only a brief summary of this complex subject is offered here, and the reader is referred to reviews in Tucker and Wright (1990) and Scoffin (1992) for further details.

Aragonite and high-Mg calcite are metastable and alter to the more stable and less soluble low-Mg calcite (or dolomite) at a rate that varies with the diagenetic environment. This is a process known as *mineral stabilization* and is governed chiefly by the composition of the water and permeability of the sediment, but dissolution is also more extensive within heavily biologically reworked sediments (Aller 1982). If grains are in contact with sea water, dissolution may take 1–3 Ma, but in freshwater (*meteoric*), complete loss of aragonite may occur in only 5–20 kyr (Humphrey *et al.* 1986). Mineral stabilization results in significant and highly complex chemical and textural changes in the limestone. In a system which involves the active pumping of water undersaturated with respect to aragonite, dissolution of aragonite will be total. Most Pleistocene and older marine limestones have stabilized to calcite: the oldest known aragonite is 350 Ma from Early Carboniferous organic-rich mudstones (Hallam and O'Hara 1962). At times in the geological past, the depth of aragonite dissolution may have been much higher: indeed, aragonite dissolution may have occurred on shallow sea floors (especially in areas of slow sedimentation) during the Jurassic and Ordovician when sea water was only supersaturated with respect to calcite (Palmer *et al.* 1988; Section 4.4.4).

Shallow-water carbonates are particularly prone to alteration by subaerial diagenesis due to changes in sea level. Rates of meteoric alteration depend primarily upon climate, the chemistry of pore-fluids, and local hydrodynamic regime. Complete stabilization of minerals occurs in only a few tens of thousands of years in humid climates, whereas in arid areas, aragonite and high-Mg calcite can persist for several hundred thousand years. Arid climates also promote sea-floor cementation, whereas humid climates favour dissolution and cementation by meteoric processes. The style of early diagenesis will affect later diagenesis: significant early lithification can prevent later compaction during burial. Burial diagenesis occurs below the depth where sediments are affected by marine and meteoric processes.

4.4.3 Medium-term geological change (10–100 kyr)

Environmental change in the order of 10–100 kyr may be due to changes in orbital forcing (known as Milankovitch cycles), which are probably responsible for the many metre-scale, shallowing-upwards cycles on carbonate platforms, and also for the major ecological disruption of reef communities.

Solar radiation

The Earth's mean surface temperature depends upon the luminosity of the Sun and the planet's distance from the Sun. Predictable changes in Earth's orbit around the Sun have resulted in recurring changes in global climate, and have been implicated in initiating major glacial–interglacial cycles. As Earth moves closer to the Sun, the climate warms, resulting in the melting of ice-caps and glaciers such that sea level rises. In contrast, during cool periods, sea level falls. With transgressions and regressions of the sea, reefs which lived near sea level would be flooded or exposed—so producing cycles of abandonment and recolonization. Such climate-driven fluctuations in sea level have occurred through geological time, but appear to have become more pronounced after the break-up of the single supercontinent Pangaea 250–200 Ma.

Earth's mean surface temperature also depends upon planetary albedo (surface and cloud reflectivity), determined by the distribution of continents, and on the composition and dynamic interaction of the atmosphere and hydrosphere. The Earth's dominant greenhouse gas is CO_2, which prevents heat loss from the surface of the Earth by reradiation trapping. The CO_2 content of the atmosphere has changed through time in response to changes in tectonic activity, which cause areas of the sedimentary reservoir to become heated to such a degree that the carbon is converted into CO_2 gas, which is then released back to the atmosphere by volcanic activity (Frakes *et al.* 1992). It is these processes related to the carbon cycle which lead to variations in the Earth's climate over geological time.

Sea level and climate

Our record of the last 150 kyr based on uplifted Pleistocene reef terraces in New Guinea shows that sea level has fluctuated dramatically, with oscillations through 100 m and rates of change that have exceeded 10 metres per 100 year (Chappell 1983; Fig. 4.5). For example, 18 ka, sea level was approximately 120–135 m lower than today, varying geographically with local tectonic regimes and isostatic movement. These oscillations have been interpreted as a low-frequency cycle of major glacial and interglacial periods with a periodicity of more than 100 kyr and a smaller higher frequency oscillation of 20 kyr (Chappell 1981). Palaeoclimatic records through this interval also show considerable variability in SST considered to have resulted, at least in part, from orbital forced variations in trade wind and monsoon strength (A. McIntyre *et al.* 1989).

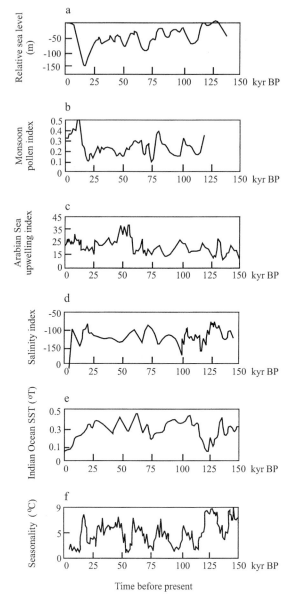

Fig. 4.5 The record of the last 150 kyr shows that sea level has fluctuated dramatically. Palaeoclimatic records through this interval also show considerable variability in sea surface temperatures (SST), salinity and monsoon strength. (Modified from Brown 1997a, by kind permission of Academic Press.)

In the eastern Pacific, Glynn and Colgan (1992) have estimated that during the past 6 kyr between 12 and 30 ENSO events of severe magnitude have caused a temporary cessation in reef-building. Ball *et al.* (1967) have estimated that between 160 000 and 320 000 hurricanes have hit the Florida Keys during the last 2 Myr.

Biotic response

Such radical oscillations must have severely disrupted shallow marine reef habitats, particularly by the smaller fluctuations during the interglacial–glacial cycles of the late Quaternary. But exactly how reef communities responded to these oscillations is much debated. Ecological disruption was considerable, however, as changes in local environments must have accompanied major glacioeustatic changes.

The Indo-Pacific centre of coral diversity was strongly affected by these changes; Western Indonesia and Japan were both part of the Asian continent, and Australia and Papua New Guinea were also joined. The northpolar ice-cap extended far south to northern North America, and the Antarctic sea–ice barrier lay at about 70° of latitude, much closer to the equator than at present. Many seas were landlocked, and many currents were displaced away from present coastlines to beyond the continental shelves. Populations probably respond to such change by a latitudinal shift in abundance or geographic range boundaries. This would have major effects on the stability and composition of communities; species previously disconnected would come into contact, and co-occurring forms might become separated. At least for the Pleistocene, where the evidence for a highly fluctuating climate is strong, we should interpret biological patterns within this highly dynamic framework (Roy *et al.* 1996).

Veron and Kelley (1988) have calculated that rates of origination and extinction in the Indo-Pacific over the past 2 Myr have been similar, and so conclude that these climatic fluctuations and sea-level changes had little effect upon the diversity of reef corals. Most important reef-building families (e.g. the Pocillioporidae, Faviidae, Siderastreidae, and Agariciidae) survived the Pleistocene environmental oscillations, with the possible exception of the Acroporidae. Their study emphasizes, however, that while diversity remained constant, reefs were not always present through this interval. Coral distribution and diversity, therefore, are probably poor indicators of palaeoclimatic patterns: reef distribution is a more reliable indicator of the relative climatic stability of tropical shelf environments conducive to carbonate production for periods exceeding a few thousand years (Veron 1995).

Potts (1983, 1984) has argued that the main outcome of fluctuations on 10–100-kyr scales was that their frequency in relation to the long generation times of corals provided insufficient time to respond in an evolutionary sense. He proposed a model in which coral communities were continuously and severely disrupted, mainly due to smaller fluctuations during the latter part of the Quaternary glacial–interglacial cycles of the last 2 Myr. Many of these fluctuations (involving sea-level changes of approximately 25 m) exceeded the bathymetric zone supporting most coral growth (less than 20 m), such that only relatively short periods of time (approximately 3.2 kyr) compared to the long generation times of large, long-lived corals was available for an evolutionary response. Such profound disturbance may have favoured the evolution of extreme intraspecific variability found in Indo-Pacific corals, and the paucity of living endemic species in the region.

This record of the Indo-Pacific is in contrast, however, to that of the Caribbean, which shows marked faunal changes over this period (Budd *et al.* 1993, 1998). During the Plio-Pleistocene some 4–1.5 Ma, a major turnover of reef corals and molluscs occurred. The coral genera *Pocillopora*, *Stylophora*, and *Goniopora* became regionally extinct, and *Acropora* became more abundant. Despite this faunal turnover, species richness changed little and reef communities remained common throughout the Caribbean. The change does not appear to have been abrupt as predicted by the theory of 'coordinated stasis', but may have taken 2 to 3 Myr at any one geographic location. In a case study from Curaçao, members of the post-turnover fauna were well established in the species pool before members of the pre-turnover fauna disappeared: faunal change was slow and not coordinated (Budd *et al.* 1998).

However, the patchy distribution patterns characteristic of many reef-building communities at small spatial scales means that most species occur at relatively few localities and so tend to be uncommon or rare, even though they are often widely distributed geographically across a region (Budd *et al.* 1998). On this local scale, high observed variability between different localities suggests that species associations and biological interactions are not tightly bound, and that the observed faunal change may have proceeded by varying dispersal and colonization rates at different spatial and temporal scales. Moreover, this patchiness would also explain how reef faunas appear more or less 'stable' at scales less than 1 to 2 Myr (Jackson 1992; Pandolfi 1996).

While conventional arguments seek to find a correlation between major faunal change and changes in climatic or oceanographic factors, this relative stability in Caribbean coral faunas argues against a primary evolutionary role for either global cooling or eustatic sea-level change as, like the Indo-Pacific, faunas were stable throughout the high-frequency, sea-level and temperature fluctuations of the Pleistocene. Instead, Budd *et al.* (1993) argue that regional physicochemical factors—such as altered ocean circulation patterns associated with the closure of the Isthmus of Panama some 3.5 Ma—may be ultimately responsible. However, areas such as the tropics where climatic change would have been greatly reduced, still show major faunal turnover. How do such minor climatic changes then (perhaps of only a few degrees Celsius)—which might only be predicted to compress latitudinal ranges of species—bring about extinction and radiation, especially since subsequent fluctuations of increasing magnitude appear to have had so little evolutionary effect (Budd *et al.* 1994)?

Metapopulation theory may provide some explanation (Nee and May 1992). The initial rapid fall in sea level would have drastically decreased the available habitat area and the number of suitable patches for all species. Declining temperature would further decrease the number of patches available to non-cold tolerant species causing local extinction at the extremes of tolerance. For species with low colonization potential, the decrease in patches would exceed their capacity to survive the usual cycle of local extinction and recruitment of individuals. This would lead to very patchy communities, and the result would be massive regional extinction (Tilman *et al.* 1994). Speciation would lag behind local extinction, with life-history characteristics determining the magnitude of the lag. So, the life history of species which survive turnover are predicted to have features which allow differential survival—that is they are an accidental side-effect of features evolved for other reasons. Acroporoids grow up to ten times faster than any other corals and are so able to keep pace with sea-level changes (Chappell and Polach 1991), and this group both radiated and increased in abundance during the turnover event. Budd *et al.* (1993) propose that once the benthos had changed, all that were left were eurytopic species—those with relatively wide environmental tolerances such as acroporoids.

What is clear from the Pleistocene record, is that coral reefs continued to exist even when sea levels dropped by over 100 m and mean temperature fell by 8 °C. Reefs flourished again during the warmer periods of higher sea level (Wilkinson and Buddemeier 1994). These oscillations occurred in concert with near-doubling of CO_2 concentrations, and were certainly associated with 'greenhouse' climate changes such as shifting rainfall patterns, storms and currents, and cloud cover. These rates of change probably exceeded those that coral reefs are predicted to experience during ongoing, anthropogenic climate change (Callander 1995).

4.4.4 Long-term geological change (10 Myr)

Physical processes that exert control globally, or at least over very large areas, over millions of years are movements of the Earth's crust (tectonics), which will also affect global changes in sea level and climate. These will, in turn, determine mean temperatures, salinity, patterns of water circulation, nutrient concentration and flux, tidal current strengths and storm frequency, and wave energy.

The influence of global tectonics on the record of reef-building

The large-scale differences apparent between reefs of the Caribbean and eastern Australia are, in part, a response to differences in tectonic setting. The Great Barrier Reef grows on a passive continental margin which provides a broad, uninterrupted shelf for the development of a long, continuous barrier reef system. The Caribbean is a tectonically active area characterized by continual movement during the Tertiary. There, the fragmented crust has formed a jigsaw of often steep-sided islands each with their own history, rather than a continuous broad shelf with smaller reefs closer to shore. However, where broader shallow platforms have formed, e.g. the east coast of Belize, the growth of a near-continuous barrier reef has been possible.

Barrier reef systems which produce a pronounced environmental gradient have had an episodic geological history, as their distribution is dependent upon the availability of high-energy shallow margins in settings where organisms can withstand the hydrodynamic regime conducive to rapid $CaCO_3$ production (see Section 4.1.1). When these conditions are not met, no barrier reefs form but high-energy carbonate facies such as oolite and grain-stone shoals develop instead on shelf margins. The apparent abundance of barrier reef complexes in modern seas as compared with the ancient record has two possible explanations: first, it may be due to the common occurrence in modern seas of subsiding coastal platforms as a consequence of Pleistocene sea-level fluctuations, or second, it may be a reflection of the inability of many ancient reef organisms to grow into shallow, highly agitated waters (see Section 6.5).

The importance of the hydrodynamic regime was stressed by Wilson (1975), who divided platform margins into quiet, quiet–moderate, and rough sea types. He proposed that in tranquil settings, sand shoals develop in the most agitated areas with mud-mounds developing downslope. In quiet–moderately energetic settings, discrete reefs form in the most shallow areas, but only in the most energetic seas do reefs actually form at the platform margin to produce a continuous barrier. Tectonic cycles consist of phases of widespread rifting, followed by quiet phases, which create different types of continental margins. These are closely related to changes in sea level, which directly affect climate and so both the global distribution of reefs and the type of reefs formed. Two case studies from the Early Carboniferous (Case study 4.2) and the Jurassic (Case study 4.3) illustrate this phenomena.

Case study 4.2: The Early Carboniferous record

The Early Carboniferous has often been quoted as a time of reef 'devastation' following the widespread loss of large reef-building skeletal metazoans (stromatoporoids and corals) in the end-Devonian mass extinction, which supposedly restricted reef-building to deep-water settings resulting in the formation of Waulsortian mud-mounds (see Section 3.5). The subsequent pattern of Waulsortian mud-mounds evolving into diverse, shallow-water, heterotrophic metazoan reefs is also cited as reef 'recovery'. These phenomena are, how-ever, simply distributional artefacts (Horbury 1992): it is now clear that the distribution and type of reefs developed during the Early Carboniferous was controlled by regional tectonic settings, depositional environments, and local faunal provinciality (Webb 1994).

During the Tournaisian ramps dominated many continental margins, and sea level was globally very low. Shelves world-wide were often clay-rich, such that carbonate production

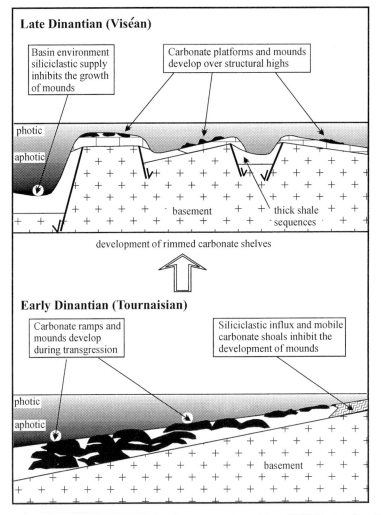

Fig. CS 4.2 The tectonic development of shelf platforms and intrashelf ramp environments during the Early Carboniferous of Britain influenced the development of different reef types. (Modified from Bridges *et al.* 1995.)

was restricted to areas far removed from heavy terrigenous input. This resulted in the displacement of carbonate production downslope leading to the formation of often deepwater, low-energy, Waulsortian mounds dominated by microbial communities and few encrusting metazoans. But by the late Tournasian to early Viséan, these ramps evolved into shallow carbonate shelves and fault-bounded platforms due to aggradation and progradation of detrital carbonates (Wright and Faulkner 1990; Ahr *et al.* 1991). As a result, deepwater reefs disappeared and shallow, high-energy shelf reefs were able to develop on fault-controlled highs or, upon shelf-margin sand shoals. Although such reefs were widespread, many regions were isolated and so supported diverse reef communities of largely endemic encrusting metazoans such as corals, bryozoans, and sponges bound by rigid microbial communities. No single reef community with cosmopolitan elements is apparent at this time (Webb 1994).

The fact that Waulsortian-like mounds occur in ramp-like settings later in the Carboniferous, and indeed again in the Permian, supports the contention that tectonics plays a controlling role on the distribution of this reef style.

Case study 4.3: The Jurassic record

A variety of reef-types formed during the Jurassic, variously composed of corals, siliceous sponges, and thrombolites. These reefs grew under different environmental conditions and their distribution and abundance largely reflects changes in shelf morphology, ocean circulation, sea level, and climate during the break-up of the super-continent Pangaea (Leinfelder 1994).

Lower Jurassic reefs are globally scarce, as terrigenous rather than carbonate sedimentation dominates most shallow marine shelves at this time. Corals may have survived on seamounts and isolated platforms away in the Pacific remote from terrigenous input. On the northern shelf of Tethys, reefs did not reappear until the mid- to late-Jurassic when carbonate sedimentation became widespread. In contrast, there were reefs thriving on the southern margin of Tethys by the Early Jurassic.

The development of the southern tethyan shelf was governed by tectonic controls that maintained shallow carbonate shelves during the Early Jurassic, but drowning removed them during the Middle Jurassic and tectonic movements uplifted some blocks during the Late Jurassic. Mixed coral–sponge reefs grew on steepened slopes.

On the stable northern shore of Tethys, gently inclined to moderately steepened ramps developed, which consist of coral-debris sands and extensive siliceous sponge communities. These sponge communities show only isolated occurrences on the southern shore during the Early Jurassic. Reefs which generated and shed considerable amounts of debris are characteristic of steepened slopes, caused mainly by the presence of bypass margins.

The temporal reef distribution on the southern Tethys shelf reflects the successive drowning of carbonate platforms during the early and mid Jurassic, and local uplift during the Late Jurassic. Reefs expanded on the northern shelf due to increased habitat-availability caused by sea-level rise which accompanied tectonic activity.

Climate

While reconstructions of long-term climate change show smooth and gradual changes in parameters such as temperature and CO_2 concentrations (Wilkinson and Buddemeier 1994), this may be a misleading impression based on poor stratigraphic resolution and incomplete records (Buddemeier and Hopley 1988).

Climate is the basic control on the distribution of reefs, as temperature gradients control the distribution of carbonate production (Section 4.1.1). That climate is a major determinant of species distribution is also beyond doubt, as supported by a diverse and vast literature.

Earth's climate has evolved through time on a global scale, passing through alternating phases of cooling and extreme glaciation to those of aridity and warmth, as recorded by palaeontological and sedimentary evidence (Fig. 4.6). For most of the Proterozoic, warm climates ('greenhouse') prevailed, as demonstrated by the widespread occurrence of shelf carbonates. However, two cool phases ('icehouse') are indicated by glacial features and deposits around 2.3 Ga and 860–600 Ma. During the Phanerozoic, four cycles of cool–warm phases of varying intensity are established, lasting on average approximately 150 Myr (Frakes *et al.* 1992). The transition from a warm to cool phase may be slow, in the order of 50 Myr.

Climate states are governed by changing continental distributions, where during cold phases the presence of substantial continents at high latitudes provides sites for the growth

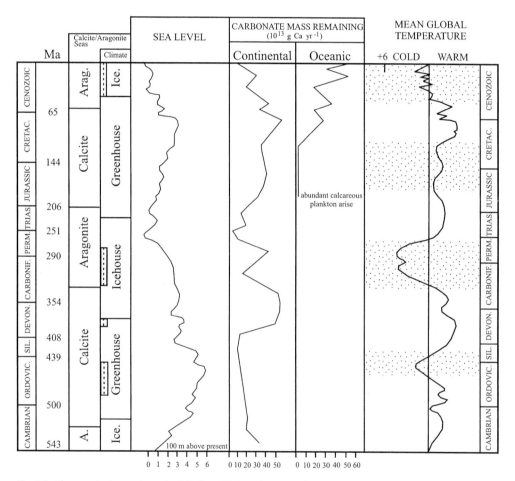

Fig. 4.6 Phanerozoic changes in sea level (Hallam 1984), marine water chemistry (Sandberg 1983), mean global temperature with the warm and cool modes (stippled) of Frakes *et al.* (1992) and global climate of Sandberg (1983), and preserved carbonate mass (after Wilkinson and Walker 1989). (Modified after Webb 1996).

of glaciers. During any one cold episode, glaciation only took place in a hemisphere where large land masses were present, and over a period when relatively large amounts of carbon was locked within sedimentary or biological reservoirs due to high productivity. Large, elevated masses of high-latitude continental ice also increases albedo which intensifies global cooling trends. During warm periods, high sea levels would lead to the sequestration of substantial amounts of organic carbon (on land, within shallow shelf sediments, and possibly oceanic sediments). This may have resulted in a gradual increase in biological productivity during cooling, eventually leading to the development of glaciers and large falls of global sea level (Frakes *et al.* 1992). Cooling may have been further enhanced by changes in orbital influences. The onset of warm phases may have been far more rapid, and although the mechanisms for this are not fully understood they may, in part, be driven by an increase in sea-floor spreading, volcanism, and increased CO_2 output. The end of cool modes is characterized by decreasing levels of productivity and/or the accumulation of organic matter. This suggests the release of organic carbon, and the initiation of substantial carbonate production.

Sea level

The level of the ocean is controlled by climate and tectonics: climate controls the volume of water in the oceans by influencing glacial ice volumes; plate tectonic movements affect the volume of ocean basins. Sea-level fluctuations due to changes in ocean-basin volume are controlled by rates of sea-floor spreading to the order of 0.01 m 1000 year^{-1}. Using seismic profiles (which display the two-dimensional distribution of strata) from different ocean basins, it is possible to correlate recurring depositional patterns with changes in sea level that relate to changes in polar ice volume (glacioeustatic changes). This enables the reconstruction of *global sea level* through time (e.g. Hallam 1984; Haq *et al.* 1987), and the differentiation of more local tectonic influences which are responsible for *relative sea level*. When global sea-level changes are superimposed upon local tectonic history this can lead to very individual reef depositional histories.

Over the Phanerozoic, global sea level has changed greatly, on a variety of scales: two first-order cycles occur through the Phanerozoic, driven by supercontinent fragmentation, sea-floor spreading, and construction with the resultant formation and closure of major oceans; second-order cycles are thought to be the result of passive margin subsidence; other cycles are driven by global sea-level changes, local subsidence rates, and tectonic regime.

Global sea level is a major influence on reef distribution, as the height of sea level determines the availability of shallow sea area and the accommodation space available for carbonate production. More shallow marine carbonates were deposited at times of higher sea levels and continental submergence than at times of continental emergence (Fig. 4.6). In the Late Cretaceous for example, nearly 40% of the continents were flooded, leaving only 18% of the Earth's surface as land (Howarth 1981). At this time, extensive epeiric seas were common, and carbonate production rates were high. Dolomite production also shows two maxima that correspond to the high sea-level stands of the Phanerozoic (Given and Wilkinson 1987), which is considered to be due to increased continental flooding, leading to higher atmospheric $p$$CO_2$, lower oceanic carbonate concentration and/or lower calcite saturation resulting in increased rates of dolomitization of carbonate sediments. Also, these maxima may reflect the increased flow of sea water through carbonate platforms by ocean current pumping: processes which will be more widespread at times of high sea level. We might also predict that more substantial shelf margin reefs will form due to higher rates of cementation at these times, independent of the relative hydrodynamic resistance of the organisms present.

If the accumulation rate of carbonate is related to ambient sea-water composition, then the latitudinal limits of ancient carbonate accumulation can be used to derive gradients of sea-water temperature/saturation state. For example, using palaeolatitude data on more than 2500 Phanerozoic carbonate sections in the continental United States, Opdyke and Wilkinson (1993) have shown that many Cretaceous carbonates (both pelagic and shallow-water sediments) were deposited 10° more poleward than the latitudinal limits of modern shallow-water carbonate accumulation.

These data indicate that the mean annual Cretaceous sea-water surface temperature gradient between the equator and 38° north latitude was about at least 5 °C lower (warmer) than today. Higher atmospheric pCO_2 most likely existed at this time (Berner 1994) due either to increased volcanic outgassing or by changes in continental weathering patterns caused by high sea levels. As a result, higher temperatures and sea levels would have been necessary to balance the enhanced rates of weathering of calcium carbonate into the global oceans. Indeed, the demise of numerous Cretaceous guyots in the Pacific at low latitudes suggests that the tropics have not always been the refuge for healthy, isolated carbonate-platform growth that they are today (Wilson *et al.* 1998).

Bosscher and Schlager (1993) have estimated carbonate-platform accumulation rates over the Phanerozoic. They have shown that while there are no apparent long-term trends, there are periods of lower accumulation rates, such as in the Early Devonian, latest Devonian–Carboniferous, and Early Jurassic (Fig. 4.7). These authors suggest some coincidence between these low rates, mass extinction events, and the availability of large reef-associated skeletal metazoans (but see Section 5.5).

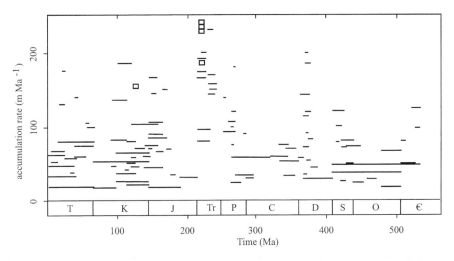

Fig. 4.7 Accumulation rates for Phanerozoic carbonate platforms; □ indicates rates based on cyclostratigraphy. (Replotted from Bosscher and Schlager 1993, by kind permission from the University of Chicago Press.)

Surface circulation

Global circulation patterns are determined by atmospheric circulation and the tectonic placement of land masses. The climatic and oceanographic patterns on Earth are driven by excess heat at the tropics and a deficit at higher latitudes. As equatorial water warms, the air expands and rises and is replaced by cooler air from the north and south, forming the trade winds. As the earth spins on its axis from west to east, the trade winds move westwards, which in turn force the major equatorial currents westwards. Equatorial currents turn away from the equator when blocked by continents, moving clockwise in the Northern Hemisphere and anticlockwise in the Southern Hemisphere to produce the Coriolis effect. The Coriolis effect creates the broad patterns of nutrient input onto shallow shelves: on the western sides of continents, shallow shelves receive concentrated nutrient input by upwelling deeper waters from below the photic zone. On the eastern sides, nutrients and sediment are shed from the continents via rivers. Such large-scale oceanographic processes in nutrient distribution result in profound differences in the patterns of coral reef growth, including life-history traits of the dominant organisms, i.e. predation patterns.

Supersaturation levels

We have established that gradients in temperature/supersaturation are very important in determining not only the broad distribution of carbonate production, but also the distribution of modern coral reefs. Some authors have suggested that saturation levels have changed through geological time, not only in concert with climate change, but also as a result of the evolution of new groups of organisms capable of producing calcareous skeletons.

The supersaturation of the world's oceans may have undergone a series of dramatic declines, with concomitant changes in the nature of carbonate-producing biota. Knoll *et al.* (1993) argue that the ubiquity of stromatolites (and an absence of calcified cyanobacteria) in the Archaean and Proterozoic was probably aided by very high, calcium carbonate saturation levels. This would have induced the abundant generation of tiny carbonate crystals in the water column and on the sea floor. These crystals would serve as preferential nucleation sites for inorganic carbon cements, such that cyanobacteria could only calcify in conditions that excluded them. However, from the Early Cambrian, the appearance and diversification of calcareous skeletal benthos would have sequestered calcium carbonate: as a result, competing crystals were absent and so stromatolites declined and calcified cyanobacteria became common (Fig. 4.8).

Gebelein (1976) has suggested that a further decline in supersaturation to reach modern levels occurred during the Jurassic–Cretaceous due to the rise of calcareous plankton, which, in turn, preferentially sequestered carbonate. Indeed, Wilkinson and Walker (1989) have documented an exponential shift in carbonate accumulation from predominantly shallow, continental settings prior to the Cretaceous to deep, pelagic settings in the Recent (Fig. 4.6). This decrease may explain the absence of Tertiary and Recent calcification of cyanobacteria in marine environments of normal salinity. But what additional data might support this hypothesis? If saturation levels were higher before the mid–late Mesozoic, what further changes might we predict in the style of shallow marine carbonates?

There is some anecdotal evidence of very rapid growths of cements within many Palaeozoic reefs, especially within reef crypts. That these cements grew rapidly and during

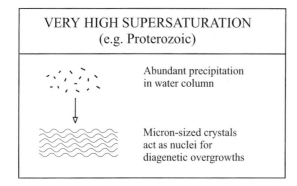

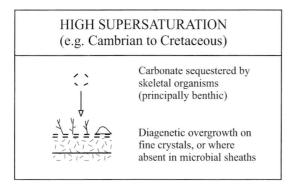

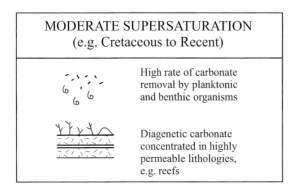

Fig. 4.8 Schematic diagram illustrating the different rates and styles of carbonate precipitation in the Proterozoic, Palaeozoic to Cretaceous, and post-Cretaceous. (Modified from Knoll *et al.* 1993, by kind permission of SEPM Society for Sedimentary Geology.)

the lifetime of the living reef is shown by the fact that some cryptic organisms abut against and become distorted by the growth of cements (Fig. 4.9). This raises the possibility that Palaeozoic reef substrates may have lithified more rapidly that post-Jurassic substrates in the same environments, and that the growth of reef-builders was aided by the stability of rapidly lithifying sediments. This would suggest the operation of important feedback mechanisms for ancient reef formation, which have yet to be explored.

As well as cement-rich reef frameworks, microbialites were certainly more widespread in ancient reefs, especially prior to the Cretaceous (Webb 1996). The geological distribution of microbialites may therefore be controlled by physicochemical factors including the satura-

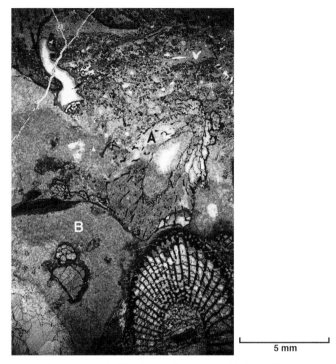

5 mm

Fig. 4.9 Distortion of a cryptic archaeocyath (A) by the growth of a cement botryoid (B).

tion state of sea water driven by changes in $p\mathrm{CO}_2$, supersaturation, or Ca/Mg ratios, as well as global temperature distribution. The decline in abundance of reefal microbialite after the Jurassic may be the result of the relatively reduced saturation state of sea water which may have lowered supersaturation levels to a threshold for abundant microbialite formation, so restricting microbialite formation to cryptic reef habitats where abnormal chemistries can develop. Such a scenario might also be related to the absence of stromatactis in the Mesozoic.

Nutrient levels

Thayer (1983) has suggested that the coincident rise of land plants and deposit-feeders was related perhaps by the increased input of nutrients to shallow shelf areas. Likewise, the rise of bioturbators may have released more nutrients into the ocean water column as it is well established that benthic recycling of nitrogen may exceed input from freshwater. These nutrients may, in turn, have been utilized/taken up by a new planktonic biomass—perhaps in the form of phytoplankton, such as dinoflagellates, coccolithophorids, and diatoms which all appeared and diversified during the Mesozoic.

Secular variation in mineralogy

While aragonite and high magnesium calcite are the principal inorganic carbonates formed in modern seas, calcite alone seems to have dominated the Early Ordovician to the Middle

Carboniferous, and the Early Jurassic to the Late Cretaceous (P.A. Sandberg 1985*b*; Wilkinson *et al.* 1985). This oscillating trend between 'Aragonite' and 'Calcite' seas—sometimes known as the Sandberg curve—has been attributed to changes in pCO_2 or the Mg/Ca ratio of sea water, and hence to changes in global tectonic activity.

The almost complete absence of radiaxial calcite from Quaternary limestones but abundance in Palaeozoic reefs suggests some changes in sea-water chemistry. It has been proposed that when sea water is undersaturated with respect to aragonite, active circulation during shallow burial processes and further fluctuations in the degree of saturation with respect to calcite may control radiaxial calcite formation (Saller 1986). Aragonitic botryoids appear to be restricted to the Early Cambrian, mid-Carboniferous to Early Jurassic, and the mid- to Late Cenozoic, whereas ooids are conspicuously reduced in abundance in the Mid-Ordovician to Lower Carboniferous and Upper Cretaceous to mid-Tertiary sediments. It has been suggested that these absences might be explained by the fact that pCO_2 levels reached sufficiently high values to inhibit physicochemical precipitation by lowering the degree of carbonate saturation (Wilkinson *et al.* 1985).

While these secular changes are apparent in the history of inorganic cements, there is also some suggestion that they may have controlled the dominant mineralogy of organisms (Harper *et al.* 1997). Bivalves may broadly follow these secular trends, with those forming aragonite shells being at a selective disadvantage during times of highly corrosive 'calcite' seas. This might explain the dominance of rudist bivalves with outer calcitic layers, rather than wholly aragonitic corals during the Cretaceous. The dominant mineralogy of some reef-building organisms also mirrors these changes: aragonitic calcified sponges are known only from the Late Carboniferous to the end of the Triassic, and from the Tertiary to Recent (Wood 1987).

4.5 Carbonate and the carbon cycle

Calcifying organisms are active participants in the carbon and CO_2 cycle. As the surface ocean is usually close to equilibrium with atmospheric CO_2, supersaturated with respect to calcium carbonate and buffered to a pH of 7.8 to 8.3, this ensures that most dissolved inorganic carbon is in the form of bicarbonate ions.

There is a widespread misconception that reefs and their associated carbonate platforms are sinks for atmospheric CO_2. But carbonate precipitation lowers pH and converts bicarbonate to CO_2 gas, so that carbonate formation is actually a source of CO_2 release to the atmosphere. One mole of marine carbonate yields approximately 0.6 mol of CO_2 due to the buffering action of the ocean (Ware *et al.* 1992). Field measurements of the partial pressure of CO_2 on reef flats have since confirmed that these sites release CO_2 to the atmosphere, except where the reefs are stressed (Kayanne *et al.* 1995). Calcification thus releases both CO_2 to the environment and at the same time removes carbonate ions, such that the process is both a carbon sink and a source of CO_2. This apparent paradox is resolved by recognizing that the source of both carbonate and CO_2 is the marine alkalinity reservoir, which contains more exchangeable carbon than the atmosphere (Broeker and Pang 1982).

Atmospheric CO_2 concentration is the result of many processes operating at different timescales. Shallow-water carbonate production is an important part of the global carbon cycle (Milliman 1993), and patterns of carbonate deposition have been used as proxy indicators of the rate of weathering and volcanic output of CO_2 into the atmosphere (e.g. Berner

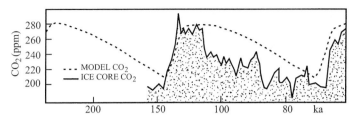

Fig. 4.10 Calculated atmospheric CO_2 change as a consequence of changing the shelf carbonate-deposition rate using a dissolution rate constant of 0.0002 [g cm^{-2} year^{-1}]/[mmol m^{-3}]. The overall pattern and amplitude of changes compares well to the data from the Vostok ice core (shaded curve, after Barnola *et al.* 1987). (Modified from Opdyke and Walker 1992, by kind permission of American Journal of Science.)

1994). Estimates suggest that coral reefs release between 0.02 and 0.08 Gt C (gigatonnes of carbon) as CO_2 annually (Ware *et al.* 1992). Notwithstanding its relative volumetric unimportance compared to pelagic carbonate production in the Holocene, shallow-water carbonate is now also thought to be a sedimentary sink of quantitative importance (Opdyke and Walker 1992). Reefs are estimated to be sinks for 111 million tonnes C year^{-1} (Kinsey and Hopley 1991), accounting for approximately 6% of Earth's present carbonate production.

Changes in the volume of shallow carbonate deposited are involved in a complex and dynamic internal cycle of deposition and erosion which can be related to changes in atmospheric CO_2. It has been argued that during glacial–interglacial sea-level changes, rising sea level shifts the location of carbonate deposition from the deep sea to shelves, such that deposition occurs on the shelves when sea level is rising and CO_2 production is high. When sea level is falling or when sea level is low, shelf carbonates undergo net dissolution, so depressing CO_2 levels. Differences in the rate of coral reef carbonate deposition due to sea-level change from the Pleistocene to the Holocene can account for broad changes in Quaternary atmospheric CO_2 concentrations, as recorded in Greenland ice cores (Opdyke and Walker 1992; Kleypas 1997; Fig. 4.10), although such modelling does not account for the many interrelated feedback mechanisms, such as the lowering of CO_2 by deep ocean circulation through dissolution of carbonates near the CCD. Coral reef carbonate production may, therefore, have been an important mechanism forcing Quaternary climate fluctuations, although it is not clear to what degree increased reef growth is a consequence, rather than a cause, of CO_2 increase, and it was also probably important further back into geological time.

The levels of primary production on coral reefs are high, and perhaps even higher than reported for any other ecosystem (Kinsey 1991). However, almost all of this production is consumed by community respiration. Net production has been estimated to be only 3% of gross primary production, amounting to ~0.1 g C m^{-2} d^{-1}—a value that is remarkably close to that of typical nutrient-poor open oceans (Kinsey 1991). This places severe limitations on the sustainable yield of coral reefs, and on their management for human benefit (Wilkinson and Buddemeier 1994).

4.6 The role of catastrophes

Extreme, seemingly chance, catastrophic disturbances can radically restructure a reef community, as evidenced by the response of reefs globally to severe hurricane damage

(Knowlton *et al.* 1990), high or cold temperature events (Glynn and Colgan 1992), widespread coral disease (W.D. Gladfelter 1982), bleaching (Brown and Ogden 1992), outbreak populations of voracious predators and their subsequent mass mortality (T.P. Hughes *et al.* 1987), and overfishing (Hay 1984*b*).

All these events have caused profound changes in the distribution and abundance of dominant reef species. Species appear to dominate in any particular environment until removed by catastrophic disturbance, after which a different species may come to dominate. This may be because established organisms resist invasion of newcomers in a frequency-dependent way, often independent of their life history or morphology (Jackson 1988).

Disturbance is thus an important process on coral reefs, but the role of disturbance in maintaining diversity may vary greatly in magnitude over time, and the impact of disturbance processes may depend much on their chronology and may interact in surprising ways (T.P. Hughes 1989). Multiple stresses act synergistically, with reefs often showing very poor recovery following second sources of stress (T.P. Hughes 1994).

4.7 Summary

Reefs appear to be robust to environmental change on geological and evolutionary timescales, but individual reef communities can be transient and fragile—constantly shifting and accommodating to unpredictable and local environmental change (Buddemeier and Smith 1992). Modern corals show a formidable array of acclimatization processes, adaptations at the population level, and changes in community composition over longer time periods (Kinzie and Buddemeier 1996). But while the tremendous variability apparent within the shallow tropical coral reef environment may explain the ability of corals to survive major change, this may not be apparent using standard taxonomic methods and so many fossil corals may belie this variability—appearing as stable form-species (Buddemeier 1997).

Our understanding of how environmental variation interacts with reef communities is still in its infancy. Moreover, our powers to deduce exactly how living coral reefs respond to such change may have been further complicated by the fact that because reef degeneration is so widespread due to the activities of our society—particularly by overfishing—that when we describe the functioning of modern coral reefs we are in fact describing ecosystems that are already far from pristine (Jackson 1997).

Anthropogenic stress, particularly those cumulative aspects of global change, clearly present the most immediate threat to coral reefs and reef corals, and it is the potential synergism of these factors that makes the current period of environmental change so different—and so potentially disastrous—to those experienced by reef communities in the past.

5   Mass extinctions: collapse and recovery

The previous chapter emphasized the importance of disturbance for maintaining the high diversity of modern coral reef communities. The response of reefs to events which cause catastrophic disturbance is, however, highly complex and depends on the timing, intensity, and history of events. Indeed, some combinations of disturbances can act synergistically to cause seemingly permanent shifts in the composition of individual reef communities. But it is now clear that while reef biotas (and therefore ecosystems) may be robust to environmental variability over both ecologically and geologically significant timescales, reefs themselves are relatively fragile, transient phenomena. And we have seen that over even longer timescales, a succession of different ecosystems have formed reefs over the past 3.5 Ga, and that the 'incumbency' of many of these long-lived ecosystems was terminated by mass extinctions.

Origination and extinction of species have occurred throughout the history of life, with the continual increase of diversity indicating the predominance of origination. The only major reversals to this trend have occurred during mass extinctions. *Mass extinctions* can be defined as the extinction of a significant proportion of the Earth's biota over a geographically widespread area in a geological insignificant period of time, often one that appears instantaneous when viewed at the level provided by the geological record (Fig. 5.1). Although relatively continuous levels of background extinction accounts for all but a few per cent of all species extinctions (Raup 1991), mass extinctions are disproportionately significant in that they record crises that—by virtue of their speed, unpredictability, and magnitude—are capable of removing dominant taxa and their habitats, which can lead to the collapse of whole ecosystems.

Research into the cause and role of mass extinctions in the history of life was tremendously invigorated by the publication of a seminal paper by Alvarez *et al.* (1980), which suggested that a giant bolide impact was largely responsible for the major extinction event at the end of the Cretaceous. This hypothesis is now widely accepted, and the considerable body of data generated since 1980 on the End-Cretaceous and other mass extinctions, has highlighted their potentially devastating effect and led to an apportioning of special status within the study of evolution. The removal of evolutionary resilient 'incumbent' groups during mass extinctions can allow for the subsequent radiation of remaining, or new, taxa.

While the many long-term trends evident in the evolution of life are probably due to biological interactions among species, the synchronous extinction or turnover of species in the geological record points to the operation of extrinsic environmental factors. Mass extinctions appear to be due to sudden and often unstable fluctuations in the environment caused by changes in sea level, climate, or oceanographic characteristics. Rapid environmental change may therefore underlie the global discontinuities in the fossil record that mark major boundaries of the geological timescale (Jackson 1995).

The fossil record shows that most marine species appear to originate abruptly and persist unchanged for millions of years, but may go extinct as co-ordinated ecological units. Once a

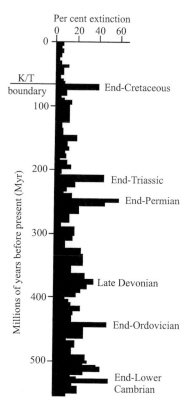

Fig. 5.1 The extinction intensity of all marine genera measured for each stage of the Phanerozoic. Although the major mass extinction events only account for a few per cent of all species extinctions, their unpredictable and global nature makes them of special importance in explaining the history of life. (Modified from Sepkoski 1995, by kind permission of Springer-Verlag.)

shift in a biota has occurred, the replacement biota will persist until the next extinction event. The mathematical predictions of relative species abundance developed by Hubbell (1997) can explain this evolutionary pattern of 'sequential incumbency'. His theory demonstrates how only modest rates of migration of new species are sufficient to yield regional stability in the composition of communities for geologically significant periods of time and over wide areas. This is especially true for species with adults or larvae capable of dispersal over long distances.

Many authors have suggested that reef communities are particularly susceptible to mass extinction, and that recovery occurs more slowly than in other communities, with reefs often taking 2–10 Myr to reappear in the geological record (e.g. Copper 1988; Talent 1988; Kauffman and Fagerstrom 1993; Jablonski 1995; Erwin 1996). Implicit within many of these analyses, is the belief that reefs take longer to re-establish not because of the continuation of environmental stresses hostile to reef formation, but because of the time necessary to restore the complex ecological characteristics (particularly mutualisms) that characterize some reef communities.

This chapter will concentrate on four main issues concerning mass extinction and the evolution of reefs:

1. Are reefs more susceptible to mass extinctions than other marine communities?

2. Is the loss of reef biota, or continued environmental perturbation, responsible for the absence of reefs after mass extinction events?

3. Do reef communities recovery at a different pace from other communities?

4. What effect have mass extinctions had on the evolution of reefs?

5.1 Problems of measuring extinction in the fossil record

Mass extinctions in the fossil record—as other evolutionary events—are usually analysed by changes in biodiversity. This requires sound biostratigraphy and correlation, as well as a secure taxonomy. Although skeletal marine organisms provide the best record, this is still very incomplete at the species level and subject to considerable sampling and preservational biases. For example, taxa compiled in global diversity data might be biased towards the most abundant, widespread, and geologically long-lived species, where the best time resolution possible is usually limited to 10^3–10^4 years (Jablonski 1995). While global compilations of data can identify mass extinction episodes, they do not reveal details of timing or local geographic information (Jablonski 1995). Poor phylogenetic knowledge of clades that survive extinction can also lead to an underestimation of the number of surviving taxa, which may inflate not only the apparent severity of the extinction event but also estimates of 'endemic' extinctions (A.B. Smith and Jeffrey 1997). Moreover, change in diversity is not always a straightforward expression of the magnitude of ecological change (Droser *et al.* 1997), as compilations of the numbers of taxa may reveal little of ecological importance (Jackson 1988). Mass extinctions are complex phenomena, and involve the preferential elimination of some parts of the biota—such as phytoplankton at the base of the food-chain—which will cause a variety of cascading ecological effects.

Range charts that record the stratigraphic distribution of fossil species are frequently used to assess extinction rates. However, as the most recent stratigraphic occurrence of a species is unlikely to represent the last individual of that species, ranges of extinct species are likely to be artificially truncated. This sampling bias will have the effect of making even sudden extinctions appear more gradual, a phenomenon known as the *Signor–Lipps effect* (Signor and Lipps 1982). Indeed, it is likely that only through using a variety of statistical methods (e.g. C.R. Marshall 1994), will we be able to estimate the true range of extinct species and so gain insight into whether apparent patterns of extinction are genuine or mere artefacts.

Notwithstanding these problems, careful documentation of the fossil record can provide quantitative information on ecological selectivity in species that survive extinction, such as differences in geographic range, feeding preference, niche breadth, larval and adult lifestyle, and relative abundance.

5.2 Terminology

Particular terms have been established to describe recurrent biological phenomenon associated with mass extinctions. These are introduced in the following section.

Extinction events can be divided into three phases: extinction, survival, and recovery (Fig. 5.2). Taxa that outlive the majority of their clade are known as *holdover taxa*, and those

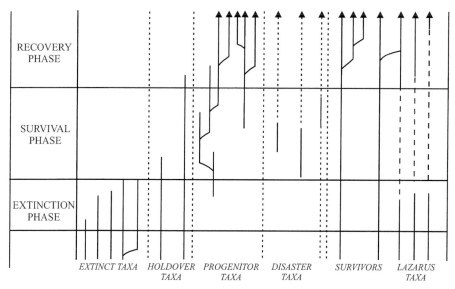

Fig. 5.2 Generalized model of a mass extinction showing typical phases and major biotic response. (Modified from Kauffman and Erwin 1995.)

which appear during the extinction, or subsequent survival phase, and then radiate during the recovery phase are known as *progenitor taxa*. Progenitor taxa are often adapted to the extreme environmental conditions of the extinction phase. The survival phase may also be characterized by *disaster taxa,* which are usually long-ranged species of opportunists, whose presence in vast abundance is a clear sign of considerable environmental perturbation. Disaster taxa are typically small and morphologically simple, and the ecological preferences of such taxa can provide clues as to the nature of the environmental conditions during extinction and survival phases.

The recovery phase is characterized by holdover taxa and a radiation of progenitors, as well as the reappearance of taxa that had disappeared from the fossil record during the extinction phase several million years earlier. These are known as *Lazarus taxa,* and are either genera or families; no Lazarus species are currently known. Lazarus taxa are presumed to have weathered the extinction event, and their temporary absence in the fossil record is due either to survival in isolated (relict) communities or to sampling or preservational bias.

5.3 Mass extinction events

Mass extinctions may have occurred throughout the history of life. While there is some evidence for mass extinctions in the Late Proterozoic (Vendian), five major mass extinction events have been identified from the Phanerozoic geological record (end Ordovician, Late Devonian, end Permian, end Triassic, end Cretaceous), and no fewer than 27 further minor extinction events (Sepkoski 1986). For a detailed synthesis of our current understanding of the causes of Phanerozoic mass extinctions, the reader is referred to the comprehensive review provided by Hallam and Wignall (1997). In the following section, the causes of these five major extinction events and their effect on reefs is discussed, with the addition of the Lower Cambrian extinction event which had a major effect on reef communities (Fig. 5.1).

5.3.1 Late Lower Cambrian (~ 525–520 Ma)

Lower Cambrian reefs formed subtidally by the intergrowth of archaeocyath sponges and calcified cyanobacterial thrombolites. The Archaeocyatha appeared on the Siberian Platform at the base of the Tommotian (530 Ma) and radiated spectacularly during the Early Cambrian reaching ~180 genera by the Botoman 1: accounting for approximately half of described metazoans at this time. Many other sessile metazoans—most of problematic affinity—grew within these cyanobacterial-archaeocyath reef communities, including radio-cyaths, cribricyaths, and coralomorphs.

This reef community persisted until the virtual demise of the archaeocyaths and the complete extinction of associated reef-dwelling metazoans and some calcified cyanobacteria at the end of the Early Cambrian (Fig. 5.3). Toyonian reefs are known only from Labrador and Altai-Sayan-Tuva. Archaeocyaths underwent a precipitous decline in diversity from the early Botoman onwards, with only 35 genera remaining by the Toyonian 1 and none by the Toyonian 3. Two archaeocyath-like genera are known in post-Lower Cambrian sediments from Antarctica (Debrenne *et al.* 1984; Wood *et al.* 1992*b*). Other benthos such as trilobites (especially redlichiids), hyoliths, anabaratids, and tommotiids were also affected (Zhuravlev and Wood 1996). This resulted in the extinction of the distinctive Lower Cambrian reef and level-bottom faunas by about 520 Ma (Bowring *et al.* 1993).

Causes

The late Early Cambrian extinction can be resolved into two distinct events: the late Botoman to early Toyonian Hawke Bay Regression Event and an earlier, and more severe disruption, during the early Botoman. This earlier event has been named the Sinsk Event, during which the shallow-water benthos of the so-called Tommotian fauna, together with archaeocyaths and some trilobites, underwent a rapid decline (Zhuravlev and Wood 1996). All areas except Australia experienced a substantial reduction in archaeocyath diversity after this event.

The Sinsk Event, estimated to have lasted some 50–300 kyr (Zhuravlev 1996), is characterized by the significant accumulation of non-bioturbated, well-laminated black shales in tropical shallow waters, especially on the Siberian and Yunnan platforms (Fig. 5.4). Lamination is due to the fine alternation of clay- and organic-rich laminae, with calcite-rich laminae containing abundant monospecific acritarchs. These shales are enriched by pyrite and elements typical of anoxic conditions and support a benthic biota of dysaerobic character.

This suggests that extinction during the early Botoman was caused by extensive encroachment of anoxic waters onto epicontinental seas, associated with eutrophication and resultant phytoplankton blooms. Stratigraphic data have not yet, however, established whether the Sinsk Event was truly global.

The Toyonian Hawke Bay Regression Event would have resulted in the extensive reduction of shallow marine habitat area in epicontinental seas. This crisis also correlates with a time of supposed reduced organic productivity (Brasier 1995), and an estimated gradual decline in CO_2 levels (Berner 1994).

Reef Benthos	Nemakit-Daldynian	Tommotian 1 2 3 4	Atdabanian 1 2 3 4	Botoman 1 2 3	Toyonian 1 2 3

```
Non-calcified
microbes:                      <-------------------------------------------------------------------
Calcified cyanobacteria:
    Korilophyton    X------------------------------X
    Angulocellularia X--------------------------------------------------------------------------
    Botominella           ?------------------------------------------------------------------
    Renalcis        X--------------------------------------------------------------------------
    Tarthinia       X--------------------------------------------------------------------------
    Girvanella      X--------------------------------------------------------------------------
    Obruchevella    X--------------------------------------------------------------------------
    Epiphyton            X-----------------------------------------------------------------
    Tubomorphophyton        X------------------------------------------------------------
    Gordonophyton           X------------------------------------------------------------
    Kordephyton             X------------------------------------------------------------
    Bija                                                   X------------------------
    Chabakovia                                             X------------------------
    Wetheredella                                                       X---------
"Encrusting microfossils"*                    X-------X
"Calcareous microspheres"*                    X-------X
?Fungi              X------------------------------------------------------------------------
Archaeocyaths:      X------------------------------------------------------------------------
Cribricyaths                        X------------------------------------------X
Coralomorphs:
    Cysticyathus   X-X
    Hydroconus                  X----------------------X
    Khasaktia                   X--------------X
    Rackovskia                         X
    Aploconus                                      X--X
    Tabulaconus                                     X--X
    Labyrinthus                                               X--X
Radiocyaths                         X------------------------------------------X
Corals                                                 X        X
"Stromatoporoids"                                      X
Microburrowers  X-------------------------------------------------------------------------
Siliceous sponges X-----------------------------------------------------------------------
Calcarean sponges       X----------------------------------------------------------------
Stenothecoids                                  X--------------------------------------
Archaeotrypa                                                       X---------
Unidentified borings  ?-----------------------------------------------------------------
Microborings                        X------------------------------------------
```

Fig. 5.3 The stratigraphic range of Lower Cambrian reef biota shows that a major extinction event took place during the Botoman–Toyonian. (Modified from Zhuravlev and Wood 1996.)

A further scenario explaining the Lower Cambrian extinction might be that enhanced ocean thermal transport cooled the tropics (the engine of global heat transfer), shrinking the latitudes at which carbonate production and reefs could develop and increasing phosphate levels by the emplacement of cold, phosphate-rich oceanic upwelling (Copper 1997). Such a mechanism has been suggested for the collapse of rudist communities during the Late Cretaceous (C.C. Johnson *et al.* 1996; see Section 5.3.6).

Recovery

Despite common reference to the demise of reefs after the extinction of the archaeocyath sponges (e.g. Fagerstrom 1987), reefs in fact continued to flourish for the remainder of the

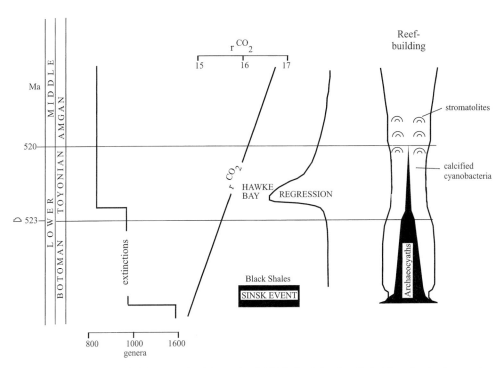

Fig. 5.4 Changes at the end of the Lower Cambrian and early Middle Cambrian show that while regional archaeocyath species' diversity plummeted during the Sinsk and Hawke Bay events, reefs continued to be built by calcified cyanobacterial communities and stromatolites. (Extinctions: Zhuravlev and Wood 1996; rCO_2 estimate curve: Berner 1994.)

Cambrian (Fig. 5.4). Middle and Late Cambrian reefs were built by a variety of thrombolites and stromatolites, and calcified cyanobacteria in shallow subtidal areas, and other microbes in deeper waters. Indeed, the largest Cambrian reefs known are recorded from the Middle Cambrian of Laurentia and on the Siberian Platform. So while reefs were much reduced in biodiversity after the Early Cambrian extinctions, they were not in either size or apparent abundance.

Although a few new calcified cyanobacteria appeared during the early Middle Cambrian, their role in reef-building was relatively insignificant, and global diversity continued to decline after the extinction events. However, lithistid demosponges — which had been rare components of late Lower Cambrian (Toyonian) reefs — became prominent in some reef communities.

Coincident with the decline of calcified cyanobacteria, was the rise of thrombolites and stromatolites, and huge Late Cambrian stromatolite complexes are known from supratidal to subtidal settings in Laurentia, South America, the Siberian Platform, and Australia (see Zhuravlev 1996 for a summary). These might represent disaster floras, and it is possible that the Lower Cambrian extinction events removed micrograzers that had formerly promoted the growth of more diverse, calcified cyanobacterial communities. Various microbial communities, together with lithistid sponges, continued to dominate reef-building until the early Middle Ordovician.

5.3.2   Late Ordovician (~ 440 Ma)

The Ordovician diversification was a major adaptive radiation, with peak diversity occuring during the mid-Caradoc when sea level appears to have been at an all-time Phanerozoic high. Some continents supported endemic biotas that undoubtedly contributed further to high global biodiversity. Both platform carbonates and basinal black shales were deposited over vast areas at this time (e.g. Brenchley 1989).

The end of the Ordovician (Ashgill) is marked by the first major Phanerozoic extinction: 26% of families and 60% of genera are estimated to have gone extinct (Sepkoski 1995). The graptolites suffered near-total extinction during the late Ashgill, losing all characteristic Ordovician biserial forms with only a few species surviving (Fortey 1989). Conodonts were also badly hit, with only 20 of the approximately 100 species known globally from the Ashgill surviving into the Silurian (C.R. Barnes *et al.* 1995). Plankton (acritarchs and chitinozoans), bivalves and echinoderms also suffered a crisis. Brachiopods and trilobites both show two phases of extinction separated by an interval dominated by higher-diversity, cosmopolitan taxa. All trilobites that were either pelagic or had planktonic larval stages went extinct (Chatterton and Speyer 1989).

Bryozoans suffered only modest extinction rates by comparison with the trilobites and brachiopods, although Llandovery faunas are impoverished and most genera of cystoporate and cryptostome orders are absent from the fossil record at this time; these Lazarus taxa reappear in the late Llandovery and early Wenlock (P.D. Taylor and Larwood 1988). Rugose and tabulate corals suffered 70% extinction of genera (Kaljo and Klaaman 1973; Kaljo 1996), but apparently this had little effect upon overall standing diversity (Scrutton 1988). In general, reefs were not as severely affected as level-bottom communities during the late Ashgill crisis (Copper 1997), and many families of reef metazoans survived to repopulate Early Silurian reefs. Indeed, trilobites that lived associated with mud mounds do not show generic-level extinction (Brenchley 1989), yet few reefs are known from the last 0.5–1 Myr of the late Hirnantian, and globally carbonate platforms are not widespread. Carbonate production rates, however, do not appear to have been severely reduced by the event (Bosscher and Schlager 1993).

Causes

The late Ashgill crisis was probably caused by several environmental perturbations. The double, mass extinction event probably relates to the sudden onset—followed by sudden termination—of an intense but short-lived Gondwanan glaciation (Fig. 5.5). Most recent work concludes that this occurred during the million or so years of the Hirnantian Stage (e.g. Brenchley 1989). The initial phase of extinction that caused the demise of graptolites and other pelagic groups points to a fundamental change in ocean circulation and chemistry, as well as cooling and regression (although most extinctions probably occurred prior to the main regression). Trilobites show extinction patterns that are consistent with some deleterious change in water quality (Chatterton and Speyer 1989). This crisis amongst pelagic life may have been triggered by a substantial decline in productivity when deep oceans became oxic with the onset of glaciation. The second phase of extinction also indicates a rapid change in oceanic state, probably marking a return to *euxinic conditions* (that is with free hydrogen sulphide in the lower water column). The concentration of these extinctions among mid- and outer-shelf faunas sug-

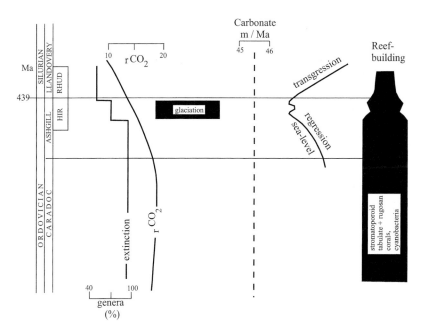

Fig. 5.5 Notwithstanding the 70% loss of coral genera, reef-building continued across the end of the Ordovician extinction event, which is related to Gondwanan glaciation and sea-level change. (Estimated carbonate production: Bosscher and Schlager 1993; sea-level curve: Brenchley 1989; rCO₂ estimate curve: Berner 1994.) (HIR: Hirnantian; RHUD: Rhuddanian.)

gests the upwelling and spread of bottom-water anoxia during the Llandovery transgression which eliminated mid- and outer-shelf habitats (Hallam and Wignall 1997).

Both reefs and coeval level-bottom communities on Anticosti Island show evidence of this double extinction, with the most marked event at the end of the Richmondian, and a lesser event at the Ordovician–Silurian boundary (Copper 1997).

Recovery

Graptolites and conodonts show spectacular and rapid radiations after the Ashgill extinction events. For example, more than 200 graptolite species are known by the end of the Silurian (Berry and Boucot 1973). For conodonts, the extinction was concentrated on shelf faunas, and the recovery appears to have involved the invasion and subsequent radiation into these vacated shallow areas by deep-water or bathyal stocks (Armstrong 1996). Brachiopods also radiated relatively rapidly. Although only 57 genera are known in the Rhuddanian (early Llandovery)—of which over half were Lazarus taxa (Cocks 1988)—a radiation was evident by the late Rhuddanian with the diversification of athyrids, pentamerids, and spirifids (Harper and Rong 1995).

As all trilobites which were either pelagic adult phases or planktonic larval stages went extinct, the radiation of trilobites in the late Llandovery was derived, at least in part, from trilobites that had lived in Ordovician reef habitats of shallow, inshore clastic settings (Fortey 1989). Like brachiopods, trilobites had become widespread by the Llandovery (Chatterton and Speyer 1989).

In North America, the extinction of rugose corals was coincident with a regression, and taxa from continental margins appear to have survived preferentially, perhaps due to their broad geographic ranges and/or ability to withstand relatively cool waters (Elias 1989). Rugose survivors were very small, and a radiation did not become fully underway until the Late Llandovery. While the Ordovician events appear to have had little effect on reefs (Fig. 5.5), Latest Ordovician reefs are composed of 'Silurian' taxa (Copper 1997). The beginning of the Silurian is marked by widespread transgression, and Early Silurian reefs are limited in both size and occurrence (P.M. Sheehan 1985; Copper 1988) and populated by cosmopolitan coral taxa (Elias 1989). When reefs became widespread in the Wenlock, they retained the same ecological cast as the pre-extinction communities.

5.3.3   Late Devonian (~370–360 Ma)

The crisis at the Frasnian–Famennian boundary is considered to be one of the major extinction events of the Phanerozoic, with 57% of genera going extinct (Sepkoski 1995). This extinction can in fact be resolved into three distinct events during the Givetian, Frasnian–Famennian, and Late Famennian, known as the Taghanic, Kellwasser, and Hangenberg events, respectively (Fig. 5.6). Of these, the Frasnian–Famennian event (FF) represents a significant period of extinction comparable in magnitude to the End Cretaceous event

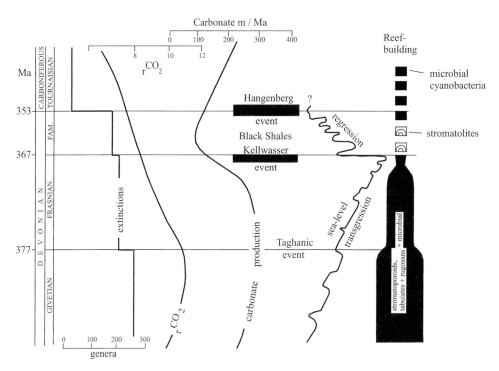

Fig. 5.6   Collapse of reef communities during the Late Devonian, showing a stepped series of extinction events. The reduction in reefs is probably related to the loss of carbonate environments due to regression and anoxic events. (Carbonate production curve: Bosscher and Schlager 1993; sea-level curve: J.G. Johnson and Klapper 1992; and rCO$_2$ estimate curve: Berner 1994.) (Modified from Copper 1997.) (FAM: Famennian.)

(Sepkoski 1995). Most extinctions occurred during two intervals—the *gigas* and *trianglaris* conodont zones of the late Frasnian—and these are known as the early and late Kellwasser events (Buggisch 1991). Extinction is notable in both benthic and pelagic groups.

Brachiopods—which were diverse and abundant members of Devonian benthic communities—suffered some extinction during the mid-Givetian, but numerous brachiopod families were eliminated during the Frasnian, most notably from tropical (Copper 1997) and mid- and outer-shelf settings (Bratton 1996). The Orthacea, Pentameracea, Atrypacea, and Stropheodontidae became extinct. The early Famennian contains an impoverished brachiopod fauna, with several holdover taxa.

Many trilobites also became extinct throughout the Late Devonian, but with little discernible selectivity (Briggs *et al.* 1988). As a group, they never again achieved their Devonian abundance and diversity.

Bryozoans suffered substantial extinction at the Givetian–Frasnian boundary, but were little affected by the later Devonian crises. Diversity rose markedly during the Early Carboniferous due to the radiation of fenestrates into shallow marine and reef habitats (Fagerstrom 1994). Large and complex foraminifera appear to have gone preferentially extinct during the FF crisis, which has been suggested to be due to the loss of reef habitats.

Corals suffered major extinction. Rugosan diversity plummeted during the mid-Givetian event, but recovered rapidly to diversify in the Frasnian to yield cosmopolitan faunas (Scrutton 1998). This radiation was abruptly and severely curtailed during the Kellwasser event: of the 43 Frasnian rugosan genera, only 14 are known from the Famennian. The Kellwasser appears to have had a differential effect on rugosan species, removing nearly all shallow-water but only some deep-water forms (Sorauf and Pedder 1986). Rugosans diversified in the Famennian, mainly from deep-water Frasnian species that invaded vacated shallow environments; colonial taxa had re-appeared by the late Famennian (Pedder 1982). However, the Hangenberg event probably represents a further major crisis for rugose corals. For example, although rugose distribution is not well documented globally across this boundary, Poty (1996) shows that no rugose species cross the Devonian–Carboniferous boundary in Belgian sequences. Tabulate corals also suffered major extinction during the Kellwasser event, with the extinction event removing perhaps as many as 80% of genera (McGhee 1996). All Famennian survivors, however, crossed the Devonian–Carboniferous boundary (Scrutton 1988).

Stromatoporoid sponges declined rapidly in the lower *gigas* Zone of the Frasnian, and by the end of the Frasnian had lost 50% loss of generic and familial diversity (Stearn 1987). Only the labechiid stromatoporoids show significant recovery after the Kellwasser crisis (Stearn *et al.* 1987), but this group became extinct in the late Famennian (Fagerstrom 1994). In contrast to stromatoporoids, hexactinellid sponges appear to have proliferated during the FF interval in sites occuring close to the Devonian equator (McGhee 1982).

Although the FF mass extinction removed a substantial proportion of invertebrate reef-builders and -dwellers, diversity had begun to decrease over a long interval from a maximum in the mid-Devonian (Fagerstrom 1994) and did not terminate until the Devonian–Carboniferous boundary. Many reefs in Europe drowned during the major early Frasnian transgressions, but a global decline in the abundance of reefs occurred rapidly in the lower *gigas* Zone of the Frasnian (Fig. 5.6). Microbial and calcareous algal communities continued to build substantial reefs in the early Famennian in the Canning Basin, Australia, and a few other localities. Surprisingly, however, most calcareous reef algae went extinct in the mid- to late-Famennian—a time when other groups were radiating—with only *Renalcis* surviving into the Tournaisian (Chuvashov and Riding 1984).

One group that has received detailed scrutiny across the FF extinction is the extinct crico-conarids. These problematic organisms, which may be molluscs, are common in Frasnian strata and probably had a pelagic lifestyle (Hallam and Wignall 1997). All four families went extinct during the Kellwasser crisis in a stepped extinction pattern. Other microplankton such as acritarchs also suffered substantial extinction during the Late Devonian, but it is not clear to which events these extinctions are related. However, exceptional abundances of tasmanaceans (other microplankton) followed by a single species of chitinozoan in the FF boundary at Coumiac, France suggest the proliferation of a disaster biota (Paris *et al.* 1996).

Detailed changes in plant communities during the Late Devonian are based mainly on the miospore record. The number of miospore species reduced from more than 50 in the Early Frasnian to less than 30 at the FF boundary (Boulter *et al.* 1988). The miospore record of the Hangenberg crisis also shows major extinctions and floral turnover, suggesting that a significant terrestrial crisis was coincident with the marine event (Bless *et al.* 1992).

Causes

The Late Devonian extinctions are clearly complex, with biotas responding in a variety of ways. As a result, several hypotheses have been promoted to explain the extinction events.

Some authors favour a bolide impact as the cause of extinction (e.g. McGee 1996). The most persuasive evidence is the presence of shocked quartz, a modest iridium anomaly, and in particular a spectacular 70-m thick breccia bed in early Frasnian sequences from Nevada and adjacent States (Leroux *et al.* 1995). This proposed impact, however, does not appear to coincide with the major extinction intervals.

At the end of the Givetian, a precipitous drop in atmospheric CO_2 began which is coincident with the Taghanic crisis (Fig.5.6). The principal victims were biotas from low-latitude, shallow marine waters (brachiopods, rugose corals, and ammonoids) which declined while pelagic biotas (ostracods and some conodonts) radiated. This drop in CO_2 continued until ~320 Ma, marking a dramatic reduction in global temperatures. Sea level rose during the Late Devonian, but the Frasnian–Famennian boundary is marked by a shift to high-frequency, sea-level fluctuations, possibly regressive cycles that exposed large continental areas (J.G. Johnson and Klapper 1992). In particular, the Kellwasser events may have been deepening events (C.A. Sandberg *et al.* 1988). After the FF interval, carbonate production declined by >70% (Bosscher and Schlager 1993) (Fig. 5.6).

The extinction of shallow marine taxa in a series of step-down extinctions accompanied this dramatic climate change. The early and late Kellwasser events are associated with a sea-level rise and may have been related to the upwelling of cold, oxygen-deficient and nutrient-rich waters onto shallow carbonate shelf settings (Buggisch 1991). This is evidenced by the development of anoxic/dysoxic sediments (including black shales), which some believe show a close correspondence to the extinction events (Hallam and Wignall 1997). In several areas of the Canning Basin, Australia, there is a transition from reefs to deeper water facies—often thinly bedded, ammonoid-bearing limestones interbedded with breccias (Becker *et al.* 1991). This is coincident with the Kellwasser horizons in German sections.

Although the mechanism to explain the widespread generation of oxygen-poor shelf waters is hotly debated, such a scenario would explain the preferential loss of warm-water taxa (for example atrypid and pentamerid brachiopods), the preferential survival of dysoxia-tolerant benthos (ostracods and bivalves), and the proliferation of rugose corals, and hexactinellid sponges—normally encountered in deep, cool waters—near the Devonian equator. Only the labechiid stromatoporoids show significant recovery after the Kellwasser crisis, perhaps due to their preference for cold waters (Stearn *et al*. 1987). The stepped extinction pattern of cricoconarid families during the Kellwasser crisis might indicate progressive deterioration in water conditions, although little is known of the ecology of this extinct group. Tasmanaceans are usually associated with anoxic and/or cool waters, and chitinozoan blooms may likewise be triggered by cool or nutrient-rich waters. Anoxia is also the favoured candidate for the Hangenberg extinctions, which also coincide with the global development of black shales (Hallam and Wignall 1997).

Recovery

In striking comparison to the widespread distribution of Frasnian reefs, estimated to cover some 5 million square kilometres (Copper 1994), Famennian shallow-water reefs are relatively rare, known only from the Canning Basin, Alberta and Omolon, and north-eastern Russia, and probably covered less than 1000 square kilometres (Copper 1994). This collapse occurred over a period of 1–4 Myr. However, the effect of the Frasnian extinction on reefs may be overemphasized, as stromatoporoids and corals were not affected on the Russian platform (Ulmishek 1988), and stromatoporoid-bearing reefs are known from the Famennian of Canada (Stearn *et al*. 1987), and Poland and Moravia (Buggisch 1991). Moreover, microbial and calcareous algal communities continued to build substantial reefs with associated lithistid sponges in the early Famennian wherever environmental conditions allowed. Notwithstanding the near-complete loss of large, heavily calcified metazoans, reef development was continuous across the FF boundary in some parts of the Canning Basin, Western Australia. There, reef growth was dominated by microbialites, which had been major framework builders during the Frasnian. Deep-water mud mounds and shallow-water stromatolites are more widespread, built mainly by calcified cyanobacteria (*Renalcis*) and other microbial communities. Lazarus tabulate corals were almost exclusively creeping or branching auloporids, the small colonial michelinids and the dendroid pachyporids; larger, massive colonial species went extinct (Copper 1994).

Microbially-dominated reefs persisted into the Late Carboniferous and Permian, particularly on the northern margin of Pangaea (G.R. Davies *et al*. 1989). Many groups diversified during the Permo-Carboniferous—a ~100 Myr interval that lacks any major crises except for a modest Mid-Carboniferous extinction coinciding with glaciation. Radiating groups included the rugose corals, crinoids, gastropods, productid brachiopods, and foraminiferans. However, most Early Carboniferous corals maintained a deep-water habitat (Sando 1989). This suggests that the distribution and ecology of Early Carboniferous reefs was not controlled by the availability of sessile skeletal metazoans or microbial communities, but rather by extrinsic environmental factors.

Labechiids may represent disaster taxa as they disappear by the Carboniferous. But small solitary rugosans, as well as sphinctozoan and lithistid sponges, michelinid and auloporid tabulate corals, chlorophytes and rhodophytes survived the end-Devonian extinctions. Cool-adapted brachiopods (rhynchonellids, terebratulids, athyrids and spiriferids), and ammonoids may represent opportunists that moved onto carbonate shelves from deeper water or temperate settings (Copper 1997). By the later Famennian, a rapid recovery was underway with the radiation of spiriferids and rhynchonellids—the two orders least affected by the Kellwasser events—as well as the productids (Ji 1989).

5.3.4 End Permian (245 Ma)

The extinction at the end of the Permian has long been recognized as the greatest of the Phanerozoic (e.g. Phillips 1841), which probably eliminated 82% of all genera (Sepkoski 1995) and 80–95% of species in the oceans (Erwin 1994). Permian levels of species diversity were not achieved again until the Cretaceous.

Notwithstanding its obvious importance, the end-Permian extinction is probably the least well understood of the five major mass extinctions (Erwin 1993), as its study has been plagued by poor preservation of biotas, a paucity of successions across the interval, and difficult stratigraphic correlation. Although there is still little consensus as to the rate and pattern of extinction (Erwin 1994), it is now clear that the supposed long, Late Permian decline in diversity in fact consists of two distinct extinction episodes separated by an extended period of recovery and radiation (e.g. S.M. Stanley and Yang 1994). The first extinction occurred in the Late Maokouan, the second and more disastrous appears to be entirely confined to the *changxingensis* Zone (representing perhaps less than 1 Myr) in the best-dated sections (Fig. 5.7). The intervening Lopingian radiation may have lasted as long as 16 Myr.

Numerous excellent Permian–Triassic (PTr) boundary sections have now been documented from China, including the finding of abundant, substantial latest Permian (Lopingian)

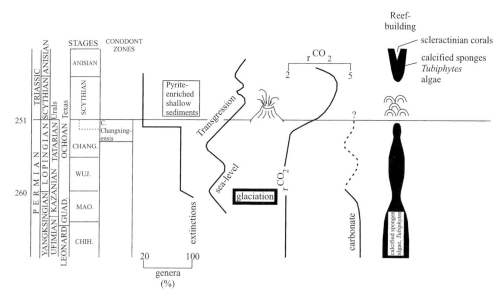

Fig. 5.7 It is now clear that the supposed long Late Permian diversity decline in fact consists of two distinct extinction episodes separated by an extended period of recovery and radiation. The first extinction occurred in the late Maokouan, and was probably caused by climatic cooling induced by glaciation and an attendant, and possibly related, major global major regression. The regression terminated reef formation in the Guadalupian sections of Texas and New Mexico, and caused the loss of vast areas of other shallow-marine tropical carbonate habitats elsewhere resulting in the extinction of endemic biotas. A modest radiation has been recorded from the Lopingian. The second and more disastrous crisis is probably to be entirely confined to the *changxingensis* Zone (representing perhaps less than 1 Myr) and appears to have been the result of global warming—perhaps, in part, triggered by the eruption of the Siberian traps—and marine anoxia. (Carbonate production curve: Bosscher and Schlager 1993; sea-level curve: Hallam and Wignall 1997; rCO₂ estimate curve: Berner 1994.) (CHANG: Changxingian; WUJ: Wujiapingian; MAO: Maokouan; CHIH: Chihsian.)

reefs that contain an hitherto unappreciated diversity of invertebrates. Lopingian reefs are known throughout Tethys. They show increasing size and biodiversity towards the end of the Changxingian (Reinhardt 1988; Flügel and Reinhardt 1989), including a rapid and marked increase in the diversity of sphinctozoan sponges (Rigby and Senowbari-Daryan 1995). In addition to calcified sponges, the reef biota includes *Tubiphytes*, various algae, as well as brachiopods, crinoids, non-fusuline foraminifera, and six families of bryozoans (Fan *et al*. 1990). Reef-building terminated abruptly in the late Changxingian; in Sichuan the reefs are overlain by deeper water wackestones only a few metres below the *parvus* Zone (Wignall and Hallam 1996).

Most tabulate corals went extinct in the late Maokouan and only a few michelinids persisted to the end of the Changxingian (Federowski 1989). Rugosans underwent a similar decline in diversity at this time, but show a modest radiation in the Wujiapingian. The demise of the Rugosa has been traditionally viewed as a progressive decline, starting with the loss of larger colonial forms, then fasciculate forms, leaving only solitary, deep-water corals by the end of the Permian (Flügel and Stanley 1984). This sequential loss has been well documented in Transcaucasia and Iran, where it is closely associated with the development of deep-water sediments in the Changxingian (Ezaki 1993). By contrast, in South China, where shallow-water sediments are found right up to the end of the *changxingensis* Zone, the disappearance of colonial forms is far more abrupt. Such forms could clearly thrive almost to the end of the Permian given the persistence of the right habitats.

The supposed latitudinal contraction of rugosans is also no longer tenable as it ignores the occurrence of corals in the uppermost Permian of Spitzbergen, ascribed a Tatarian age, and also a fauna from north-east Greenland of probable Late Changxingian age (Ezaki *et al*. 1994). Over 30 genera of bryozoans are recorded from the Spitzbergen sections.

Foraminifera offer one of the best studied and most informative fossil records across this extinction event. The record shows that the extinction was highly ecologically and environmentally selective. Foraminifera suffered two extinction events, the first in the late Maokouan which primarily affected fusulinids—particularly those with distinctive honeycomb wall structures—and the second and more severe crisis in the latest Changxingian which affected all groups (S.M. Stanley and Yang 1994). The radiation of many foraminiferas that occurred in the intervening Lopingian produced larger and more complex forms, which supports a scenario of crisis followed by recovery and radiation to fill the ecological niches vacated by the fusulinids. The latest Changxingian mass extinction was highly selective, affecting primarily tropical, architecturally complex forms, including those which had radiated in the Lopingian and those with calcareous microgranular tests (Brasier 1988).

On land, the extinction is marked by the abrupt demise of the cold-adapted glossopterid flora in high southern latitudes (Retallack 1995), and the disappearance of diverse tetrapod communities involving the loss of 21 families (Benton 1995). A 'fungal spike' occurs in the *changxingensis* Zone in the best-dated sections (Visscher and Brugman 1988), which has been speculated to have resulted from the widespread proliferation of fungi due to the abundance of rotting vegetation and the extinction of many insect groups (Hallam and Wignall 1997).

Causes

The late Maokouan extinction was probably caused by climatic cooling and an attendant, and possibly related, major global regression. This extinction preferentially affected tropical

Tethyan biotas, as temperate and polar faunas escaped unharmed (Jin *et al.* 1994). Evidence for glaciation is well established in Siberia and Gondwana for this time (Caputo and Crowell 1985). The crisis removed many endemic biotas, and the regression terminated reef formation in the Guadalupian sections of Texas and New Mexico and caused the loss of vast areas of shallow-marine tropical carbonate habitats elsewhere.

Contrary to the popular belief that a further major regression occurred at the very end of the Permian which reduced shelf areas and so eliminated many shallow shelf habitats, current evidence points to the occurrence of a major transgression across the PTr boundary (Hallam and Wignall 1997). It has now been established that the huge eruptions of flood basalts in Russia known as the Siberian traps exactly coincided with the End-Permian (latest Changxingian) mass extinction. Furthermore, it has been proposed that these eruptions caused global warming by providing a major source of CO_2, along with that generated by the uplift and oxidation of coal formations during orogeny of the margin of Gondwana in the Late Permian (Faure *et al.* 1995). Several lines of evidence support the contention that global warming occurred that also reduced environmental heterogeneity on land: the extinction was not as marked in the tropics as at high latitudes; the abrupt demise of the cold-adapted glossopterid flora in high southern latitudes was replaced by remarkably uniform warm, temperate floras; near-Tundra peats were replaced by warm, temperate palaeosols in the Triassic (Retallack 1996); and the formerly diverse tetrapods were replaced by low-diversity, pandemic forms, in particular the dicynodont *Lystrosaurus*.

That marine anoxia was also responsible for the Late Permian extinction is supported by both palaeontological and geological evidence. The only benthic groups to weather the latest Changxingian crisis were all Palaeozoic dysaerobic taxa, including 'paper pecten' bivalves e.g. *Claraia*. Postextinction biotas of foraminifera have been proposed to consist preferentially of infaunal detrital feeders (Tappan and Loeblich 1988) and forms adapted to dysaerobic conditions (Wignall 1990). Ostracods with an ability to withstand dysoxia also appear to have preferentially survived the PTr crisis, and some sediments contain enormously abundant arcritarchs. Marine nektonic predators — including many fish groups — also preferentially survived, for reasons as yet poorly understood, although interestingly the survival of such metabolically active organisms is predicted by a model involving hypercapnia (death by CO_2 poisoning) (Knoll *et al.* 1996.)

Boundary sections commonly show the disappearance of diverse Permian assemblages coincident with a change from burrowed aerobic sediments to laminated, pyrite-rich, anaerobic strata (Hallam and Wignall 1997). On shelf-settings, micritic limestones are common which show the preservation of high-quality organic carbon, indicative of anoxic depositional conditions. These observations are consistent with the proposal that the world's oceans became euxinic (with free H_2S) at this time.

Recovery

The Scythian was characterized by highly distinctive, very low-diversity (depauperate) assemblages which were remarkably cosmopolitan (Erwin 1993; Schubert and Bottjer 1995; Hallam and Wignall 1997). In the sea these included *Lingula*, 'paper pecten' bivalves, agglutinating foraminifera, palaeopsychrospheric ostracods, and abundant acritarchs; on land, biotas were dominated by the dicynodont *Lystrosaurus* and only three plant genera. Many of these groups were evolutionary dead-ends. Schubert and Bottjer (1992) suggest that stromatolites became abundant in normal marine shallow seas following the extinction, forming a disaster flora in response to the new availability of such habitats after the removal of herbivores and superior benthic competitors. Pratt (1995), however, suggests that some of these stromatolites grew in deep-shelf settings. All these features are highly suggestive of widespread and prolonged, stressful conditions, and are all the more remarkable as Late Permian faunas, such as the bivalves, can be readily divided into tropical, temperate, and high-latitude realms. During the Scythian there is a noteworthy absence in the fossil record of numerous Lazarus taxa—gastropods, calcified sponges, brachiopods, and bryozoans—which are all benthic groups.

Although a few new groups radiate in the first 4–5 Myr after the extinction event, full recovery after the end-Permian mass extinction is generally recognized to have begun at the base of the Anisian—some 7–10 Myr after the latest Changxingian crisis—when normal marine communities reappeared abruptly (Erwin 1996). Reefs (non-stromatolite) appeared at the base of the Anisian in the mid- and deep-ramp settings of the Northern Italian Dolomites, although some Scythian reefs are found on oceanic islands in the ancestral Pacific and some possible Upper Sycthian–Anisian reefs are known from Iran (R. Brander, unpublished data, in Webb 1996) and southern China (D.J. Lehrmann and P. Enos, unpublished data, in Webb 1996). Coals also appeared at the base of the Anisian, with red radiolarian cherts reappearing later in the stage.

While the long absence of reefs from the record has been attributed to a period of 're-organisation' (P.M. Sheehan 1985), it is clear they reappeared at the same time as other marine communities. Indeed, brachiopod and bryozoan radiations did not start until the Ladinian or later; these groups were derived from formerly insignificant Palaeozoic groups. The well-known bivalve genera *Pinna*, *Chlamys*, and *Thracia*, although appearing in the Late Permian, were not seen again until the Middle Triassic.

Many of these new communities had a Permian caste: for example, normal marine gastropod communities in the Anisian and Ladinian were more like those from the Guadalupian than the Jurassic (Erwin 1996). Anisian reef communities resemble Late Permian sponge–algal reefs, consisting of calcified sponges particularly sphinctozoans, *Tubiphytes*, and algae such as *Archaeolithoporella*. With the resumption of widespread carbonate deposition they re-established rapidly once suitable environmental conditions established. Of the 30 genera of sphinctozoans known from the latest Permian reefs, only 9 crossed the PTr boundary; the survivors are apparently long-ranging conservative forms (Rigby and Senowbari-Daryan 1995). This style of reef-building continued into the succeeding Ladinian and Carnian (Fois and Gaetani 1984).

The close similarity of ecology and biota has lead to the suggestion that Middle Triassic reefs were populated by Permian Lazarus taxa. However, closer taxonomic scrutiny has failed to confirm true Permian holdovers in Triassic reefs (Senowbari-Daryan *et al.* 1993; Senowbari-Daryan and Flügel 1994), suggesting instead that Triassic examples of *Archaeolithoporella*, *Tubiphytes*, and some sphinctozoan sponges (e.g. '*Girtycoelia*') are convergent homeomorphs of Permian taxa. However, phylloid algae have been reported from the Stikinia Terrane of the Yukon (Reid 1987), one of the many displaced terranes of volcanic origin thought to have been exotic to North America during the early Mesozoic. This suggests that during the Triassic, the ancestral Pacific was strewn with volcanic islands such as those now found within the circum-Pacific region (Hallam 1986), and it lends some credence to the idea that reef biota may have survived on island refuges where reefs could have flourished away from the environmental crisis on platform margins and interiors.

Scleractinian corals appeared in the Anisian, with a high-standing diversity. This record indicates a considerable phase of preskeletal diversification, and molecular phylogenies show that the scleractinian corals probably arose as an unskeletonized group of sea anemones or corallimorpharian-like organisms, within which two subordinal groups later diverged and independently acquired calcareous skeletons (Romano and Palumbi 1996). These data strongly suggest the operation of an environmental or ecological cue in the Anisian that in some way initiated skeletonization.

It is not yet clear whether this long period of recovery was due to the profound disruption of Permian communities that required a long interval of recovery, or to continued environmental perturbation (Erwin 1996). Doubts have been raised, however, to the artefactual nature of this slow recovery due to preservation bias (e.g. Hallam and Wignall 1997): the ubiquity of Lazarus taxa in the Early Triassic is certainly suggestive of limited sampling and/or preservation. However, current evidence suggests that oxygen levels may not have improved until the Middle Triassic, until CO_2 levels were finally lowered by the combined effects of high rates of global weathering, the formation of thick carbonate sequences, and the burial of large amounts of organic carbon in the deep sea. In the aftermath of the extinction, formerly minor groups of mobile predators such as gastropods, crustaceans, and fishes radiated, as well as bivalves. This resulted in a fundamental reorganization of marine life and the dawn of typically Mesozoic biotas.

5.3.5   End Triassic (208 Ma)

Two extinction events of importance to reef communities have been documented at the end of the Triassic: the first across the Lower–Upper Carnian boundary and a second at or close to the end of the Triassic (Fig. 5.8).

Several important reef-builders declined across the Lower–Upper Carnian boundary, including *Tubiphytes*, calcified sponges, bryozoans, and rhodophyte and chlorophyte algae. This coincided with a diversification of scleractinian corals, 'spongiomorphs', and foraminifera (G.D. Stanley 1988). While these events appear broadly synchronous, they are poorly constrained stratigraphically, and it is also not clear whether this accelerated turnover was global in extent. Benton (1988) has demonstrated that four out of six families of marine tetrapods became extinct at the end of the Carnian.

The extinction that occurred at or close to the end of the Triassic is estimated to have claimed 53% of the marine invertebrate genera (Sepkoski 1995), and a continental mass extinction event is also probable. Most dramatically, no less than six superfamilies of ammonites became extinct at or shortly before the Triassic–Jurassic boundary. The subsequent spectacular radiation of the group during the Jurassic and Cretaceous is derived from a sole surviving species. Half of the known bivalve genera also disappeared.

During the Norian (including the Rhaetian) of the Northern Calcareous Alps, scleractinian corals underwent a substantial increase in diversity; they had become major components of reefs in the Carnian. But at the end of the Triassic, coral reefs suddenly disappeared. Of the 50 scleractinian genera known globally, only 11 are recognized to have survived into the Liassic (Beauvais 1984); other reef-associated organisms also disappeared, including calcified sponges. Although this extinction is considered to be global, no other localities are known which contain a comparatively rich, Upper Triassic reef biota (G.D. Stanley 1988).

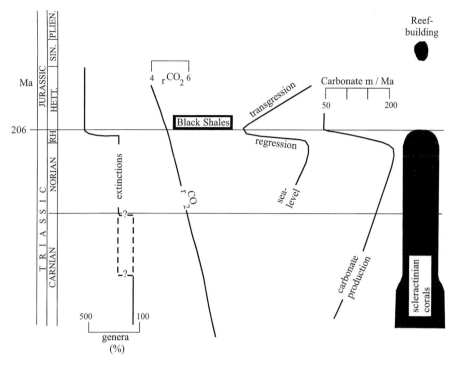

Fig. 5.8 Scleractinian-dominated reef communities disappeared at the end of the Triassic. The dramatic loss of reefs is probably related to the loss of carbonate environments due to regression and anoxic events. (Carbonate production curve: Bosscher and Schlager 1993; sea-level curve: Embry and Suneby 1994 and Brandner 1984; rCO$_2$ estimate curve: Berner 1994.) (RH: Rhaetian; HETT: Hettangian; SIN: Sinemurian; PLIEN: Pliensbachian.)

Causes

There is much debate as to the nature of the end of the Triassic extinction event. Coccolith and palynological data, like that of the late Norian reef biotas, show no reduction in biodiversity through the Rhaetian before a substantial reduction at the end of the Triassic, suggestive of a short-lived extinction event. But such data are limited to a few localities, and so the catastrophic nature of the end of the Triassic event cannot be confirmed until the compilation of more detailed studies on a global basis. Marine successions across the Triassic–Jurassic boundary are very restricted due to low global sea level.

There is little evidence to support any climatic disturbance (either cooling or warming) at the end of the Triassic. But there is good evidence for an extensive and significant regression in many localities at this time (e.g. Embry and Suneby 1994; Fig. 5.8). Analysis of the Alpine sequence shows an abrupt fall in sea level and a subsequent rapid rise (Brandner 1984). This regression led to the widespread emergence and karstification of the reef complexes (Satterley *et al.* 1994), and in the Southern Alps an episode of shallowing within the carbonate platform sequence is also evident (McRoberts 1994).

In Northern Europe, as well as evidence for a short-lived regression followed by a major Early Jurassic transgression, the basal Jurassic sequences consist of alternating marine limestones and shales that are well-laminated, organic-rich, and contain no benthos, with beds that contain abundant, low-diversity faunas suggestive of opportunistic colonization under dysaerobic conditions (Hallam 1995). These sediments may indicate the spread of anoxic or dysoxic bottom waters associated with marine transgression (Fig. 5.8). That an anoxic event was responsible for the mass extinctions at this time is the best-supported model currently available (Hallam and Wignall 1997).

Recovery

The end-Triassic event was profound, and affected both marine and terrestrial biotas. Reefs were globally rare for 4–10 Myr following the end of the Triassic (G.D. Stanley 1988), although small, low-diversity coral reefs are known. Throughout the Jurassic, reef abundance mirrored eustatic sea-level changes (Leinfelder *et al.* 1994), and by the Upper Jurassic, when substantial shallow epicontinental seas had developed, reefs formed in the Tethys stretching across Europe, North Africa, and the Middle East, and also locally in North and South America.

Approximately 18 of 67 genera of scleractinian corals survived the end-Triassic extinction. But whilst the Jurassic inheritance was depauperate, these genera were systematically diverse. At the end of the Triassic, carbonate production plummeted, and during the Hettangian–Sinemurian, there was minimal shelf carbonate development and therefore few reefs (Fig. 5.8). Sinemurian reefs are known from a displaced terrane in British Columbia, which are presumed to represent volcanic islands in the ancestral Pacific. These are constructed by Late Triassic coral holdover species such as *Phacelostylophyllum rugosum* which appear to have survived in these refugia (G.D. Stanley and Beauvais 1994). This finding has been suggested as lending support to the role of displaced terranes in providing refugia from the drastic environmental perturbations which affected Pangaea.

A substantially different reef coral fauna had emerged by the Pliensbachian. All Triassic genera became extinct by the late Early Jurassic, and beginning in the Toarcian a major faunal changeover occurred (Beauvais 1984). In these reefs, endemic taxa predominate together with a few remaining Norian taxa. This marked a substantial phase of scleractinian coral radiation: whilst only 21 genera are known from the Liassic, 100 are recorded from the Middle Jurassic, and in many regions of Europe, spectacular shelf-edge coral reefs are known (Heckel 1974). This radiation and abundance of reef-building activity may have been due to a warm climate (Beauvais 1984), to the formation of barriers produced by intraoceanic carbonate platforms (Pantić *et al.* 1983), or to the initiation of the central Atlantic seaway connecting Tethys directly to the Pacific, which would have opened up new shallow shelf habitats and created a cosmopolitan fauna. The distribution of reefs, however, appears to have been reduced during a cooling episode in the Oxfordian.

5.3.6 End Cretaceous (65 Ma)

The end of the Cretaceous extinction (known as the Cretaceous/Tertiary (KT) event) is estimated to have claimed up to 47% of marine invertebrate genera (Sepkoski 1995; Fig. 5.9). Although correlation between terrestrial and marine deposits is problematic, data suggest a coincident catastrophe on land and in the sea. Terrestrial sequences show that the last dinosaurs in North America disappeared—probably abruptly (P.W. Sheehan *et al.* 1991)—at the level of the so-called 'fern spike', which marks a drastic reduction in the angiosperm

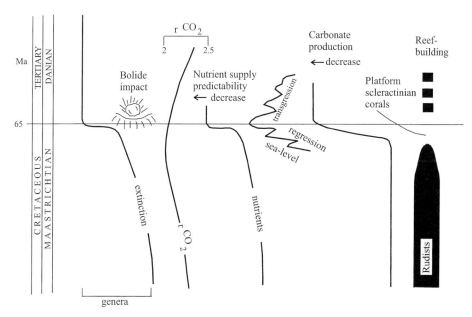

Fig. 5.9 While there is little doubt that a bolide impact caused widespread extinction at the Cretaceous/Tertiary boundary, Late Cretaceous climates were already deteriorating during the Maastrichtian and there is also evidence for unpredictable nutrient supply during this time (A.B. Smith and Jeffrey 1997). Platform-dwelling rudists became extinct 1–3 Myr before the KT boundary, but scleractinian corals persisted. Tropical carbonate platform development was markedly reduced after the KT boundary. (Sea-level curve: Rohling *et al.* 1991; rCO_2 estimate curve: Berner 1994.)

flora but a proliferation of ferns. A similar fern spike has been recognized in Japan that is coincident with marine sediments containing calcareous plankton used to locate the KT boundary. Planktonic foraminifera show a major extinction event, and some 85% of calcareous nannoplankton suffered extinction at or shortly after the boundary. Although planktonic foraminifera have been used to define the boundary, there is some debate as to whether this should be marked by the sudden extinction of Cretaceous species and the proliferation of the survivor *Guembelitria cretacica,* or the first appearance of Palaeocene species. There is also continued debate as to whether this faunal change was instantaneous or rather more gradual, because there is no unequivocal method for estimating the time involved in the KT event.

Diatoms and dinoflagellates—probably by dint of their benthic resting stages—survived the event relatively unscathed. No sudden mass extinction of planktonic foraminiferans has been recognized at high palaeolatitudes, where unspecialized, small cosmopolitan faunas and presumed eurytopic species are thought to have survived (Keller *et al.* 1993). The species that disappeared from low latitudes were large, specialized forms which is suggested to indicate greater severity of the extinction at low latitudes. However, no such selectivity has been found amongst molluscs if the exclusively tropical rudists are excluded (Raup and Jablonski 1993), or in echinoids (A.B. Smith and Jeffrey 1997), and amongst plants those extinctions in the tropics were less severe than in the northern hemisphere (Hickey 1984).

Kaiho (1992) has documented an 80% extinction rate of benthic foraminifera in the photic zone from Japan, compared with only 10% for forms living below 150 m. This is consistent with an extinction scenario that affected surface, but not deeper water, biotas. The sudden drop in diversity at the boundary section at El Kef, Tunisia, is coincident with a proliferation of opportunistic, low-oxygen-tolerant forms, and a preferential extinction of those forms with higher nutrient demands (Speijer 1994). This selectivity suggests that extinction was caused by a massive drop in pelagic primary productivity.

In deep-sea sites, the KT transition is marked by an abrupt decline in sedimentation rate, and in outer-shelf and upper-slope sequences the lithologies change from carbonate-dominated sediments to black clays. This decrease in carbonate production is seen in all environments; in pelagic environments as a result of the mass extinction of calcareous plankton, and in shelf environments as the result of an increase in clastic input due to sea-level lowstand (Fig. 5.9).

While an extinction event of catastrophic proportions occurred at the KT boundary, some major groups show a decline in diversity considerably before this time (C.R. Marshall and Ward 1996). Rudist and inoceramid communities declined such that they were nearly extinct between 1.5 and 3 million years prior to the bolide impact (Kauffman 1984; C.R. Marshall and Ward 1996; Steuber 1997); other major molluscan groups had been in decline earlier than this (Fig. 5.9). High floral turnover rates are also evident in the Late Cretaceous independent of any end-Cretaceous catastrophe (J.G. Johnson 1993). But although data is limited and stratigraphic resolution coarse, the presence of rich Maastrichtian coral faunas suggests that extinctions may have been concentrated in the late Maastrichtian (Beauvais and Beauvais 1974).

Although scleractinian coral diversity was high, coral reefs were not a feature of the shallow, epicontinental platforms of the Upper Cretaceous. One-third of coral families (about 57% of the genera) went extinct at the KT boundary. There is some indication that inferred photosymbiotic corals (based on those with complex colonial organizations) were more susceptible to extinction (70%) that inferred non-photosymbiotic forms (solitary and simple branching colonies—40%) (Rosen and Turnšek 1989). Only 13 of the 27 genera of the non-photosymbiotic Caryophyllidae survived; and 3 genera of the Rhizangiidae.

Causes

There can now be little doubt that the KT boundary was marked by a substantial bolide impact. A global iridium anomaly is well documented at the boundary (an anomaly far greater than at any other horizon in the Phanerozoic), as well as shocked quartz from the North American Western Interior, tektites from Haiti, and tsunami deposits in the Gulf Coast–Caribbean region. The impact was located on the Yucatan peninsular in Mexico, as evidenced by the presence of the Chicxulub crater, which is estimated to be between 150–300 km in diameter and occurs within sulphur-rich evaporates within a carbonate platform sequence.

Such a substantial bolide impact is predicted to have had devastating environmental consequences on a global scale. Exactly how the bolide impact caused a severe environmental perturbation leading to mass extinction is not clear, but suggestions include profound changes in atmospheric gas composition, a global drop in temperature, the formation of acid rain (from sulphur derived from evaporites at the impact site), global wildfires, or a reduction in light levels causing a nuclear winter (see summary in Ward 1995). Careful dating of the extensive and thick basalts of India, known as the Deccan Traps—previously thought also to be good candidates for causing the end Cretaceous extinction event—have now been shown to have erupted before the impact (e.g. Bhandari et al. 1995).

The plankton record at the end of the Cretaceous implies a catastrophic deterioration of ocean surface waters. This is consistent with a drop in primary productivity, which some have estimated persisted for about half a million years (Zachos *et al.* 1989). P.W. Sheehan and Hanson (1986) predicted that this would preferentially affect benthic suspension-feeders, and indeed all these groups experienced substantial extinctions. On the other hand, detritivores, carnivores, and scavengers would have been relatively resistant, as has been documented in some sequences across the boundary. However, echinoid clades with planktotrophic larvae do not experience a statistically significant higher extinction than those with non-feeding (non-planktotrophic) larvae, suggesting that scenarios which invoke an instantaneous catastrophe by bolide impact causing the wholesale selective extinction of planktotrophs may be invalid (Smith and Jeffery 1997). Indeed, there is indirect evidence that conditions for plankton were becoming less favourable immediately before the KT boundary.

The latest Cretaceous was a time of significant climatic and oceanographic change (Fig. 5.9). The bolide impact followed shortly (1–3 Myr) after two rapid regression–transgression eustatic sea- level cycles, and a major change in ocean chemistry (Ward 1995). Rudists disappeared during these regressions, perhaps linked to one phase in a series of sea-level fall and global cooling events (C.C. Johnson *et al.* 1996) which was also coincident with the major floral turnover (Barrera 1994). Numerous unrelated clades of echinoid have been documented to have switched from planktotrophic to non-planktotrophic development during the Maastrichtian, independent of any changes in palaeolatitude and water depth (Jeffrey 1997).

Enhanced ocean thermal transport may have cooled the tropics, shrinking the latitudes at which carbonate production and reefs could develop (Copper 1997). Such a mechanism has also been suggested to explain the paradox of apparently cool tropics despite high CO_2 levels at this time (d'Hondt and Arthur 1996). Whatever the cause, it seems likely that some extinction would have occurred at the end of the Cretaceous independent of any bolide impact (e.g. Hallam and Wignall 1997).

Recovery

There was a dramatic radiation of spinose globigerinids in the early Palaeocene, which may indicate either a radically altered post-extinction environment, or one vacated of competition by the extinction event. Analyses of planktonic foraminifera show two phases of indigenous Danian speciation which centred around low- and mid-latitude habitats and displaced many formerly cosmopolitan survivor species to high latitudes (MacLeod and Keller 1994). Some mollusc families also show faunas which are interpreted to be opportunistic 'blooms', characterized by rapid maturation, short lifespans, and wide dispersal capabilities which were replaced later in the Palaeocene by supposedly more specialized groups (Hansen 1988).

In a highly significant study, A.B. Smith and Jeffery (1997) used strict phylogenetic analyses to discern patterns of echinoid extinction and recovery across the KT boundary. They established that extinction rates were considerably lower than previous estimates, and also found a significant correlation between feeding mode and survival: urchins that lived on organodetritus in nutrient-poor settings without the aid of specialized feeding structures fared the worst; generalized omnivores suffered the least extinction. A pronounced drop in body size was also noted for most Danian clades that survived the KT boundary. Both these observations are consistent with the presence of an unpredictable nutrient supply, and possibly a sustained drop in primary productivity after the KT boundary. High levels of extinction continued into the Danian, suggesting that unpredictable nutrient supplies continued to limit growth.

Two temperate communities appear to have persisted through the KT boundary: the highly diverse bryozoan–bivalve communities of the Danish Chalk, and the deep-water, non-photosymbiotic coral reefs from Denmark (e.g. Birkelund and Hakansson 1982). In addition, the main episode of extinction for chalk echinoids was not at the end of the Cretaceous but at the end of the Danian when chalk deposition ceased (A.B. Smith and Jeffrey 1997). This suggests that survival of both echinoids and reef-building communities is closely related to persistence of habitat.

Globally, carbonate production diminished, lasting at least 1 Myr, but probably far longer (Zachos and Arthur 1986; Fig. 5.9). Carbonate platforms and reefs are rare in the Palaeocene (Bryan 1991), but the first are known some 2–3 to 5 Myr after the KT boundary—considerably earlier than previously realized (Rosen 1998). Many may not have occurred in the tropics, and were built mainly by sponges and coralline algae. Extensive reefs became established by the Mid–Late Palaeocene (Bryan 1991).

Detailed analysis of the recovery of early Cenozoic tropical reef biotas is hampered by problematic correlation and poor documentation, but unlike scleractinian corals, coralline algae appear to have survived virtually unscathed and indeed are responsible for the construction of some Palaeocene reefs, although most Palaeocene and Eocene shelf margins were dominated by accumulations of coralline algae and photosymbiotic benthic foraminifera (Frost 1986). Although inferred photosymbiotic coral species appear to have been more prone to extinction (Rosen and Turnšek 1989)—as might be predicted from climatic cooling and loss of habitats—nine genera of living photosymbiotic corals survived the extinction event into the Palaeocene and all either became, or were closely related to, important reef-builders in the Cenozoic (Rosen 1998). Current knowledge suggests that the rediversification of symbiotic scleractinian corals after the end-Cretaceous extinction was relatively rapid, perhaps in the order of a few million years, although some suggest that Cretaceous levels of diversity were not reached until the Oligocene (Rosen 1998). Although the taxonomic and stratigraphic framework is very poor, it appears that when coral reefs become common in the Eocene, there was no species level continuity with the Mesozoic. Palaeocene coral reefs mark the appearance of a new type of coral community, not the recovery of a previously existing one (Rosen 1998).

5.4 Are reefs more susceptible to mass extinction than other communities?

Many texts state that reefs, as specialized tropical communities, have been more susceptible to mass extinction than level-bottom communities (see summary in Hallam and Wignall 1997). Such a contention has, however, been subject to little detailed scrutiny.

The relative extinction rates of reef biotas have been compared to other marine communities by a compilation of marine invertebrate genera from the Ordovician–Recent (Raup and Boyajian 1988). These authors found that the extinction profiles of reef and non-reef genera were similar, suggesting not only the operation of common external factors, but also that extinction rates in reef taxa are, in general, no higher than other marine taxa during both background and mass extinctions (Fig. 5.10).

Such a broad analysis rests on the definition of what is, or is not, a reef biota: many organisms can occur in both reef and non-reef habitats. To assess the supposed differential

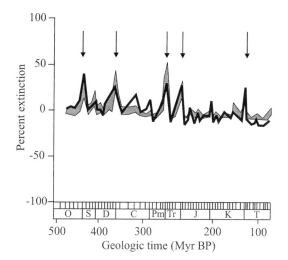

Fig. 5.10 Extinction profiles of 417 reef genera (hatched band) compared with all other genera (solid line). Arrows show the five major extinction events. (After Raup and Boyajian 1988.)

response of reef biotas to mass extinctions requires detailed appraisal of the fate of reef- and non-reef communities across each extinction boundary. The following sections examine the four hypotheses put forward (Jablonski 1995) to explain the supposed spectacular collapse and slow recovery of reef ecosystems after mass extinctions.

5.4.1 Tropical biotas are more susceptible to extinction than temperate biotas

This arguments rests on the assumption that tropical biotas are adapted to a narrow—and specialized—range of environments compared to temperate communities, which are therefore more readily perturbed by the factors that cause mass extinctions. Clearly not all mass extinctions had the same causes and effects: some were biased towards tropical benthic biotas, e.g. the end-Devonian, and late Maokouan (Late Permian) extinctions; others were either global (Cretaceous), or preferentially eliminated pelagic low-latitude (Late Ordovician), or high-latitude biotas (latest Changxingian (end-Permian)).

However, analyses of several mass extinction events have shown that while the general extinction pattern is suggestive of a tropical bias, this has yet to be confirmed statistically. In a detailed global compilation, extinction rates of marine bivalves occupying non-carbonate platform tropical settings did not suffer greater loss than non-tropical faunas during the KT event (Raup and Jablonski 1993). Likewise, A.B. Smith and Jeffrey (1997) found only a weak correlation with palaeolatitude for the extinction of sea urchins across the KT boundary.

5.4.2 Tropical biotas contain a large proportion of extinction-prone endemic species

There is some support for the contention that widespread species are more resistant to extinction than endemic species, both from ecological theories of community species abundance and community assembly, and from the fossil record.

Hubbells's (1997) theory of the relationship between metacommunity (the large source area of species) and community (local populations) predicts that common members of the metacommunity are likely to be very resistant to extinction and to persist for geologically significant periods of time. Even moderate rates of dispersal will ensure that such species occupy nearly all suitable habitats, nearly all the time. Species with larvae or adults capable of long-distance dispersal, will show particularly widespread distributions. Indeed, most living reef coral species compared to other major marine groups are widespread and show relatively low levels of endemism within geographic regions.

Some fossil data have suggested that widespread species are more resistant to extinction than endemic species (e.g. Jablonski 1989, 1995; Jackson 1995). However, A.B. Smith and Jeffrey (1997) have shown that the construction of an explicitly phylogenetic framework which identified clade survival independent of taxonomic name removed the apparent preferential extinction of endemic biotas. Such work suggests that wrongly identified postextinction taxa will artificially inflate 'endemic' extinctions, and it stresses the need for a sound phylogenetic framework when assessing extinction patterns. Detailed phylogenetic treatments are largely absent in the study of the recovery of reef biotas after mass extinctions.

5.4.3 Reef communities are structured by close biological interactions which are both more susceptible to change, and slower to reassemble, than other marine communities

Hubbell's (1997) theory also predicts that regional, long-term community stability can arise just as easily from the stabilizing effect of large species diversity, as from rules of niche assembly that supposedly limit species membership in communities.

Little is known as to the timing and origin of many apparently complex mutualisms that occur on modern coral reefs, but we have noted that the nature of biological interactions that structure coral reef communities change as community membership changes. However, in the only analysis to date, Rosen and Turnšek (1989) have shown that inferred photosymbiotic scleractinian corals were more susceptible to extinction than inferred non-photosymbiotic forms. This analysis did not consider the habitat preferences of the corals, and may simply reflect the differential extinction of shallow, tropical genera from those that inhabited deep, cold waters. There is no evidence, though, that photosymbiosis collapsed (see Section 8.5.4).

Modern reef corals live in a highly fluctuating environment—albeit within well-circumscribed environmental tolerances (Section 4.4.1)—to which they are well adapted by a considerable range of genetic and phenotypic adaptations. Even the multifarious photosymbiosis with the dinoflagellate *Symbiodinium* may have an adaptive significance. *Symbiodinium* types have different adaptive capabilities and tolerances to microenvironmental factors, allowing corals to tolerate a range of temperature and light settings. The patterns of symbiont distribution are highly dynamic, and it has been proposed that established symbioses with *Symbiodinium* may be able to respond to rapid or profound environmental change by switching partners to a more favourable combination (Buddemeier and Fautin 1993). Such continuous sampling of available algae would promote long-term stability in the face of environmental stress and change by enhancing the chances of survival of both host and alga (see Section 8.2.5).

5.4.4 The favoured habitat of tropical reef communities (low-nutrient, carbonate platforms and ramps) is itself more easily disrupted by environmental change

Although not all extinct events preferentially affected tropical areas, there are data to suggest that carbonate platform biotas are more susceptible to environmental perturbation than those occupying other habitats at low latitudes. In a global compilation, marine bivalves occupying carbonate platforms showed higher extinction rates during the end-Cretaceous extinction than those in non-carbonate platform tropical settings (Raup and Jablonski 1993). Likewise, echinoids that occupied low-nutrient, inner-shelf, carbonate platform settings experienced significantly more extinction at the KT boundary than faunas from other habitats at the same palaeolatitudes (A.B. Smith and Jeffrey 1997). Moreover, extinction within these shallow-water carbonate settings took place across all habitats, not just those associated with reefs. The contention that the degree of extinction of a carbonate biota is closely related to habitat disturbance or loss is reinforced by the observation that the main episode of extinction for chalk echinoids was not at the end of the Cretaceous, but at the end of the Danian when chalk deposition ceased (A.B. Smith and Jeffrey 1997). These authors demonstrated that in regions where carbonate environments continued uninterrupted across the KT boundary—as for the Eurasian chalk settings—there is considerably less extinction than in areas where the carbonate habitat was lost.

Shallow-water carbonate platforms are particularly susceptible to rapid sea-level change, which may eliminate endemic biotas. For example, in both the Late Devonian and Late Maokouan (Late Permian) extinctions we see that the sequential decline in diversity of colonial rugose corals, followed by branching forms, then deeper water, solitary forms is a result of the development of deeper water sediments (Scrutton 1988; Ezaki 1993).

5.5 Decoupling the disappearance of reefs and the extinction of reef biotas

The foregoing discussion has concerned the susceptibility of reef biotas to mass extinction. While there is some evidence to support the preferential elimination of endemic biotas during mass extinctions, this does not necessarily make reef biotas more susceptible—as the degree of endemism within reef biotas has certainly varied through geological time, and, indeed, reef corals today are composed of relatively widespread species. Evidence to support the contention that tropical biotas are more susceptible to some extinction–kill mechanisms appears to be based on the relatively well-documented high susceptibility of shallow-water carbonate platform and ramp biotas, rather than all biotas that occupy low latitudes. The degree of extinction in such biotas is closely related to the loss of their habitat. Moreover, extinction within all shallow-water tropical carbonate settings—not only reefal habitats—appears to be high during many mass extinctions.

Times of extensive reef-building often correlate with extensive carbonate platform development, often when sea levels were high and climates warm, and tectonic history produced extensive areas of shallow marine substrates (see Fig. 4.6). High-resolution stratigraphic analysis of Cretaceous sections of North America have established that shallow carbonate platforms are more sensitive to global perturbations than temperate habitats (C.C. Johnson *et al.* 1996). During some mass extinction events, tropical carbonate platform distribution becomes restricted.

Reefs have been suggested to disappear early in major extinction phases, and so be early indicators of environmental crisis (Copper 1994; Droser *et al.* 1997). While reefs are known right up to several mass extinction events, they appear to be reduced markedly in distribution and abundance whilst retaining complex ecologies. For example, after the anoxic Sinsk Event, subsequent early Toyonian (Lower Cambrian) archaeocyath reefs were restricted to Labrador, and Altai-Sayan-Tuva when species diversities were already low; and very few reefs are known from the last 0.5–1 Myr of the late Hirnantian (Late Ordovician). This decoupling of reefs and reef biotas is confirmed by the response of reef biotas to the dramatic climatic and sea-level fluctuations of the Pleistocene and by the distribution of reefs and reef biotas seen today.

The pitfalls of using global diversity of reef-like organisms as a measure of reef abundance is highlighted by the observation that high species diversity of reef corals does not necessarily correlate with high regional concentration of reefs in modern seas (Veron 1995). For example, notwithstanding the high diversity of scleractinian corals during the Cretaceous, such corals were platform-dwellers rather than reef-builders. Furthermore, Veron and Kelley (1988) calculated that the rates of extinction and origination of Indo-Pacific corals over the last 2 million years have been similar (about 4.4% of species per million years). These authors have argued that while major glacioeustatic changes had a catastrophic effect on the abundance and distribution of reefs, they had no major effect on the diversity of reef coral species (with the possible exception of the Acroporidae), even though there was considerable faunal turnover during this interval. Half of the genera of living photosymbiotic corals live in environments with mean annual temperatures between 14 and 18 °C, and are not associated with reefs. Living, wave-resistant coral reefs are reliable indicators of prolonged sea surface temperatures at or above 18 °C, as well as only forming in areas removed from terrigenous input. This suggests that in the modern at least, coral reef formation has a strong physicochemical basis that is likely to be constant with changing faunal compositions (Veron 1995). Coral distribution and diversity is probably a poor indicator of palaeoclimatic patterns: reef distribution is a more reliable indicator of the relative climatic stability of tropical shelf environments for periods exceeding a few thousand years, and this may explain how reefs might be sensitive to the onset of climatic or oceanographic deterioration.

It has been suggested that the remarkable stability of shallow-water mollusc and coral communities during the Pleistocene was linked to Pliocene turnover events that may already have eliminated vulnerable species (Valentine and Jablonski 1993; Budd *et al.* 1996). The continued Pleistocene cycles of low and high temperature are suggested to have had little evolutionary effect because, once through a thermal filter, only eurythermal species remained.

We have noted that carbonate platform accumulation rates plummeted after several mass extinctions (Section 5.4.4). While this has been suggested to be due to the extinction of reef biotas, if we decouple reefs from their biotas this then points to a prominent role for a loss of habitat linked to environmental perturbation, in addition to the extinction of reef biota *per se*. This reduction in habitat would also have led to the isolation of communities, perhaps promoting further extinction. Although there is evidence that mean global subsidence rates have fluctuated through geological time, there are no data to suggest that these are directly

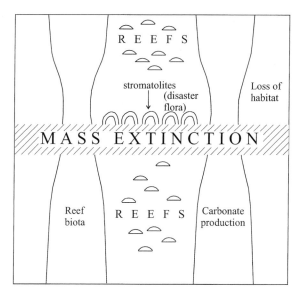

Fig. 5.11 Hypothesized relationship between the extinction of tropical reef biota, carbonate platform production, and reef appearance after a mass extinction event. Tropical reefs form only when stable carbonate platforms can develop, explaining both the early disappearance of reefs before the major phases of reef biota extinction, and the common delay between rediversification of potentially reef-forming biota and the actual appearance of reefs in the geological record.

correlated with the frequency of carbonate rocks or the abundance of reefs. For example, at the Permo-Triassic boundary, global subsidence rates were high or increasing, but there was still a dramatic drop in carbonate platform accumulation rates (Kuznetsov 1990). The same is true for the Triassic–Jurassic boundary.

These data argue for a need to decouple the formation of reefs from the availability of reef-building biota: it is reefs that have even more narrow environmental tolerances than their biotas. Figure 5.11 suggests a relationship between the extinction of tropical reef biota, carbonate platform production, and reef appearance after a mass extinction event. Tropical reefs form only when stable carbonate platforms can develop, explaining both the early disappearance of widespread reefs before the major phases of reef biota extinction, and the common delay between rediversification of potentially reef-forming biota and the actual appearance of reefs in the geological record.

5.6 Do reef communities recover at a different pace from other communities?

Given the variety of causal mechanisms, and the differences in both the duration and severity of perturbations, it seems likely that the necessary survival mechanisms will differ substantially between mass extinction events. It is clear, however, that the survival of a mass extinction is a poor predictor of later success.

There is little convincing evidence to suggest that reef biota recovery consistently occurs at a slower rate than other carbonate platform communities after mass extinctions. After the end-

Permian extinction event—the most disastrous of all mass extinctions—reef communities and reefs reappear at the same time as other normal marine communities. Relatively short-lived stromatolite or calcified cyanobacterial reefs flourish as disaster floras after several mass extinction events (e.g. Lower Cambrian, Late Devonian, and Lower Triassic (Schubert and Bottjer 1992), probably because such floras show a wide tolerance to stressful conditions including reduced nutrient levels, low or high salinity, anaerobic or dysaerobic conditions, and poor illumination. Where microbialites were already important reef-builders (e.g. Lower Cambrian, Late Devonian), reefs continued to form in isolated areas where favourable environmental conditions—that is shallow carbonate platforms—remained available. But there is also little persuasive evidence to support the contention that taxa from these relict communities reinvaded their former widespread habitats once favourable environmental conditions resumed.

Many post-extinction reef biotas are of a similar ecological caste to pre-extinction biotas. For example, notwithstanding the 7–10-Myr recovery period for the reappearance of normal marine communities after the devastating End-Permian extinction, Middle Triassic reefs retain a very similar ecology to Late Permian examples, and Silurian reefs were formed by the same subfamilies and genera of rugose corals and tabulate corals that had built reefs in the Late Ordovician (Copper 1994). In many cases, the radiation of new, or previously insignificant, components of the pre-extinction faunas into reef habitats is a long affair, with increases in diversity occurring no faster in postextinction phases than at other times. However, once carbonate platforms do become re-established, reef biotas can radiate rapidly into new habitats. Scleractinian corals did not radiate until the Pliensbachian–Toarcian—some 15 to 20 Myr after the end of the Triassic extinction event—until widespread carbonate platforms had become established. The resumption of normal marine conditions and the onset of widespread carbonate platform formation in the Anisian seems also to have provided the environmental cue for the skeletonization of two separate clades of soft-bodied hexacorals to form scleractinian corals.

5.7   Summary: what effect have mass extinctions had on reef evolution?

There appears to be no obvious correlation between the magnitude of the environmental change and the ecological or evolutionary response. Many mass extinctions are the result of a combination of interacting factors that are difficult to separate, and are poorly understood. Most mass extinctions do, however, coincide with major climatic and sea-level fluctuations that are distinctive for their rapidity, magnitude, and global extent (Hallam and Wignall 1997). These alone would most certainly have severely disrupted shallow benthic communities on carbonate platforms that are adapted to particular temperatures and light levels.

The disappearance and reappearance of reefs after mass extinction events shows only a poor correlation with the extinction of reef biotas. Even after a dramatic reduction in the diversity of reef-associated metazoans at the end of the Frasnian, calcified cyanobacterial and microbial reefs still persisted where carbonate platforms continued to form. Mass extinctions often appear to disturb carbonate platform accumulation, and widespread reef-building may therefore be possible only in times of global carbonate platform stability. Reefs disappeared as communities, while their component biotas persisted in non-reef environments to re-form after the resumption of stable carbonate platform formation.

Some assumptions of the slow recovery of reefs after mass extinctions have been based upon the mistaken identification of marine invertebrates as the main reef-builders, where recovery has been confused with a return to a substantial component of large metazoans to the reef biota.

The reappearance of stromatolites in shallow marine habitats suggests that reduced substrate competition may be lessened due to the removal of competitors. Of significance may be the observation that some cyanobacteria are able to fix nitrogen and so are not as severely nitrogen-limited as other algae (Hay 1991). This may explain their ability to live in low-nutrient environments, making them less sensitive to the sustained reductions in primary productivity associated with some mass extinctions.

Part III

Evolutionary Innovation

6 Life on a substrate: trends in form and function

The growth of sessile organisms is concerned with the successful occupation of space, and in Chapter 1 it was suggested that living reef communities grow under very specific environmental conditions that allow such growth—where suitable substrate is available, competitors are discouraged, and predators are either absent or in some way accommodated.

Far-reaching generalizations as to the distribution and abundance of sessile organisms can be made on the basis of differences in growth form and *life history*, that is the schedule of events that occurs between birth and death (Hall and Hughes 1996). Life history can control individual growth, habitat selection, and competitive ability, as well as the way in which the next generation is recruited: all features which will strongly influence the characteristics of a reef. Reefs dominated by organisms with life histories which involve relatively short lives, high fecundities, and the rapid growth of weakly calcified skeletons will produce radically different communities from those dominated by species with increased longevities, low fecundities, and heavily calcified constructions.

This chapter shows how the consideration of benthic invertebrates within a series of functional and life history types can successfully predict the ecological structure and environmental preference of a reef community. It then documents how the growth form and function of reef-building invertebrates has changed, with attendant ecological consequences, through the Phanerozoic.

6.1   Growth form, life history, and reef formation

The physical form of a reef framework is determined by differences in functional and growth morphology, and the life histories of the constructing organisms. The ability of an organism to secure living space on a stable substrate is a necessary prerequisite for the long-term persistence of any reef, and stability will be determined both by the growth form of an individual and the longevity of the substrate to which it is attached. Large, long-lived reef-builders which are elevated above a substrate allow for the coexistence of many other associated organisms, as diversity is enhanced by the degree of three-dimensionality (known as *spatial heterogeneity*) of the reef framework, itself largely a consequence of reef-builder morphology.

6.1.1   Functional morphology and growth form

Notwithstanding the great diversity of taxa that have contributed to reef-building over the course of the Phanerozoic—and irrespective of systematic position—marine invertebrates can be resolved into a limited number of functional forms according to the number and arrangement of their functional units, termed *modules* (Fig. 6.1). Modules, be they polypides (bryozoans), polyps (cnidarians), or masses of sponge or algal tissue, are generally capable of existence independent from the parent organism (Jackson 1985). Where an organism consists of modules which are also morphological individuals, the whole is known as a *colony*.

Modularity is often closely linked with *clonality*. The clonal habit is the ability to grow and asexually reproduce through a potentially unlimited production (known as *indeterminate* growth) of identical modules which are all ultimately derived from the same zygote (Beklemishev 1964; Jackson 1977, 1985). In contrast, aclonal organisms (e.g. bivalves and barnacles) consist of a single individual. This sole module will often have a genetically determined upper size limit, that is it will show *determinate* growth, and be larger and more structurally complex than those found in modular organisms. Sessile aclonal/solitary organisms bear a variety of growth forms including stalked and cup-shaped morphologies. Clonal organisms often possess runner-like, encrusting, or laminar growth forms that grow primarily over the substratum, or erect or branching forms that grow up or off the substratum.

In entirely soft-bodied organisms, modules can be reorganized on a daily basis, but the confinement of soft tissue within a rigid exoskeleton restricts module addition to areas of skeletal growth. In many skeletal modular organisms, living tissue is then effectively two-dimensional, and growth can only be achieved through volumetric increase of the skeleton. In stony corals, for example, soft-tissue modules are reflected in the skeletal modules (*corallites*) which grow divergently and branch whilst becoming incorporated within a common

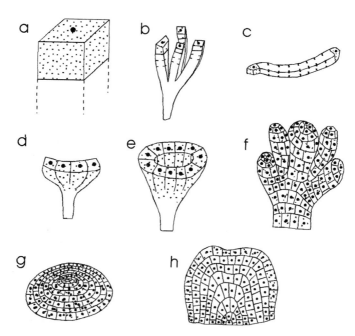

Fig. 6.1 Stylized functional organizations of epibenthic macroorganisms, showing relative integration states. Stippling shows the distribution of soft tissue during life. (a) Solitary; (b) uniserial erect; (c) uniserial encrusting; (d) catenulate; (e) cerioid. Multiserial: (f) erect; (g) encrusting; (h) massive. • Marks the centre of each module. Terminology after Coates and Jackson 1985. (Modified from Wood *et al.* 1992*a*.)

cemented matrix. Resulting growth morphologies range from single or branching chains with one functional unit isolated at the tip of each chain or branch (*uniserial* growth), to those with continuous surfaces of interconnected modules (*multiserial* growth) (Fig. 6.1).

Whether in wholly soft-bodied organisms or within a skeleton, modules that are in soft-tissue continuity are capable of translocating substances such as food and metabolites, and also nervous impulses. Indeed, new modules can only be generated if provisioned by substances translocated from their immediate predecessors. Such continuity may be expressed by the presence of connecting holes within the skeletal walls between modules, over the ends of walls, or the absence of walls entirely. The relative independence and integration of modules can thus vary greatly, from complete structural isolation to states where no separation exists between soft-tissue, skeleton, and physiological function. The term *degree of integration*, with the relative prefixes low, medium, or high, thus expresses the inferred level of modular interdependence.

Modularity offers the potential for great flexibility of growth form and organization. Individual modules may become functionally and morphologically specialized for sexual or defensive roles, a condition known as *polymorphism*. Moreover, modules can be added to various parts of an organism so that growth is possible in many directions. In this way, a tremendous array of growth morphologies can result (Fig. 6.2), with the potential to adapt to local environmental conditions. This is known as *morphological plasticity* where different growth forms of the same species are called *ecophenotypes*. For example, in scleractinian corals, colony shape is strongly influenced by wave energy and patterns of disturbance, growth rates, and light levels (Chappell 1980), and so reef corals show a distinctive pattern of zonation with increasing depth (Fig. 6.3). The highly robust, branching form of *Acropora palmata* resists all but the most severe storms, sheets and mounds are common in areas of

Fig. 6.2 The variety of coral morphology found on modern reefs showing the flexibility of multiserial growth. 1: Cup-shaped soft coral; 2: columnar; 3: free-living; 4: digitate; 5: encrusting; 6: corymbose; 7: caespitose; 8: bottlebrush; 9: massive; 10: foliaceous (cup-shaped); 11: foliaceous (forming a whorl); 12: tables and plates; 13: massive; 14: arborescent (staghorn); 15; arborescent (elkhorn). (Modified from Veron 1986; copyright, John Sibbick.)

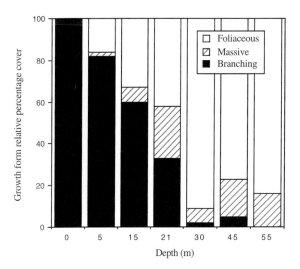

Fig. 6.3 Relative abundance of coral colony growth form at Discovery Bay, Jamaica. (Modified from Liddell and Ohlhorst 1987.)

high wave impact, and delicate branching forms are most common in less energetic waters. In shallow waters, the coral species (or species complex, see Knowlton *et al.* 1992) *Montastraea annularis* forms massive, domal colonies but deeper on the reef slope colonies become lobed, and at a depth of 35 m they grow as horizontal plates. Skeletal morphology is clearly part of the life history strategy of sessile organisms, with implications for predator resistance, substrate preference, hydrodynamic regime and coping with sediment stress.

6.1.2 Life history

The life history of most benthic invertebrates includes a relatively long-lived, sessile adult phase and a short-lived, water-borne larval phase. Aclonal benthic organisms possess very simple life cycles compared to clonal forms. In aclonal forms, population increase is achieved only through the successful recruitment of sexually produced larvae or immigration, and population decrease by death or emigration (Fig. 6.4(a)). Adults which survive to reach their upper growth limit, produce gametes or larvae, and then at some later time grow old and die. Clonal organisms, however, can increase by asexual reproduction at almost any stage in their adult life cycle, as well as being able to decrease population numbers by the fusion of previously separate clones (Fig. 6.4(b)). Colony/individual size may increase not only with growth, but also by fusion with other clones, and decrease can be achieved by fission, fragmentation, budding, or by localized non-lethal injury resulting in *partial mortality*. Partial mortality is the most common process of asexual reproduction in encrusting growth forms; fragmentation is more common in erect or free-living forms, from which fragments can regenerate into entirely new colonies morphologically similar to the parent.

6.1.3 Substrate preference

Benthic environments tend to be patchy, and favourable patches may be short-lived: habitat selection and differential survival are thus vital controls on the distribution of any benthic community. One of the most limiting resources in shallow marine environments is the avail-

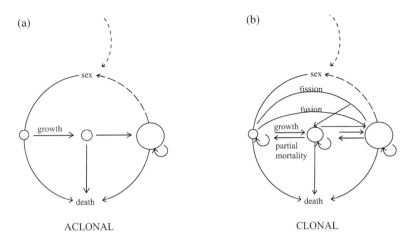

Fig. 6.4 Schematic models of the life cycles of (a) aclonal and (b) clonal benthic invertebrates. (Redrawn from T.P. Hughes 1984, by kind permission of the University of Chicago Press.)

ability of substrates sufficiently firm or stable for the attachment of benthos: any sessile benthic organism must secure enough substrate for long enough to reproduce. Substrate preferences can be profound, and the strategies employed to secure a substrate may be very different.

Substrates can range from those such as dead shells and cobbles that are small and unstable (that is easily moved or buried) and ephemeral (easily broken), to highly stable, laterally extensive areas such as reef coral skeletons and rock walls which persist virtually unchanged for decades and even centuries. The effects of physical disturbance, such as strong currents, will greatly vary with substrate size and stability: small substrates require only minor water movement to be overturned whereas large stable substrates will be unaffected even by strong currents. The distribution of substrata is thus, in part, dependent upon the ambient hydrodynamic regime.

The strategies adopted by benthic organisms for life on stable, and unstable or ephemeral substrata, are very different. Although functional types present a continuum of form, profound differences have been documented in the ecological characteristics and life histories of aclonal/solitary and clonal/modular organisms (Table 6.1). In particular, the relative abundance and diversity of clonal animals increases with environmental stability and predictability (Jackson 1985).

Solitary or low-integration modular organisms tend to have ephemeral life histories compared to high-integration forms which are more persistent (Coates and Jackson 1985, 1987; R.N. Hughes 1989; Wood *et al.* 1992a; Wood 1995). They are often *opportunistic* (rapid colonizers with high growth rates and short life cycles), and occur preferentially on unstable, soft substrates or on short-lived, patchy hard substrates (Jackson 1977, 1983). Such organisms therefore tend to be restricted to either marginal habitats such as the intertidal zone where strong environmental gradients exist, or to fugitive patches of relatively small areas of substrate which are free from superior clonal competitors.

Due to the rapid ability of large modular organisms to cover an area by clonal propagation, encrusting clonal/modular organisms such as sponges are competitively superior on

Table 6.1 Generalized comparison of the environmental preferences and ecological characteristics of reef communities dominated by aclonal/solitary and clonal/modular epibenthic invertebrates (modified from Wood *et al.* 1992a (by kind permission of SEPM Society for Sedimentary Geology); Wood 1995)

Ecological strategy	Primitive → Aclonal/solitary	Derived Clonal/modular
Environmental characteristics		
Nutrient levels	High	Low
Substrate preference	Soft or small areas of hard substrate, short-lived	Extensive hard substrate, long-lived
Level of disturbance	High	Low
Ecological characteristics		
Growth form	Stalked, cup-like	Encrusting, massive, branching
Individual colony size	Small	Large
Larval recruitment rate	High	Low
Mechanism for close packing	Larval aggregation + synchronous settlement	Competition for limited/patchy substrate
Individual/community longevity	Short	Long
Topographic relief	Low	High
Niche heterogeneity	Low	High
Community diversity	Low	High
Individual reef size	Small	Large
Wave resistance	Low	High

long-lived hard substrates, usually in subtidal environments, where environmental gradients are far less pronounced. Bryozoans, although modular, may be more common on small or ephemeral hard substrata.

A modern coral reef is primarily a hard substrate environment where modular forms are three times more common (in terms of surface area) than solitary organizations (Jackson 1977). Any area of an available hard substrate—be it geological, organic, or discarded human detritus—is rapidly colonized by a rich variety of encrusting organisms, in marked contrast to surrounding soft-substrate areas which support little sessile benthos (Figs 6.5(b)). This explains why, on any scale, the availability of hard substrate determines the resultant shape of a modern coral reef (see Section 2.1.3), and why the use of artificial substrata can be an effective method for re-establishing reef growth.

6.1.4 Clonality and size

Although many clonal/modular organisms are slow-growing, they live longer than solitary forms, although this may vary greatly from a few months (temperate hydroids, bryozoans) to several centuries (corals and clonal trees). Indeed, some may show no *senescence* (that is no increased rate of mortality, or decreased rate of growth or reproductive output, with old age). As modular organisms often grow indeterminately they can also achieve very impressive sizes, and encrusting forms may persist for as long as their substrate survives, which

(a)

(b)

Fig. 6.5 Modern soft- and hard-substrate reef communities. (a) Aclonal, monospecific aggregation: sabellariid reef, south Wales. Lens cap = 50 mm. (Photograph: J.A.D. Dickson.) (b) Clonal aggregation: a 20-year-old truck tyre provides a benthic island of hard substrate which has been colonized by multiserial coral colonies. Note that the tyre is surrounded by barren, constantly shifting, soft substrate. (Photograph: J. Warme, courtesy of N.P. James.)

may be centuries (T.P. Hughes and Jackson 1980). Age and size is typically confounded in modular organisms; for example, analyses of coral populations on the Great Barrier Reef showed that both young and old colonies were present in every size-class (T.P. Hughes and Connell 1987).

Colony size correlates with many determinants of 'success'. First, colony size relates to the ability of organisms to acquire and maintain living space, and numerous experiments have shown that the relative size of combatants can directly affect the outcome of competitive conflict (see Section 6.2.1). Second, predation on small colonies tends to be an all-or-nothing affair; they either wholly escape injury, or die completely, whereas many large colonies may only suffer partial mortality (T.P. Hughes and Jackson 1980). Clonal organisms show tremendous powers of regeneration from partial mortality, which gives them greater immunity against many causes of total mortality (except disease and catastrophic physical disturbance; see Woodley *et al.* 1981). As with large aclonal animals, there is often an 'escape in size'.

Third, colony size can determine *fecundity* (reproductive output). Unlike solitary/aclonal animals where fecundity often reaches a plateau or declines in old age, many clonal organisms show increased fecundity with colony size. In corals, fecundity is proportional to the number of polyps, so increasing with both colony size and polyp density (Hall and Hughes 1996). Growth morphology, in particular surface area, of organisms thus plays a crucial role in determining reproductive output; massive colonies with large polyps of low density may have a lower output than branching forms with multiserial morphologies and high polyp densities, even though branched forms will be more susceptible to damage. Coral colony reproductive output is not simply a multiple of constituent polyp fecundities, however, as it increases indeterminately with colony size—larger colonies have disproportionately higher fecundities (Hall and Hughes 1996). As a consequence, large colonies make a vastly disproportionate contribution to the total gene pool on coral reefs. For example, it has been calculated that 25% of the annual egg production of the coral *Goniastrea aspera* was generated by the largest colonies on Lizard Island in the Great Barrier Reef, which numerically accounted for only 3% of the population (Babcock 1984; Fig. 6.6).

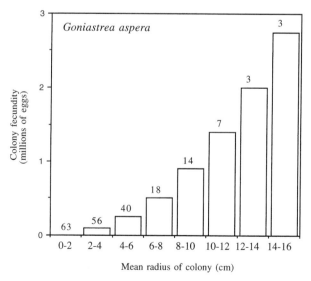

Fig. 6.6 Fecundity of the braincoral *Goniastrea aspera*, showing how after a 3–5-year prereproductive period, coral fecundity increases dramatically with colony size. (Modified from Babcock 1984, by kind permission of Springer-Verlag.)

These advantages predict that the evolutionary benefits of achieving large size in sessile marine organisms will be substantial. Not surprisingly then, coral reefs are typically dominated and structured by large, long-lived organisms. The benefits of larger colony size gained by fusion may, therefore, be of great importance in juvenile colonies, as newly established recruits are particularly susceptible to many agents of mortality. But the predominance of large organisms in a reef means that highly complex and intricate mechanisms are required to avoid the competitive dominance of any one species (Section 6.2), and the costs of potentially unlimited growth on a substrate makes encrusters in particular literally sitting targets for agents of destruction, such as sedimentation, overgrowth, and predation.

6.1.5 The phenomenon of aggregation

Although there may be considerable costs such as increased competition for food, some protection may be gained from biological and physical disturbance in sessile organisms by the adoption of closely aggregating growth. Research has shown that dense populations are less susceptible than isolated individuals to overgrowth from competitors, larval invasion and infestation, and attack by predation (Jackson 1983). Many achieve this increased resistance by modifying their feeding behaviour, such as presenting an outer 'colony' consisting almost entirely of feeding apertures, as found in vermetid gastropods (R.N. Hughes 1979). Other benefits may accrue from intimate association, such as the enhanced growth and decreased loss rates demonstrated experimentally in individual reef demosponges growing within dense, intertwined, multispecies associations (Wulff 1997).

The proximity of *conspecifics* (members of the same species) will also enhance the potential for the local proliferation of a species, and aggregations can exclude mobile bioturbators and so help to stabilize and bind soft substrates (Thayer 1983). Interconnected, cemented, or otherwise locked individuals—i.e. those forming a rigid reef framework—will gain further stability by producing topographic relief thereby creating a biologically controlled substrate suitable for the fostering of future generations.

Just as the ecological and environmental requirements of different reef communities are profoundly different, so are the mechanisms they employ to achieve sufficient close-packing to form a reef.

Aclonal/solitary organisms

Sessile aclonal/solitary organisms are particularly susceptible to overgrowth by modular, encrusting competitors, to smothering by sediment, and to attack by predation. Rapid growth rates to achieve either large individual size or dense aggregation offer the only means of escape. To achieve this, high rates of larval recruitment and aggregative settlement in a localized area are necessary to create large, monospecific populations of densely intergrown, mutually supporting, individuals.

Several groups of modern organisms aggregate to form tightly packed, monospecific groups, growing in various marginal habitats (see Table 1.1). Such reefs can be extensive, e.g. sabellariids (tube-dwelling polychaete worms) form structures up to 6 m thick by sand-trapping which cover over 30 km^2 in energetic temperate or subtropical settings (Fig. 6.5(a)) (Multer and Milliman 1967; Gram 1968; Kirtley and Tanner 1968), and may achieve sufficient rigidity and topographic elevation to restrict estuarine tidal flow (Kirtley 1994).

To achieve such dense, monospecific stands, solitary organisms must employ strongly density-dependent recruitment, that is larvae must selectively choose areas where the density of new recruits is already high, or where spawning and larval behaviour involves short-distance dispersal and so settlement close to the parent (*philopatry*). Larval recruitment rates tend to be extremely low in low-nutrient, tropical environments, but in temperate, productive waters seasonal recruitment rates commonly reach $10\,000\ \mathrm{m}^{-2}\,\mathrm{day}^{-1}$ (Jackson and Winston 1982). Densely aggregated communities are thus usually restricted to unpredictable soft substrates in high-nutrient settings.

Clonal/modular organisms

Unlike the rapid rates of colonization and growth shown by opportunistic aclonal/solitary forms, clonal organisms may be relatively slow to colonize a given area, as larval recruitment rates are usually far lower. Clonal organisms can be simultaneous hermaphrodites, or individual modules may be sequential hermaphrodites, and although larval production rates vary enormously between different species, sexual reproduction can be infrequent and relatively few larvae may be produced: some two-thirds of coral species on the Great Barrier Reef spawn synchronously once a year (Harrision *et al.* 1984). The dispersal of larvae in many clonal organisms including ascidians, bryozoans, sponges, hydrozoans, and some corals may also restricted by a variety of mechanisms, including the brooding of larvae (for between 1 to more than 100 days) until they are almost ready to settle (known as '*brooders*', as opposed to those which release long-lived and widely dispersed larvae after 4 to 8 days after fertilization known as '*broadcasters*') (Jackson 1986; Jackson and Coates 1986). Such traits in many sessile, modular organisms may be the result of strong selection pressures, as restricted larval dispersal presents abundant opportunities for philopatry, gregarious settlement, and the fusion of colonies. Indeed, philopatric dispersal of both sexual and asexual propagules is one of the most striking features of clonal benthic organisms compared with aclonal species (Jackson and Coates 1986). Philopatric larval behaviour allows inbreeding, which facilitates epistatic genetic interactions (where one allele will prevent the expression of other alleles) that are favoured in spatially heterogeneous environments such as reefs.

In some clonal organisms, new colonies/individuals form as frequently from fragmentation and fission as they do by sexual recruitment. Moreover, in some of the most important living reef-building corals, clonal fragmentation is of paramount importance. For instance, dense, monospecific thickets of loose-branching staghorn (*Acropora cervicornis*) and elkhorn corals (*Acropora palmata*) are created in shallow, wave-exposed, and moderately energetic parts of reefs, respectively, by rapid growth and colony breakage whereby fragments establish new colonies. Sexual recruitment of these species is quite rare: here the relative fragility of erect growth has been turned to considerable advantage.

Asexual reproduction is common in many reefal organisms, including various coralline algae, echinoderms, sponges, and polychaetes. This leads to the long-term persistence of many genetically identical individuals or colonies, individually known as *ramets* and collectively known as a *genet* (T.P. Hughes and Jackson 1985), whose prevalence on modern coral reefs has only recently become appreciated.

The advantages of aggregation and asexual (clonal) reproduction are many. First, asexual reproduction enables the survival and rapid re-establishment of normal population density after severe disturbance, including catastrophic events such as hurricanes—although many individuals may not survive, the genet probably will. Second, species which reproduce mainly either by asexual means or are able to self-fertilize do not need to be close to other conspecifics. However, increasing evidence does suggest that larvae of benthic colonial animals settle near their point of origin, resulting in the close proximity of relatives or genetically identical larvae and high levels of inbreeding (e.g. Jackson 1986). All these features allow reproduction to be independent of population density such that very small populations of a species are still viable. (This may, in part, explain the widely spaced nature of some clonal, modular coral species on reefs and may also provide one mechanism whereby the local domination of one species is prevented.)

Although reliance upon asexual reproduction might at first appear to restrict natural selection by mutation, in fact somatic (body) mutations may be heritable in clonal organisms as these undifferentiated cells are able to develop into gametes, i.e. become part of the germline (Buss 1987). Unlike most aclonal organisms, germ cells are not sequestered in early development. A long clonal life span also provides the opportunity for many such somatic mutations to arise, and for advantageous mutations to increase. It also seems increasingly likely that at least some coral colonies are in fact *hybrids* of two or more distinct genotypes (Resing and Ayre 1985) and that sibling species may be far more common on modern reefs than is currently recognized (Knowlton and Jackson 1994). Indeed, for *Porites* from the Pacific, the level of intraspecific variation has been demonstrated by electrophoretic data to be higher than that reported from most other invertebrates (Potts and Garthwaite 1991). That coral evolution progresses by such hybridization is a strong possibility (Veron 1995), and this may be one result of the mass spawning events described from many reefs.

The life history of clonal, modular organisms has been summarized in the Strawberry-Coral model (Williams 1975). In this model, sessile organisms gain and exploit limited resources by clonal expansion, but are able to achieve long-distance dispersal by large numbers of sexually produced (or in some cases asexually produced) larvae, which are often produced synchronously on a reef. Vacant substrate is colonized simultaneously by many diverse propagules, which thus ensures intense competition and presumably results, on average, in the survival of the more fit. These are able to occupy substrate through clonal growth and to dominate the population until population restructuring by the next catastrophe. While it is now clear that clonal organisms do not often employ long-distance dispersal of large numbers of larvae, the model does, however, offer an explanation for the retention of costly sexual reproduction within clonal organisms as an adaptation to the intense competition among sibling recruits to a local population (Jackson and Coates 1986). Moreover, environments change greatly over short distances so that among clonal organisms occupying hard substrata, wide dispersal is not necessary for encountering new environments. Indeed, the evolutionary trend in many independent clonal clades is to decrease the dispersal distance of their sexual propagules (Jackson and Coates 1986). It may be possible that the proportion of investment of sexual versus asexual dispersal may depend on the relative availability of parental habitat elsewhere: if it is common and predictable in distribution, then asexual dispersal may be favoured (Hamilton and May 1977).

This combination of infrequent larval production and short dispersal distance, with frequent asexual reproduction, yields local populations of high genetic relatedness. That parents, siblings, and clone mates can all be mixed together within small areas has been confirmed by electrophoretic analyses of some benthic clonal populations (e.g. Ayre 1984; Stoddart 1984). Such interrelatedness also decreases gene flow between populations and increases the chances of both speciation and of evolutionary adaptation to local environmental conditions, as well as maintaining high levels of genetic variation. The advantages of inbreeding should be greatest for long-lived organisms with low fecundity—all characteristics of clonal organisms (Shields 1982). This high level of genetic variation has profound consequences for our interpretation of the evolutionary history of a group. Such variation may leave no trace in the palaeontological record, as fossilizable skeletons may not reflect all-important behavioural, reproductive or soft-part morphological differences between genets. As a result of these biological insights, serious doubts are raised as to the validity of fossil species and their temporal stability.

6.1.6 Implications for reef community structure

The persistence of dense assemblages, a necessary prerequisite for the formation of reefs by macroorganisms, thus depends either on continual aggregative larval settlement and rapid growth, or upon clonal growth with partial mortality or fragmentation (Jackson 1977).

Aclonal/solitary organisms are relatively small and short-lived, and often show *constratal growth* (that is they do not project substantially above a substrate), so producing little topographic relief and minimal primary cavity development (Fig. 6.7). As they tend to be prone to frequent physical disturbance, the communities they maintain are relatively ephemeral, usually forming aggregations in marginal, soft-substrate environments; they do not often form dense aggregations subtidally except in crypts (Jackson *et al.* 1971). Moreover, the very dense packing of many solitary aggregations provides little spatial heterogeneity, allowing only minimal development of an associated or differentiated biota. Thus such reefs often have low overall biodiversity (Table 6.1). Many Recent solitary organisms also commonly occur in cryptic habitats (Jackson 1977).

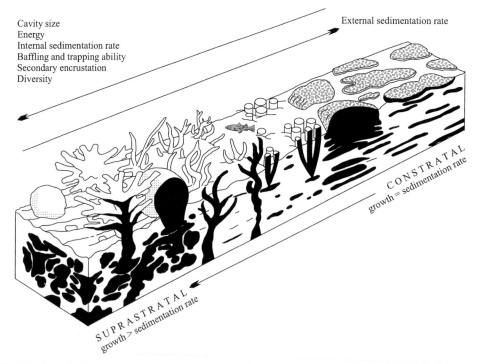

Fig. 6.7 Generalized environmental preferences and ecological consequences of constratal and suprastratal growth.

The longevity and vast size achieved by some clonal/modular organisms can lead to a very different style of reef. When heavily calcified, the secure, permanent attachment to extensive subtidal hard substrates of modular, encrusting organisms (such as many massive scleractinian corals and coralline algae) can create very persistent communities. Although some clonal organisms are constratal, others adopt *suprastratal growth* which produces elevation above a substrate, with resultant considerable topographic relief and large, primary cavity formation by the roofing-over of primary framework builders (Fig. 6.7). Such topographic relief and spatial heterogeneity continues even after death. This explains why modern coral reefs have the capability to reach vast sizes if hard substrate availability allows.

Without topographic relief, high spatial heterogeneity and the cryptic refuges present in such reefs, many herbivorous food webs and those involving suspension-feeders would simply not exist; nor would cryptic communities (Jackson and Winston 1982; Adey and Steneck 1985). Very small areas of coral reef can support enormous numbers of individuals and species. In particular, cryptic niches are very important in modern reefs, as many organisms are far more abundant in crypts than on open surfaces; indeed many may be obligate crypt dwellers. For example, 300 species of cryptic organisms were noted on the undersurfaces of foliaceous corals in just one habitat in Jamaica (Buss and Jackson 1979). Likewise, a single coral skeleton (weighing 4.6 kg) taken from the Great Barrier Reef contained 1441 *endocryptic* (organisms living within a hard substrate) polychaete worms belonging to 103 species, as well as amphipods, isopods, oligochaetes, crustaceans, and ophiuroids (Grassle 1973). Reef rock can support an equally abundant motile cryptofauna (Klumpp *et al.* 1988). Crypts thus house a significant proportion of the overall biotic diversity of a modern coral reef.

The striking differentiation of modern reef communities into open surface and cryptic communities is a consequence of differing competitive abilities of organisms to secure limited hard substrata. Crypts offer niches which are protected from direct exposure to open-surface hazards such as wave scour, intense predation, and high levels of irradiation (Table 6.2). Whilst rapidly growing *phototrophic* (those which obtain energy from sunlight)

Table 6.2   Differences between the environmental and ecological characteristics of open-surface and cryptic communities in modern coral reefs

	Open surface	Crypt
Environmental		
Irradiance	High	Medium–non-existent
Hydrodynamic energy	Medium–high	Low
Predation pressure	High	Low
Ecological		
Main energy source	Light	Suspended organic matter
Dominant biota	Phototrophs (coralline algae + photosymbiotic metazoans)	Heterotrophs (e.g. sponges, bryozoans, brachiopods)
Dominant morphologies	Clonal/modular, massive, encrusting	Solitary/aclonal
Dominant growth direction	Upwards	Laterally + downwards
Size	Large	Small

organisms dominate exposed surfaces (Fig. 6.8(a)), *heterotrophs* (filter- and suspension-feeders) flourish within hidden, cryptic niches (Fig. 6.8(b)). Solitary organisms such as ser-pulids, crinoids, foraminiferans, and brachiopods are conspicuous within crypts but occupy

(a)

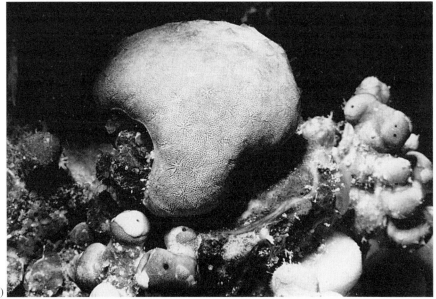

(b)

Fig. 6.8 Community differentiation of coral reefs. (a) Open-surface phototrophic community (note corals and the photosymbiotic giant clam, *Tridacna gigas*). (Photograph: F. Talbot.) (b) Cryptic communities dominated by calcified demosponges: *Astrosclera* (centre) and the sphinctozoan *Vaceletia* (bottom and right). Lizard Island, GBR, Australia. (Photograph: J. Reitner.)

little space; encrusting ectoprocts and particularly sponges are abundant as they appear to be the best overgrowth competitors (Jackson 1977; Jackson and Winston 1982).

6.1.7 The role of disturbance

We have seen that reef environments vary greatly in disturbance. While deeper water habitats appear to show remarkably constant conditions, modern shallow marine tropical environments are subject to considerable daily, seasonal, and annual fluctuations (Section 4.4.1), and intermediate levels of disturbance are of fundamental importance in structuring many aspects of coral reef communities.

Modern coral reefs are a mosaic where organisms show the whole range of life histories designed to cope with fundamentally differing degrees of disturbance. Short-lived opportunistic forms are able to invade newly available patches of substrate. They may dominate the areas of a reef which are subject to frequent and intense physical disturbance, such as the shallow wave-crest environment. Here, few corals survive for more than 10 years, and the periodic removal of competitively dominant and fecund branching forms, which are more susceptible to hurricane damage, may help to maintain high diversity in these areas (Connell 1978). The long-lived (potentially infinite), large encrusting forms that are highly resilient to disturbance and are able to repair damage rapidly will determine the general physical structure of the reef, and these populations show adult-dominated size distributions as befits their longevity. Indeed, although both are modular, erect, and encrusting morphologies they may have subtly different life histories. Encrusting growth makes organisms vulnerable to all the vagaries of life on a substrate, such as sedimentation, overgrowth, and predation; erect morphologies can escape from potential sources of mortality or partial mortality. Erect forms—whilst gaining mainly advantages connected with high, soft-tissue surface area: volume ratios and life away from a substrate—must contend with the problems of increased susceptibility to breakage. Erect growth can also set overall size constraints relative to mechanical support and the strength of the attachment site, which may lead to determinate colony growth unless fragments are able to regenerate independently (Woodley *et al.* 1981). Encrusting forms, whilst being more stable, may suffer increased partial mortality as a result of frequent biological attack, but predators apparently have a far greater negative effect on the survival of erect growth forms (Table 6.3).

Some reef organisms are long-lived: several members of the fish families Lethrinidae, Lutjanidae, and Serranidae have an average life span of 19 years (Loubens 1980), and it seems likely that many reef organisms show constant population size, as noted in populations of massive corals (Shinkarenko 1982), as well as the giant clam *Tridacna maxima* (McMichael 1974) and even starfish (Glynn 1982). However, underlying this apparent picture of stability at the population level, is the constantly changing status of individual ramets as colonies recruit or die, grow or shrink, and divide or fuse. During a three-year study of Jamaican foliaceous corals, although the population size and structure remained the same, 36% of the original colonies died and 57% suffered fission through partial mortality producing 21% further colonies. Sexual recruitment also added 23% while fusion removed 5% (T.P. Hughes and Jackson 1985). Reef populations are in a continuous state of flux.

Predation and disturbance can act so as to maintain the densities of potentially competing species, to a level so low that competition through direct contact between structurally or numerically dominant taxa is reduced or prevented (see Section 7.1.10). This was first

Table 6.3 Relative advantages and disadvantages of erect versus encrusting multiserial growth

	Advantages	Disadvantages
Erect growth		
Disturbed regimes	Escape vagaries of life on a substrate, including encrusting competitors, some predators, and sedimentation	Stringent mechanical demands required for colony strength and attachment site
	Increased tissue surface area and volume increase feeding and reproductive capacity per area of substratum	More vulnerable to grazing and predation
	Increased access to food (including light) in the water column	
	Decrease fitness of encrusting competitors by shading	
	Fragmentation frequently leads to new colonies	
	Fast growth rates	
	Clones can disperse over wide area	
Encrusting growth		
Stable regimes	Earlier onset of reproduction	More susceptible to overgrowth, sedimentation, and predation
	Difficult to break, manipulate, and eat	
	Better competitor for space	
	Superior resistance to larval invasion	
	Potentially unlimited growth over substrate	

suggested by Paine (1966) in his predation hypothesis, which holds that predation is directly related to the prevention of resource monopolization. This hypothesis is also known in modified form as the *intermediate disturbance hypothesis* (Connell 1978), which states that intermediate levels of disturbance (physical and biological) lead to the highest standing diversities; diversity will decrease if competitive subordinates are the preferred prey, but where predators prefer the competitive dominant, diversity will be highest at intermediate predation levels.

Models have been used to simulate the community changes in species abundance and community composition following major disturbance such as a hurricane (Tanner and Hughes 1994). Competitively inferior species did not disappear in these simulations because routine mortality ensured that space was always available for colonization. The length of time required to reach a climax community (>20 years) was far greater than the observed interval between major disturbances, so also supporting the non-equilibrium theories of coral reef communities. However, diversity remained at very high levels at equilibrium (that is long after a major disturbance), which does not support the intermediate disturbance hypothesis.

Since the highly variable factors of recruitment and disturbance have become appreciated, the view that niche diversification is an important factor in promoting coral reef diversity has become widely discredited. Regional enrichment via dispersal of larvae from outside a local reef community is probably an important process in the maintenance of diversity (Cornell and Karlson 1996).

6.1.8 Life history, ecological succession, and diversity

Ecological succession is a well-described phenomenon on modern coral reefs. *Succession* refers to the sequence of community changes initiated when new substrate becomes available, *primary succession* applies to newly available substrate which has never been colonized, such as the emplacement of submarine lava flows. Modern reefs show a succession from low diversity ('pioneer') communities of rapid establishing, often solitary heterotrophs to high diversity ('climax') communities dominated by modular coralline algae and corals. Succession then occurs due to differences in life-history traits, according to the differing abilities of benthic organisms to colonize newly available substrate (recruitment), propagate clonally and grow, and to compete for space. The ratio of modular to solitary species has thus been proposed to be a function of substrate longevity (Jackson 1985). The rates of colonization and primary succession may be highly variable in different reef settings; lava flows on Gunung Api, Banda Islands (Indonesia), have only taken some 5 years to reach climax stages with 124 species (Tomascik *et al.* 1996; Figs 6.9(a), (b)), whereas climax communities may take several decades to form in the Eastern Pacific (Glynn 1997*a*).

Similar concepts of community succession have been applied to ancient reefs. However, in most documentations of supposed ancient reef succession, it is virtually impossible to eliminate the role of extrinsic change, and the nature of the preservation of fossil communities only rarely allows recognition of ecological succession upon a single substrate.

6.1.9 Summary: the importance of the clonal/modular habit for reef-building

We have seen that the flexibility of growth conferred by a clonal/modular (including colonial) organization is especially advantageous to organisms that remain permanently attached to one site. This is confirmed by the observation that the vast majority of living modular organisms are sessile. The functional morphology and life history of reef-building organisms determines many aspects of reef style and environmental preference, and in modern seas, the possession of a clonal/modular habit is a vital prerequisite for the formation of a wave-resistant reef in energetic regimes, and indeed for colonization of the resultant spatially heterogeneous framework. This is because:

● Modularity can confer almost unlimited growth, and hence size and longevity. Modularity also allows fusion to produce a rapid size increase. Size correlates with many measures of success and fecundity, and longevity of individual reef-builders confers longevity to the reef. Once a clonal/modular organism has achieved a certain size, it becomes virtually immune from normal predation.

● Modular growth allows for polymorphism and morphological plasticity.

● Modularity enables the possession of an encrusting habit. When combined with an attachment mechanism, this allows secure, permanent attachment to hard substrates and growth

Fig. 6.9 Primary succession. (a) Lava flows only 5-years-old on Gunung Api, Banda Islands, Indonesia; (b) these have been rapidly colonized by up to 124 species of scleractinian corals. (Photographs: P. Copper.)

in energetic settings. Large, skeletal reef-building organisms which rise above the substrate provide considerable spatial heterogeneity to the reef framework, even after death.

● A clonal organization offers defence from partial mortality as new modules can regenerate from surviving parts. By compartmentalizing damage and increasing regenerative ability, modularity facilitates cloning by either fragmentation or fission. The separation of a genet into ramets spreads the risk of mortality, enhances colonization ability, and preserves successful genomes.

- In modular organisms, the relative allocation of resources to sexual reproduction or asexual growth, and regeneration can be adjusted according to environmental circumstances.

- Modularity permits self-fertilization such that sexual reproduction can occur independent of population density.

- Clonality allows for a large amount of genetic variation and the heritability of somatic mutations, which may result in considerable genetic variability even within a single large colony.

- A clonal organization is favoured for endosymbiont acquisition, including multiple occupation (see Section 8.1.3).

- The high surface area to volume ratio present in multiserial modular organisms favours light, prey, and organic matter capture.

6.2   Interspecific competition

In interspecific competition—competition between species—the reproductive success of interacting individuals is decreased because all require the same limited resource. This means that the removal of one combatant will increase the survival and reproductive rate of individuals of the remaining species. However, for species to coexist (that is avoid or reduce interspecific competition) they must differ in the way in which they use such limiting resources. This is know as the *competitive exclusion principle* (Gause 1934), and these differences may often be seemingly subtle, such as variation in size, chemistry, or behaviour. Phenotypic evolution shaped or maintained by interspecific competition is known as *ecological character displacement*. Interspecific competition thus produces selection for the diversity of coexisting species.

Community composition on large, stable substrata such as coral reefs is controlled by interactions among established organisms because such substrata are a limited resource. For encrusting species, this is most dramatic when growth is halted by the presence of another encrusting organism (Jackson 1983). This control is in contrast to communities living on small or ephemeral (soft) substrata which is determined more by rates of recruitment and larval habitat selection than by postrecruitment interactions (Section 6.1.5).

If interspecific competition is an important control in structuring reef communities, then we would predict that two species which coexist have more restricted niches than if they were found singly. This is indeed the case, as zones in Caribbean reefs are typically dominated by only one or a few species, even though most species of corals occur over a very wide range of depths even at single localities (Goreau and Wells 1967). Modern coral reefs show an abundance of finely tuned displacements, some of which are documented below. However, to prove that ecological character displacement was important in the process of speciation requires historical knowledge; to rule out chance, evidence must be mustered to demonstrate that divergence in morphology or resource use has occurred repeatedly in independent clades.

6.2.1   Spatial competition

Encrusting, clonal species are likely to come into contact with other encrusters growing on the same substrate which may have superior competitive capabilities. These may include the capacity to overgrow as well as numerous anti-fouling mechanisms (those which prevent the growth of rapidly establishing, often small, sessile biota). Skeletonized, clonal organisms (such as corals, coralline algae, and bryozoans) are typically overgrown by forms which generally do not possess a rigid skeleton (demosponges and ascidians). Many clonal organisms show complex forms of aggression towards other species (Fig. 6.10(a)) or in some cases other clones of the same species (Fig. 6.10(b)), including direct contact overgrowth

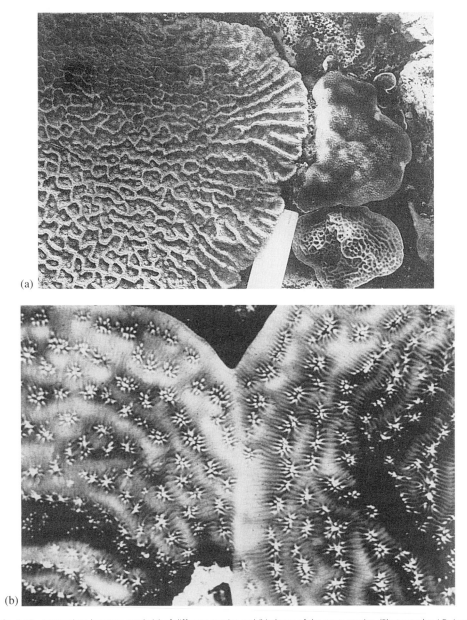

(a)

(b)

Fig. 6.10 Interactions between corals (a) of different species and (b) clones of the same species. (Photographs: J.D. Lang.)

and overtopping (shading overgrowth) (Connell 1973), extra-coelenteric digestion by mesenterial filaments (Lang 1973), sweeper tentacles (Chornesky 1983), nematocyst attack (Richardson *et al.* 1979), immunological responses (Potts 1976), and the secretion of allelochemicals (Jackson and Buss 1975) and water-borne toxins (Rinkevitch and Loya 1983).

Variation in the relative position of colonies can also affect the competitive ability of encrusting organisms (Buss 1986). Such differences are often due to settlement-induced patterns—which influence density, colony size, and elevation—or to chance encounter symmetries (such as the angles of contact; Jackson 1979). Large reef corals can recover, retaliate, or grow away from areas of contact, but small (2–3 cm diameter) colonies often die when they are placed adjacent to larger colonies (Sheppard 1985). The periphery of a colony may be specialized for defence against predation or aggressive encounters with other competitors, and damage caused by the attacking mesenterial filaments of the solitary coral *Scolymia* has been shown to vary with polyp size (Bak, in Lang and Chornesky 1990). A significant correlation between colony size and competitive abilities against macroalgae has also been noted on Jamaican reefs. In one survey only three of the relatively small colonies of *Agaricia agaricites* of the 205 surveyed in 1983 survived to 1987, in contrast to all ($n = 22$) the large colonies of *Colophyllia* and *Montastraea* (T.P. Hughes 1989). Colony size is thus an accurate predictor of competitive ability in many types of interactions.

Many of these competitive reactions are impossible to detect in the fossil record, because they either involve toxins, soft tissues, or spatial relationships that are not preserved. However, some competitive reactions which involve changes in skeletal morphology induced by the close presence of another reef organism can be recognized in the Ancient (see Section 6.4).

6.2.2 Competitive networks

Intricate competitive relationships can exist between species within a sessile community, especially between those which compete for the same substrate resource. Competitive ranking shows great variability both between communities and between species within a community. Although most rankings are *transitive*, i.e. the ranking is hierarchical and fixed, they can, however, be far more complex and form a network of competitive relationships, known as *competitive networks*, that are *intransitive* (Jackson and Buss 1975), where no one species is competitively dominant over all others. This may involve the direct action of predators, which can modify existing relationships acting to decrease the incidence of transitive competitive relationships (see Section 7.1.10). In this way, the complicated effects of direct competition can prolong the existence of diverse communities.

6.3 Trends in growth form and function

As a highly integrated, modular organization is of such clear benefit to sessile marine organisms, it is not surprising that clonal/modular representatives have evolved many times through the Phanerozoic in unrelated clades, especially in organisms associated with reefs. Major groups, such as the Cnidaria, display the entire range of aclonal and clonal life histories found in all known epibenthic invertebrates (Jackson 1985). The distribution of growth forms thus expresses patterns of functional or adaptive significance rather than evolutionary relationships (McKinney and Jackson 1989), and the temporal distribution of functional organizations in fossil groups, irrespective of their systematic placing, can therefore offer a guide to changing ecological adaptation and also record how the interactions within reef communities may have changed through the Phanerozoic. Indeed, the relative form and function of solitary versus modular organizations represents one of the earliest recognized macroevolutionary patterns in metazoan evolution (Boardman *et al.* 1973).

Patterns of aclonal/solitary and clonal/modular organism distributions suggestive of competitive exclusion are apparent ever since metazoans have been associated with reefs. The earliest sessile metazoans known were aggregating, solitary organisms, such as the globally distributed Neoproterozoic *Cloudina*. Likewise, the Lower Cambrian archaeocyath sponges were also highly gregarious, but their communities were often dominated by dense clonal populations of branching growth forms (Wood *et al.* 1992*a*). By the Ordovician, modular sponges were abundant on patch reefs, and for most of the middle Palaeozoic, reefs were dominated by large skeletal modular organisms such as tabulate corals and stromatoporoid sponges. Such skeletal organisms could reach very dense accumulations, dominating back-reef, lagoonal, and reef-margin settings.

Solitary, aclonal organisms were important reef-builders only when they were either large and/or formed dense aggregations. For example, dense accumulations of erect probable gastropods are known from the British Lower Carboniferous (Burchette and Riding 1977); spine-bearing, conical richthofeniid brachiopods up to 5 cm in diameter and 10 cm high formed small bioherms in the Late Permian (Senowbari-Daryan and Rigby 1996); and rudist bivalves that achieved lengths of up to 2 m formed reefs of low topographic relief during the Upper Cretaceous (Fig. 6.11). These dense monospecific stands, often consisting of several hundred individuals (Schumann 1995), suggest that philopatric larval settlement was important in structuring these communities, either as a result of rapid colonization of new substrates by a limited number of larval spat falls, or by preferential settlement upon adults of the same species. The high-abundance but low-diversity nature of these communities, in addition to the general dearth of associated biota, also suggests that they occupied restrictive environments that excluded most other organisms. On the basis of well-preserved growth bands, it has been estimated that such rudists had a life span of between 20 and 50 years (Schumann 1995).

Not only have modular representatives appeared within many clades of sessile organisms, but progressive *trends* (that is non-random, directional changes over time) of increasing modular integration have been documented in several groups of reef-building or associated clonal organisms. In all cases, the solitary condition is considered as the primitive state and modular as derived. (In corals, the modular habit is thought to be acquired by paedomorphosis of bud development (Rosen 1986)). These trends can be demonstrated statistically, are *polyphyletic* (that is they show multiple ancestry and are so are independent of systematic origin), and are not due to phyletic design (that is confined to particular inherited morphological organizations) or to chance. Such trends can thus be best explained as progressive adaptation to the environment, to physical processes, and/or increased success in biological interactions (McKinney and Jackson 1989; Jackson and McKinney 1991; Wood *et al.* 1992*a*).

These trends do not necessarily imply that geologically more recent organisms are in any way better adapted than their predecessors, but they do suggest that adaptation has proceeded in a measurable, directional fashion. If these progressive trends are not uniform, but

Fig. 6.11 A substantial monospecific aggregation of the large Cretaceous rudist *Vaccinites vesicularis*. The view probably shows a succession of communities. Individual size is remarkably uniform within each community suggesting that it grew as a consequence of colonization of adults of the same species by larval spat-falls that showed philopatric behaviour. On the basis of well-preserved growth bands, it has been estimated that such rudists had a life span of between 20 and 50 years. Upper Cretaceous (Campanian), Central Oman. Hammer = 32 cm long. (Photograph: D. Schumann.)

show periods of rapid change, then these might be indicative of causative changes in the environment.

The following section examines temporal trends in the development of the modular habit, attendant ecological consequences, and considers the relationship between the ecology and modular status in ancient, sessile reef organisms.

6.3.1 Porifera

Living poriferans (sponges) are suspension-feeders that passively or actively pump water through a ramifying network of canals and pores known as an *aquiferous filtration system* to gain nutrition. Although clearly individuals and not colonies (Hartman and Reiswig 1973), sponges may be clonal and show subdivision of their aquiferous system into functional modules (Fig. 6.12). Each module consists of a number of associated inhalant pores (*ostia*) which lead to a layer of choanocyte cells from which excurrent canals converge towards a single large exhalant pore (*osculum*) (Simpson 1973; Fry 1979; Wood 1987). A solitary sponge will therefore possess one osculum, a modular form many oscula. Although these modules may be highly interdependent and transitory entities, sometimes even re-organized on a daily basis, the skeleton of a sponge serves only to support the aquiferous system, and will therefore reflect its organization when the soft tissue has gone (Fry 1979; Wood 1987).

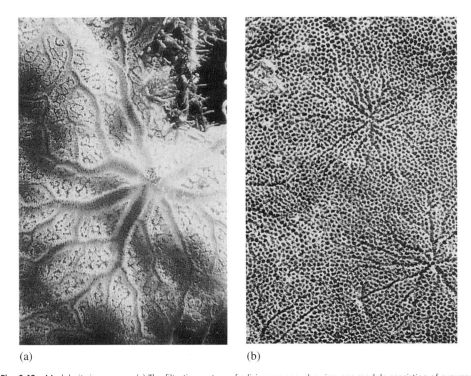

(a)   (b)

Fig. 6.12   Modularity in sponges. (a) The filtration system of a living sponge, showing one module consisting of excurrent canals converging towards a single, central osculum (photograph reproduced by permission from Glenat, *Mediterranée vivante*). (b) Traces of these features preserved on the surface of a Devonian stromatoporoid skeleton, showing several modules. (Photograph: J. Reitner.)

The vast majority of living sponges (over 95%) are modular and soft-bodied, and most of these show high integration organizations. This is in marked contrast to the earliest known complete fossil sponges from the ?late Proterozoic and Early Cambrian (hexactinellids, demosponges, and archaeocyaths), which were mostly solitary. This suggests that the solitary habit is a primitive state, retained by few living sponges, and the modular habit advanced.

Sponges, in particular calcified sponges, have been associated with reefs since the late Lower Cambrian (see Chapter 3). The Phanerozoic history of calcified sponges (Fig. 6.13) shows skeletally complex forms (archaeocyaths) followed by more simple organizations (stromatoporoids, chaetetids, and sphinctozoans). The arrangements of the aquiferous filtration system which typify each calcified sponge grade (Table 6.4), but are not exclusive to each, can be viewed in terms of different degrees of modular integration and are probably therefore ecologically determined (Wood 1990, 1991*a*). Can varying modular integration then explain the differing reef-building abilities and possibly temporal distribution of the grades of calcified sponge?

Archaeocyatha

Development of the modular habit in the highly gregarious Lower Cambrian archaeocyaths has been considered by Wood *et al.* (1992*a*). Depending upon the disposition of modules, as

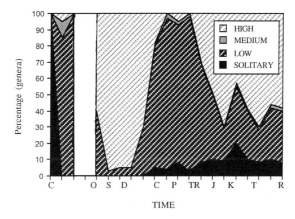

Fig. 6.13 The broad evolutionary trends within calcified sponges through the Phanerozoic, expressed as the percentage of genera that show solitary, and low, medium, and high integration. (Modified from Wood *et al.* 1992a, by kind permission of SEPM Society for Sedimentary Geology.)

Table 6.4 Range of growth forms found in calcified sponges. **X** indicates the most common growth forms of each group (modified from Wood *et al.* 1992a, by kind permission of SEPM Society for Sedimentary Geology.)

Group/grade	Archaeocyath	Stromatoporoid	Sphinctozoan	Chaetetid
Morphology				
Low integration				
1. Solitary	X		X	
2. Pseudocolonial				
(a) encrusting (uniserial)			X	
(b) branching	X	X	X	
Medium integration				
3. Uniserial erect				
(a) catenulate	X		X	
(b) pseudocerioid	X			
High integration				
4. Multiserial horizontal				
(a) encrusting (single)	X	**X**	X	**X**
(b) massive (multilayered)	X	**X**		**X**
5. Multiserial effect		X		X

defined by the placing of the central cavity within the skeleton, archaeocyaths show the modular growth forms illustrated in Fig. 6.14. Note that archaeocyaths exhibit almost the entire range of growth forms found in calcified sponges (Table 6.4).

Low-integration forms appeared first in the middle Tommotian. High-integration forms (Irregulars; see Box 3.3) followed in the Atdabanian 2, with medium-integration forms in the

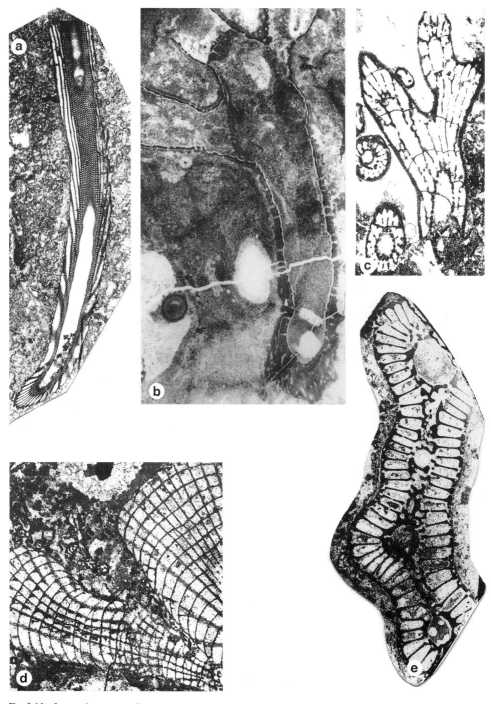

Fig. 6.14 See caption on opposite page

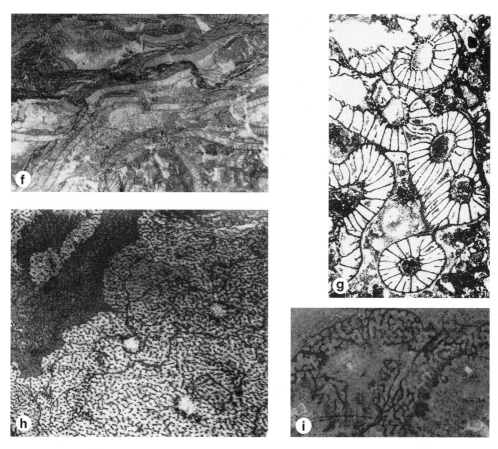

Fig. 6.14 Archaeocyath growth types. (a) Solitary, *Formosocyathus bulynnikov*, Western Sayan, USSR, Botoman. (b)–(d) Pseudocolonial, uniserial branching: (b) longitudinal subdivision, *Archaeolynthus polaris*, Zhurinsky Mys, middle Lena River, Siberian Platform, Russia, Tommotian; (c) interparietal budding, *Cambrocyathellus tuberculatus*, Zuune-Azts Mts, Mongolia, Atdabanian; (d) external budding, *Claruscoscinus billingsi*, Western Transbaikalian, Baikal Ranges, Russia, Toyonian. (e) Catenulate, *Densocyathus sanashticolensis*, Western Sayan, USSR, Botoman. (f) Pseudocolonial, uniserial laminar, *Erismacoscinus calarlue* (Born), Nebida Formation, Sardinia, Botoman. (g) Pseudocerioid, *Erbocyathus heterovallu*, Eastern Sayan, Russia, Toyonian. (h), (i) Multiserial: (h) massive, *'Agastrocyathus' grandus* (Yuan and Zhang), Yindingshan, Zunyi Province, China, Botoman; (i) encrusting, *Retilamina amourensis*, forming a characteristic stacking arrangement, Forteau Formation, Labrador, Canada, lower Toyonian. (From Wood *et al.* 1992a, by kind permission of SEPM Society for Sedimentary Geology.)

Atdabanian 3. Massive forms, which form the bulk of high-integration forms (Irregulars), made only a brief appearance as they were not known beyond the Botoman 2 zone.

Presumed archaeocyath species were commonly facultatively modular, displaying solitary forms in crypts and tranquil, non-biohermal settings, and modular, branching phenotypes in biohermal niches. This observation alone strongly supports the suggested adaptive significance of modularity for reef-building. Modular Irregulars also dominate reefal facies volumetrically. A study of some Botoman archaeocyath reefs in Mexico (Debrenne *et al.* 1989) showed that the bioherms were dominated by only two to four modular taxa, which occupied up to 97–99% of the total archaeocyath volume.

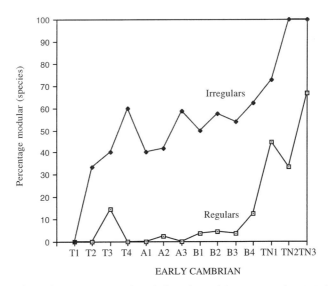

Fig. 6.15 Percentage of modular genera per zone through the Early Cambrian. T, Tommotian; A, Atdabanian; B, Botoman; TN, Toyonian. Irregulars: % = 14.605 ± 5.3859 (time), r = 0.767, p = 0.01. Regulars: % = −13.465 ± 3.5591 (time), r = 0.535, p = 0.33. (From Wood *et al.* 1992*a*, by kind permission of SEPM Society for Sedimentary Geology.)

Prominent trends are evident when the distribution of modular and solitary genera through the Early Cambrian is expressed as a percentage of modular forms per age-interval (Fig. 6.15). The dominant trend in both groupings appears to be towards increasing modularity and is especially linear in Irregulars right up to the final extinction of the group. In Regulars, a single modular genus, *Archaeolynthus*, appears in the Tommotian 3 which coincides with the emergence of vast bioherm massives on the Siberian platform, but further modular Regulars do not appear until the Atdabanian 2. After this time numbers of modular genera increase to reach 67% of Regulars by the middle Toyonian. In Irregulars, the first modular form appears in the Tommotian 2, with a subsequent general trend towards increasing modularity reaching 100% by the middle Toyonian.

Solitary archaeocyaths possess intricate skeletons with septa, inner and outer walls with complex porosities, as well as complex developmental sequences from the juvenile to adult portions of the skeleton (Fig. 6.14(a)). Attendant with the acquisition of modularity and increasing integration, is a decrease in the size of modules, a simplification of skeletal organization, and an increase in the importance of secondary-infilling structures, such as tabulae. These skeletal changes can be correlated with an inferred shift of soft-tissue distribution, from a wholly tissue-filled intervallum in solitary Regulars producing an internal, inflexible arrangement, to a highly flexible, surficial veneer in high-integration, massive forms which was almost wholly external to the skeleton (see Fig. 6.16). In many cases, massive forms have lost their inner or outer walls and there may be little or no discernible sequence of skeletal development. The bulk of such skeletons are constructed by taenia and secondary filling tissue (vesicles), which have therefore become the dominant skeletal features (Fig. 6.14(h)).

Dramatic differences are noted in the maximum size for archaeocyath species which were facultatively modular (Fig. 6.17). Modular individuals were always far larger in both dimensions than their solitary counterparts, even though the diameter of the modules in modular

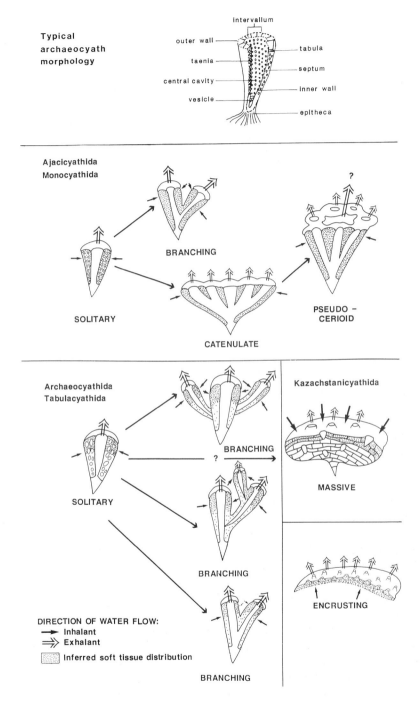

Fig. 6.16 Schematic illustrations of modular archaeocyath types based on module organization and mode of proliferation, with possible evolutionary pathways. Regulars are represented by the orders Ajacicyathida, Monocyathida, and Tabulacyathida, and Irregulars by the orders Archaeocyathida and Kazachstanicyathida. (Modified from Wood *et al.* 1992a, by kind permission of SEPM Society for Sedimentary Geology.)

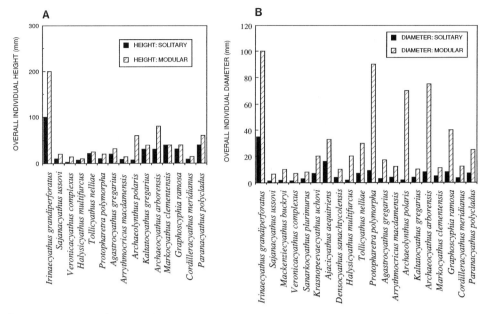

Fig. 6.17 Comparison of maximum dimensions between solitary and modular ecophenotypes of the same species. (a) Overall individual height; (b) overall individual diameter. (From Wood *et al.* 1992a, by kind permission of SEPM Society for Sedimentary Geology.)

forms is always smaller. The range and mean overall height of representatives of the various integration states are also compared in Fig. 6.18. Solitary forms show the greatest height range, but increase in average individual height with increasing integration is not apparent in the Archaeocyatha. Most notable is the far greater average height of low-integration forms compared with all other integration states, which show remarkably similar mean heights.

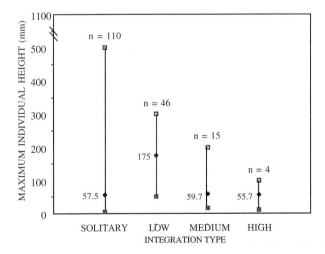

Fig. 6.18 Comparison of the range and mean overall height of archaeocyath species from different integration states. n: number of species. (From Wood *et al.* 1992a, by kind permission of SEPM Society for Sedimentary Geology.)

Such patterns of aclonal/solitary and clonal/modular distribution are suggestive of competitive exclusion comparable to that found in Recent biotas. A detailed analysis of the changing biological interactions associated with the acquisition of modularity in the Lower Cambrian Archaeocyatha is presented as a case study in Section 6.4.

Stromatoporoids and chaetetids

Living stromatoporoids (*Astrosclera*, *Calcifibrospongia*) and chaetetids (*Acanthochaetetes*, *Ceratoporella*) often have a cryptic habitat (Fig. 6.8(b)), but most pre-Cenozoic representatives lived on the open surface. Only the mid-Palaeozoic stromatoporoids built extensive reefs, particularly during the Devonian, when they displayed a great variety of multiserial modular growth forms, including foliaceous, laminar, domal, branching, and those with highly complex combined columnar (pyriform) and platy morphologies (see Fig. 6.19 and Fig. 3.16). Large multiserial, domal forms tended to dominate subtidal settings; those with fine-branching, uniserial organizations formed extensive thickets in the back-reef and lagoonal areas. Many multiserial stromatoporoids, as well as chaetetids, colonized small, ephemeral patches of hard substrate, e.g. brachiopod shells or crinoids, from which they extended rapidly over unconsolidated substrate. Others were able to encrust more extensive areas of hard substrate, such as other large skeletal organisms or microbialites.

Modularity imparted the ability to achieve great size and tremendous powers of regeneration after partial mortality in some multiserial representatives. Stromatoporoids over 5 m in diameter are known from the Devonian reefs of the Canning Basin, western Australia, and the existence of such large examples provides compelling evidence for the considerable longevity and persistence of these clones. Such examples often show multiple phases of growth, regeneration, and regrowth. For example, Fig. 6.20 illustrates a tracing of a single, laminar stromatoporoid individual which was able to produce frequent lateral outgrowths that domed over sediment, often in response to other parts of the stromatoporoid surface being smothered in sediment. These lateral outgrowths were produced by rapid growth of the basal layers of the stromatoporoid, while measurements of growth rate on the basis of seasonal banding in upwards, accretionary growth suggests growth rates of only 3 mm year^{-1} (Gao and Copper 1997). Such elevated growth away from the substrate surface is commonplace, suggesting that episodic sedimentation was the greatest source of partial and total mortality for such sponges.

Other modular, encrusting stromatoporoids were, successfully and repeatedly, able to overgrow other large skeletal organisms such as tabulate corals and bryozoans (Fritz 1977; Stel 1978), and often show regeneration from very small areas of remaining living tissue to re-cover the formerly encrusted area.

The new clades of modular stromatoporoid and chaetetid sponges that appeared during the mid-Mesozoic did not make major contributions to the energetic parts of reef complexes. But many produced monospecific aggregations of branching forms in shallow, unconsolidated substrates that formed either low-relief patch reefs (Fig. 6.21) or thickets that sometimes covered substantial areas of sea floor. These communities grew in very shallow, lagoonal conditions as evidenced by chalky marl lithologies and the presence of gymnocodiacean algae (see C.S. 3.13). Individual stromatoporoid size and shape is remarkably uniform within each community, but may differ markedly from neighbouring patch reefs. This suggests that each community formed as a consequence of localized substrate colonization by a very limited number of larval spat-falls with philopatric behaviour that produced a single generation of organisms. The monospecific and 'single generation' nature of these reefs suggests that they occupied marginal

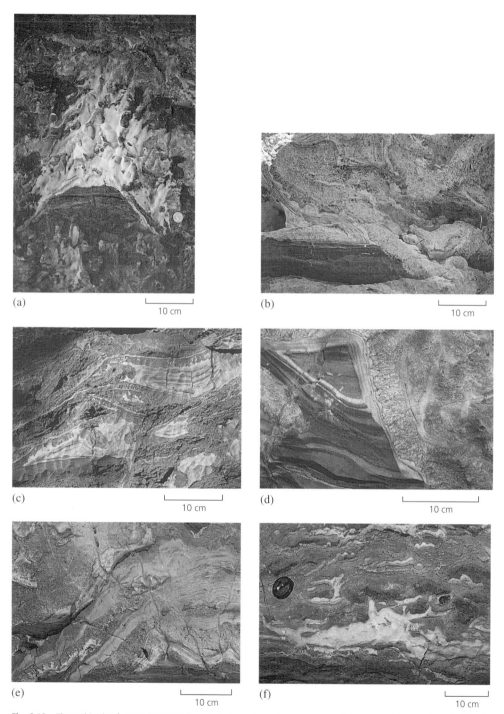

Fig. 6.19 The multitude of Upper Devonian (Frasnian) stromatoporoid sponge growth forms. (a) Columnar form (pyriform); (b) foliaceous; (c) thin-laminar; (d) branching; (e) massive with arching lateral outgrowths; (f) laminar, with development of multiple 'chimneys'.

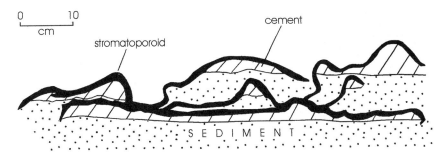

Fig. 6.20 Tracing of a stromatoporoid showing the regeneration of the skeleton after partial mortality, probably due to sediment smothering. Note how both the original laminar growth and the lateral outgrowths dome over the sediment to form open cavities which are now cement-filled. Upper Devonian (Frasnian) Pillara Limestone, Glenister Knoll, Bugle Gap, Canning Basin, Western Australia.

Fig. 6.21 A monospecific patch reef formed by aggregation of the branching Jurassic stromatoporoid *Shuqraia*. The uniformity of individual size within this community suggests that the substrate was colonized by a single phase of larval spat that showed philopatric behaviour. Hammer = 32 cm. Makhtesh Hagadol, Israel.

environments which excluded most other organisms. Other examples of densely aggregating, branching calcified sponges include the Upper Devonian stromatoporoid *Amphipora*, which can comprise up to 80% of back-reef rock volume (Playford and Lowry 1966).

Sphinctozoans

The living, chambered (sphinctozoan) sponge *Vaceletia*, is found only in reef crypts or deep-water environments (Fig. 6.8(b)). Some Ordovician sphinctozoans and Silurian aphrosalpingids (which resemble chambered archaeocyaths), as well as chambered archaeocyaths and some Permian sphinctozoans, were also common crypt dwellers (Fig. 6.22). Indeed, these groups of Palaeozoic, chambered calcified sponges, previously interpreted as erect reef framebuilders, appear to have been preferential cryptic dwellers for much of their long history (Wood *et al.* 1994, 1996; Wood 1997).

5 cm

Fig. 6.22 Cryptic sphinctozoan sponges within the Permian Capitan Reef, Mckittrick Canyon, New Mexico, USA.

Chambered sponges exhibit predominantly solitary and low-integration, branching morphologies with limited attachment sites. As we have seen, in the Modern, solitary organisms tend to be poor space competitors on hard substrates as they generally have small areas of attachment and lack specific competition mechanisms (Jackson 1977, 1985). Such organizations conferred better competitive abilities in fugitive habitats, for example within crypts, than on open surfaces, where they would have been outcompeted by modular, encrusting organisms with a marked ability to cover and occupy new substrate rapidly.

The solitary or low-integration branching organizations of sphinctozoans (and archaeocyaths) would have been severely size-limiting, with the average length reached being 6–18 cm compared with Palaeozoic stromatoporoids which often achieved over 1 m in diameter. Although acquisition of a branching, low-integration organization led to a marked increase in individual size in archaeocyaths, it is noteworthy that any further size increase was not dependent upon increasing integration (Fig. 6.18). This might suggest that growth was determinate for many archaeocyath and sphinctozoan sponges.

6.3.2 Corals (Cnidaria)

The anthozoans have calcified several times, giving rise to three major coral groups, the tabulates (Cambrian–?Triassic), rugosans (Ordovician–Permian), and scleractinians (Triassic–Recent). Corals show the entire range of integration types, as illustrated in Fig. 6.23.

Palaeozoic corals

The majority of Palaeozoic corals were small, ranging from 30 to 600 mm in diameter. But colonies up to 1 m are not uncommon, and rare examples of tabulate coral colonies up to

Fig. 6.23 Growth forms in corals: (a) solitary/aclonal; (b) pseudocolonial, phaceloid; (c) uniserial erect; (d) multiserial encrusting; (e) multiserial erect; (f) multiserial massive; (g) multiserial massive—meandroid; (h) solitary/clonal (the free-living *Fungia*).

2 m have been reported (Fagerstrom 1987). Upper size limits are difficult to establish, however, especially in branching forms which are subject to breakage, and those which are closely intergrown.

Rugose corals were predominantly small and 64% genera were solitary (Scrutton 1998). Of the 104 colonial genera described, 58 (53%) showed low-integration (phaceloid) colonies, with comparatively large corallites (often 8–30 mm; Jackson and Coates 1985). However, although more complex colonial growth forms are known, only 5% of all rugosans were of high integration (Scrutton 1998). No examples of interconnecting pores between corallites are known in the Rugosa.

Tabulate corals are exclusively colonial. The vast majority show multiserial organizations, with corallite sizes often an order of magnitude smaller than most rugose corallites; 85% of all genera show some level of integration, but only 4.9% show the highest level of integration (Scrutton 1998).

Most Palaeozoic corals did not appear to recruit actively onto extensive hard substrates. Rugose corals were never major reef-builders, they occupied a range of shallow and deepwater habitats, and most lived attached on unconsolidated substrates. Solitary forms grew either rooted in soft substrate, or occupied cryptic niches; most are found prostrate. Some species had the ability to right themselves after toppling and to regenerate—and could survive repeated disturbance producing long and convoluted skeletons.

Like stromatoporoids, tabulate corals colonized small, ephemeral patches of hard substrate, or patchy areas of stable, topographic highs such as brachiopod valves or pockets of bioclastic debris, from which they extended rapidly over unconsolidated substrate. However, most species could encrust opportunistically through chance settlement onto hard substrates such as other skeletal metazoans. The presence of an epitheca or holotheca in tabulate and rugose corals, however, limited their ability to encrust hard substrates, except for some favositids, and *Alveolites* and *Aulopora*. Where attachment scars are known from tabulate corals, they are small, demonstrating settlement upon small shell fragments or sedimentary grains (Scrutton 1997*b*).

Tabulate corals were also capable of considerable regeneration after partial mortality, and an example of colony formation by fragmentation has been described from the Silurian species *Paleofavosites capax* (Lee and Noble 1990). Localized areas which were not engulfed by sediment or encrusted by another soft-bodied organism show rapid upward growth followed by subsequent lateral extension (Stel 1978). Tabulates were clearly able to redirect resources in order to rapidly cover new substrate—two distinct growth stages are apparent: lateral growth broadly parallel to the substrate surface with widely spaced corallites suggestive of rapid growth; and upward growth with more closely spaced corallites. Widely spaced corallites are also associated with the overgrowth of other skeletal encrusters that presented obstacles to growth. Like stromatoporoids, there is evidence to suggest that the major risk of total and partial mortality was due to sediment fouling.

Several workers have proposed that stromatoporoids and tabulate corals were in probable competition (e.g. Lecompte 1959; Manten 1971). Tabulate corals appear to have been tolerant of mud-rich, turbid conditions where stromatoporoids were virtually absent: stromatoporoids flourished in clear, mud-free carbonate environments. That competition rather than different tolerances to sedimentation is responsible for this distribution is suggested by the fact that stromatoporoids commonly overgrew tabulates in well-agitated, pure carbonate environments, whereas few stromatoporoids appear to have overgrown tabulates in muddy settings. However, the limited ability of tabulate corals to form secure attachment to a hard substrate may have offered only low tolerance to highly energetic environments. There is some evidence though, to suggest that the growth of massive tabulate corals may have been aided by weighting within the sediment. Rates of overturning in Late Ordovician to Early Silurian corals decreased with increased colony size among massive tabulate corals living in relatively high energies (Young and Elias 1995). Moreover, forms with domal morphologies were the most likely to remain *in situ*.

Scleractinian corals

Scleractinian corals show the entire range of integration types: 34% of all extinct and extant genera are solitary, although only about half of these are found in shallow tropical waters (Coates and Oliver 1973). About 34% of all known genera show high-integration morphologies. Unlike Palaeozoic corals, the vast majority of scleractinian corals are able to encrust hard substrates due to possession of an edge zone. Tertiary and living corals show a tremendous number of high-integration growth types (see Fig. 6.2), and colonies often achieve several metres in diameter and height. Living scleractinian corals are found in a great range of habitats: on the abyssal plains, on shelves at temperate and even polar latitudes, and in tropics to subtropics. Much of the multifarious ecology of this group has been discussed in Section 6.1.

Trends in coral modularity were considered in a study using approximately 1600 species from 120 monographs by Coates and Jackson (1985). Their analysis of the ratio of high to low integration in all corals shows a broad increase in integration through the Phanerozoic (Fig. 6.24(a)). In scleractinian corals, the diversity of both colonial reef-associated forms and solitary non-reef organisms has increased from the inception of the group. Periods when abundant high-integration forms were present correlate with coral reef-building activity, such as the mid-Palaeozoic, mid-Jurassic, and Tertiary. Of most note is a steady and marked increase in the proportion of high-integration forms since the Triassic, which appears to be uninterrupted by the end of the Cretaceous extinction event.

In contrast, the percentage of scleractinian erect species (mainly low-integration, phaceloid–dendroid growth forms) decreased until the Turonian, but multiserial branching forms increased markedly after that time (Fig. 6.24(b); Jackson and McKinney 1991). These forms can show very light, porous skeletons, rapid rates of growth, as well as being capable of regeneration from fragments (Section 6.1.5, Clonal/modular organisms).

6.3.3 Bryozoans

Bryozoans grow as colonies of minute, clonally produced, modular units termed *zooids*. The pattern of zooidal budding has been shown to determine the growth form and to some extent the life history of the colony (McKinney and Jackson 1989; Jackson and McKinney 1991). Many trends have been documented in bryozoan morphology, in particular changes in the proportion of growth morphologies and budding types.

Bryozoans show a diverse range of growth forms including erect (branching), free-living, and encrusting types (Fig. 6.25). Erect forms were overwhelmingly dominant during the Palaeozoic, especially after the Silurian (McKinney and Jackson 1989), being abundant and diverse on hard substrata in shallow epicontinental seas (McKinney 1979; McKinney *et al.*

(a)

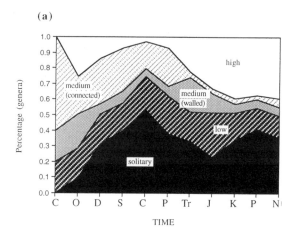

(b)

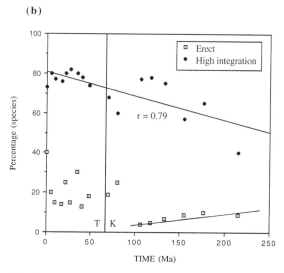

Fig. 6.24 Evolutionary trends within corals. (a) Percentage of solitary and low-, medium-, and high-integration genera in tabulate, rugosan, and scleractinian corals through the Phanerozoic. (Replotted from Coates and Oliver 1973.) (b) Percentage of high integration of corallites within the Scleractinia, Triassic–Recent regression: % = 43.7 + 0.163 (time), $r = 0.79$, $p = 0.0003$; percentage erect species, Triassic to Turonian regression: % = 13.6- 0.058 (time), $r =-0.80$, $p = 0.039$; post-Turonian regression: % = 4.87 + 0.071 (time), $r = 0.35$, $p = 0.331$. (Replotted from Jackson and McKinney 1991, by kind permission of the University of Chicago Press). (Data from sources in Coates and Jackson 1985.)

1986). Many were large, reaching 10–20 cm in height/diameter, but huge colonies up to 1–3 m are known (e.g. D.B. Smith 1981). In contrast, Mesozoic faunas show equal numbers of encrusting and erect species (Fig. 6.26). Since the Jurassic, faunas became increasingly dominated by encrusting forms, especially in the Neogene, and since the Jurassic/Cretaceous, erect bryozoans became progressively restricted to mid-shelf and deeper waters. Indeed, modern depth distributions of bryozoans show that the upper depth limit of rigidly erect species is very different from Palaeozoic distributions (McKinney and Jackson 1989). In addition, those erect forms found in shallow marine environments including reefs and hardgrounds

(a)

(b)

(c)

(d)

(e)

Fig. 6.25 Basic growth forms of bryozoans:
(a) runners (uniserial encrusting); (b) multiserial
encrusting; (c) multilayered multiserial encrusting
(massive); (d) multiserial erect; (e) uniserial erect.
(Photographs: P.D. Taylor.)

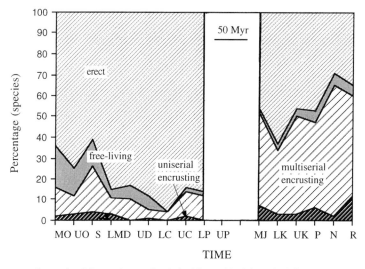

Fig. 6.26 Percentage of erect, free-living, and encrusting (uniserial or multiserial) species in bryozoan communities through the Phanerozoic. (Mann's test for trend, $T = 3.30$, $p = 0.007$; the median ratio of encrusting to erect species for the Palaeozoic is 0.10, and for the post-Palaeozoic, 0.97 (Mann–Whitney U Test, $p = 0.0009$)). O: Ordovician, S: Silurian, D: Devonian, C: Carboniferous, P: Permian, J: Jurassic, K: Cretaceous, P: Palaeogene, N: Neogene, R: Recent, L: Lower, M: Middle, U: Upper. (Data replotted from McKinney and Jackson 1989.)

became restricted to cryptic niches during the late Mesozoic, and tended to be narrow-branched, flexible, and articulated, rather than robust and rigid as shown by those which dominated shallow seas before the Cretaceous. However, as late as the Maastrichtian, and possibly the Neogene, sea-grass fronds supported dense populations of rigid, erect bry-ozoans. These fronds are now dominated by encrusting species, with multilayered species dominating on the long-lived root systems (P.D. Taylor, personal communication).

The proportion of free-living bryozoans also decreased during the Palaeozoic (Fig. 6.26). These forms were unspecialized and immobile, only able to regulate their position by current action. Free-living bryozoans, however, gained the ability to be mobile on sediments in the Late Cretaceous, and from that time the proportion of the mobile representatives increased dramatically and has remained high to present times.

Fragmentation may have been an important process of clonal propagation in populations of erect Carboniferous bryozoans such as *Archimedes* (e.g. McKinney 1979). The vast majority of branched forms can be demonstrated to have regenerated from fragments, and bases which indicate the initial attachment sites of larvae are rare.

Post-Palaeozoic bryozoans form a series of increasing skeletal defence, interzooidal integration, and zooidal polymorphism that follows their sequence of stratigraphic appearance (Jackson and McKinney 1991). All these trends are based on an increase in mean integration. As in the archaeocyaths, perhaps all these groups represent grades of organization rather than true taxa, so their diversity is a measure of comparative ecological success of differing levels of bryozoan armament and integration. In particular, there is a very striking increase in the proportion of cheilostome taxa relative to cyclostomes since the early Cretaceous (P.D. Taylor and Larwood 1988).

Species of encrusting bryozoans differ in the way in which new zooids are added to grow across a substrate, and also in their potential for vertical budding away from the substrate.

Horizontal extension can be achieved by the episodic addition of one zooid at a time, a process known as *intrazooidal budding*. But colony growth effectively may proceed continuously when a series of zooids develop (known as *zooidal budding*). These forms can achieve more rapid growth over a substratum, have greater flexibility at colony margins, and have increased regenerative capacity. Some species are also able to form multilayered colonies by frontal budding of new zooids one on top of another. The resultant thick colonies are better defended against overgrowth than single sheets, as well as from predators and grazers. Their capacity to repair injured tissue may also be markedly greater, but this remains to be proven empirically. Interzooidal communication in Palaeozoic (stenolaemate) bryozoans was apparently limited to the thin tissue at the colony surface, which would restrict regeneration of damaged colonies to budding only at the edge of any injury.

A striking pattern emerges when the budding type of the most abundant species of bryozoans occurring on a variety of substrates is compared with an inferred order of increasing stability, substrate size, and longevity (Fig. 6.27(a)). Forms with intrazooidal budding dominate small ('patchy') or ephemeral substrata because they are often small colonies of short life span, whereas species with zooidal budding, or zooidal and frontal budding dominate on long-lived substrates, including coral reef substrates where there are no abundant intrazooidally budding species. When the stratigraphic record of the appearance and subsequent abundance of budding types is considered, a marked trend is also clear (Fig. 6.27(b)). Whilst intrazooidal budding was dominant among abundant species in the Cretaceous and the Palaeocene, zoodial and frontal budding became characteristic of the most abundant species from the Eocene onwards (Lidgard and Jackson 1982).

Of these surveyed trends in bryozoan encrusting morphologies, forms with continuous budding, rapid growth, and good powers of regeneration have markedly increased since the Mesozoic, especially from the Late Cretaceous–Eocene, which was also the rapid phase of cheilostome radiation.

6.4 Ecological attributes of the modular habit—Case study: the Archaeocyatha

In the following section, I outline the ecological attributes acquired with modularity using the first reef-associated metazoans, the Lower Cambrian archaeocyath sponges (Wood *et al.* 1992*a*).

6.4.1 Community Differentiation

In an analysis of 38 assemblages, Zhuravlev and Wood (1995) showed that the Archaeocyatha differentiated both systematically and functionally early in their history into distinct open-surface and cryptic communities, which remained distinct until the extinction of the group (see Fig. 6.28). A far greater proportion of the Irregular order Archaeocyathida and the Regular orders Monocyathida and Coscinocyathida are represented in any one cryptic community than members of the orders Ajacicyathida and Tabulacyathida. Whilst open surfaces were dominated by solitary members of the order Ajacicyathida and Irregulars with branching, modular organizations, crypts preferentially housed solitary Irregulars and solitary chambered forms. Such archaeocyaths were often abundant crypt dwellers

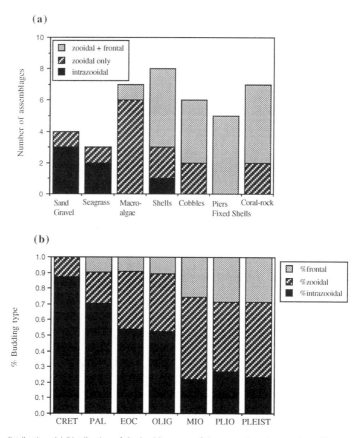

Fig. 6.27 Bryozoan distribution. (a) Distribution of the budding type of the most abundant species of bryozoans on substrates which differ in stability and longevity. (Data replotted from Lidgard and Jackson 1988.) (b) Percentages representing budding categories of the most abundant bryozoan species occurring in fossil assemblages from the Cretaceous–Pleistocene. (Data replotted from Lidgard and Jackson 1989.)

even when uncommon or absent in the open-surface bioherm community. This differentiation strongly suggests that habitat selection for small refuges was common in the Lower Cambrian.

Crypts are dominated by rapidly establishing, small, solitary organisms (Fig. 6.29(a)). The rapid growth of cements in crypts may have further reduced the time available for both colonization and growth of the cryptos, and would eventually have led to the total occlusion of crypt openings (see Fig. 4.9).

Competition for space in Lower Cambrian reefs must have been severe to produce distinct open-surface and cryptic communities. That crypt dwellers commonly formed multiple overgrowths or chains of individuals in crypts (Fig. 6.29(b)), indicates that much of the crypt surface was covered with organisms. Likewise, distortion of the undersurfaces of frame-building archaeocyaths indicates the presence of a soft-bodied cryptos that may have been better competitors for substrate space.

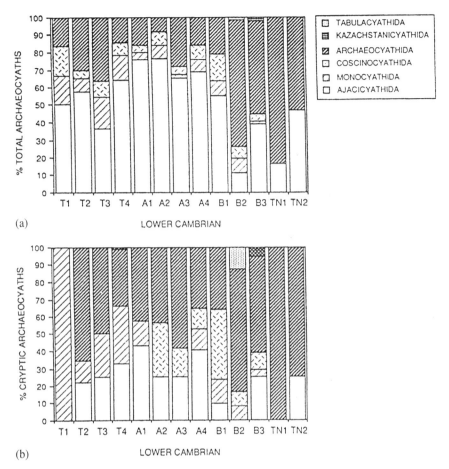

Fig. 6.28 Bar graphs showing the differentiation of archaeocyath sponges into open-surface and cryptic communities through the Lower Cambrian. The proportions of 38 assemblages have been averaged for each stage. (a) Percentage of archaeocyath species in each order within the total reef community; (b) percentage of archaeocyath species in order within cryptic communities only. T: Tommotian; A: Atdabanian; B: Botoman; TN: Toyonian. (Fisher Exact Test shows a statistically significant underrepresentation of ajacicyathids, and significant enrichment of archaeocyathids within crypts at the 5 per cent level.) (From Zhuravlev and Wood 1995.)

6.4.2 Morphological plasticity

In present-day organisms, a modular organization can lead to either specialization or generalization. In archaeocyaths, modularity appears to have conferred a more flexible response to the environment, as shown by the increased number of ecophenotypes displayed within a species. For example, *Metaldetes profundus* shows a wide range of specialized morphologies in the late Botoman–early Toyonian Forteau Formation in Labrador, Canada, from large, solitary, open bowl-shaped forms in tranquil lagoonal settings, to solitary cups in interbiohermal areas, and branching morphologies in more turbulent biohermal settings (Debrenne and James 1981; Hart 1994).

(a)   2 mm   (b)   2 mm

Fig. 6.29 Lower Cambrian cryptic reef communities. (a) Crypts dominated by a monospecific population of solitary archaeocyaths. (b) Crypt with chains of solitary archaeocyath cups.

6.4.3 Interactions

Solitary archaeocyaths were highly competitive, and show signs of profound incompatibility. This incompatibility is manifest in a wide range of skeletal reactions produced in response to the close proximity of other archaeocyath species, including the production of tersiae (calcified structures that extended beyond the outer walls of the archaeocyath cup), and growth away from, or partial atrophy on contact with, an aggressor (Figs 6.30(a) and (b), compare with Fig.6.10(a)). Such reactions that maintain spatial distance between archaeocyath individuals are clearly not conducive to the mutual encrustation necessary for reef framework formation.

In contrast, Irregular, modular forms appeared to coexist in either close proximity or actual contact with only a minimal detectable skeletal reaction, such as secondary skeletal thickening of both interacting individuals in the immediate area of contact or continued growth (Figs 6.30(c) compare with Fig. 6.10(b)). No atrophy and growth away from the aggressor are seen. Although we cannot conclude from this reduced skeletal reaction that modular archaeocyaths were in any way less aggressive than solitary forms, it is clear that these individuals were able to tolerate a closer proximity to members of other species to such an extent that formation of a skeletally bound framework resulted. This type of reaction would also be expected if these closely associated archaeocyaths, as in most dense thickets of living branching corals, consisted of only a few clones that did not exhibit intraclonal aggression.

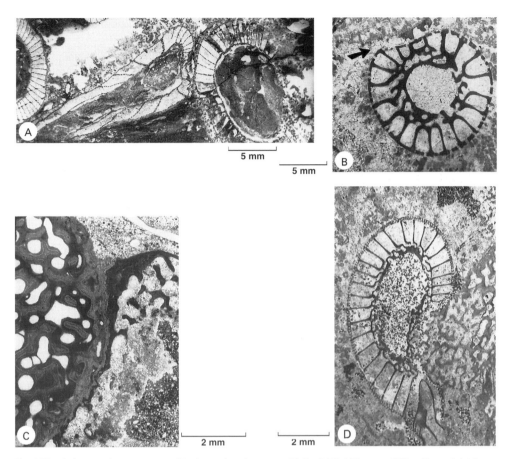

Fig. 6.30 Archaeocyath sponge competitive interactions (compare with Fig. 6.10). (a) Incompatibility with no skeletal contact—intraspecific competition between two solitary individuals, presumably different clones, of the solitary Regular *Nochoroicyathus*. Separation is maintained between these two individuals and the weaker combatant (left) shows atrophication of its skeletal element. (b) Competition (arrowed) between two solitary forms, the Regular *Parethmophyllum cooperi* (centre) and the Irregular *Markocyathus clementensis* (upper left). The elements of the Regular are being distorted and this individual maintains a constant distance from the Irregular. (c) Detail of the skeletal reaction of the encruster *Retilamina amourensis* (right) binding to the outer wall of *Archaeocyathus atlanticus*. Minor secondary thickening has occurred in both forms in the immediate zone of contact. (d) The modular Irregular *Dictyocyathus tennerimus* (right) is disrupting the skeletal elements of the solitary Regular *Inessocyathus spatiosus* due to its indeterminate rate of growth. (From Wood *et al.* 1992a, by kind permission of SEPM Society for Sedimentary Geology.)

Although far more data are needed to describe the numerous competitive reactions between modular forms of differing integration states, it would appear that acquisition of a modular habit markedly changed the nature of incompatibility between different archaeocyath taxa and possibly clones.

6.4.4 Regeneration

As in living sponges, archaeocyaths show different responses to injury according to modular status, with Regular solitary forms and Irregular modular forms showing different modes of

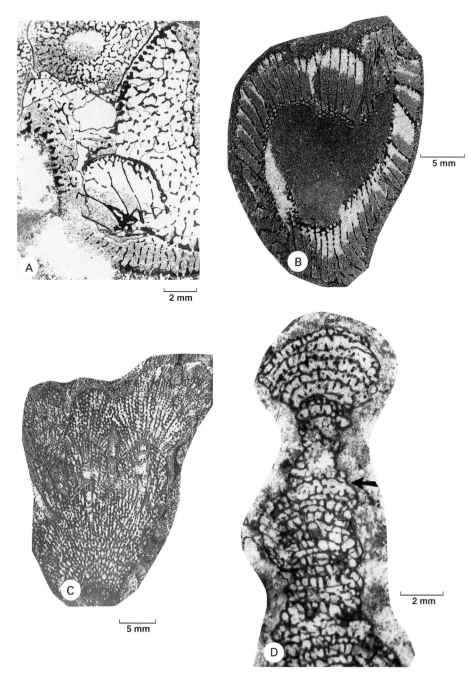

Fig. 6.31 Skeletal regeneration in archaeocyath sponges. (a) Solitary, Regular, *Rossocyathella* sp. showing disruption and gradual reorganization of skeletal elements after damage. (b) A modular form of *Metaldetes* sp. showing budding after damage. (c) Fission and fusion of branches in an individual of *Archaeocyathus arborensis*. (d) The high-integration massive form *Altaicyathus* sp. showing waxing and waning of growth; at the arrowed point, the individual was able to regenerate after the loss of about half its soft tissue. (From Wood *et al.* 1992a, by kind permission of SEPM Society for Sedimentary Geology.)

regeneration. Solitary Regulars repair the immediate area of damaged skeletal tissue. After damage, elements initially grow in a distorted fashion, but regain regularity and symmetry with further growth (Fig. 6.31(a)). Modular Irregulars bud to form completely new modules (Fig. 6.31(b)). Budding involves a total reorganization of the sponge, in addition to well-integrated soft tissue.

Modular forms show other indications of great flexibility of form. Low-integration forms may exhibit fission and fusion of their branches (Fig. 6.31(c)), and the often ragged edges of multiserial forms demonstrates their ability to cope with environmental fluctuations, such as sediment incursions, even after substantial loss of soft tissue (Fig. 6.31(d)). There is no evidence, however, that these high-integration archaeocyaths were capable of the tremendous powers of regeneration noted in mid-Palaeozoic stromatoporoids (Fig. 6.20).

Thus it appears that the thin, externalized and more flexible soft tissue of modular forms is more resistant to partial mortality. It may be that the irregular body outline and less complex skeleton allow a greater ease of regeneration involving a simpler process. Budding also involves an increase in individual size rather than simply repair of the damaged area, and may allow fragmentation of a clonal individual with the possibility of regrowth and spread of the clone.

6.4.5   Style of growth

Acquisition of modularity is predicted to confer a change in the style of growth, from determinate to indeterminate (Jackson 1985). This may be tested by considering the outcome of intergrowth between adjacent solitary and modular archaeocyaths. In all cases noted, modular forms were seen to grow into solitary forms resulting in disruption of the skeletal elements (Fig. 6.30(d)). This suggests that the indeterminate growth of modular forms conferred competitive superiority. An alternative explanation, however, might be the advantage conferred by the varied disposition of interacting growth axes. Solitary forms possess only one vertically inclined growth axis with little possibility of encountering growth into other individuals. In contrast, the inclined angle of the branches of modular forms would place growing tips against the outer wall of adjacent solitary individuals. This is especially marked in forms which branch by budding, where daughter branches have a more shallow angle of inclination than the parent branch. In either case, the characteristics of modular forms are such that they appear to have out-competed solitary forms in close interactions.

6.4.6   Size, longevity, mortality, and fecundity

Modularity in archaeocyaths conferred the predicted increase in overall individual size (Fig. 6.17), but only in low-integration, branching forms (Fig. 6.18). The main cause of mortality in archaeocyaths appeared to be due to storm or current action which either toppled the sponges or smothered them with sediment. Evidence from modern populations suggests that catastrophic physical disturbances, such as storms, kill individuals of any size (although branching growth forms may be more susceptible than massive morphologies), but there is insufficient data to determine whether the probability of catastrophic death may decrease with size. In archaeocyaths, secureness of attachment to the substrate would appear to be a major factor in determining mortality in such circumstances. Exothecal skeletal structures (see Box 3.3) may limit the risk of physical damage by providing a greater strength between interconnected individuals. Such structures offer one of the few means available for archaeo-cyaths to achieve extra stability in solitary forms in high-energy environments. Secondary calcification of the basal, abandoned parts of the skeleton, which was especially developed in modular Irregulars, might also be interpreted as an adaptation for increasing stability. Erect cheilostome bryozoans also show such branch thickening, which confers greater resistance of colonies to breakage (Jackson and McKinney 1991). Therefore, forms with only holdfasts would be less secure than forms with additional secondary calcification, which would, in turn, be less stable than encrusting forms, i.e. *Retilamina* (Fig. 6.14(i)). Modular forms, especially those which bear the extra weight of considerable secondary thickening, do tend to remain in life position more commonly than solitary forms in high-energy biohermal settings. *Retilamina*, is nearly always *in situ*, and appears to have stabilized soft sediment and encrusted toppled archaeocyaths (Hart 1994).

6.4.7 Case study summary

Archaeocyaths show the predicted ecological changes with the appearance of a modular habit. However, only genera possessing porous septa show any development of modularity, suggesting that the possession of well-integrated, soft-tissue communication is a prerequisite for acquisition of the habit. This is even true when closely related forms are compared, e.g. *Anthomorpha* is non-porous but solitary, while the sister taxon is a branching porous modular form, *Tollicyathus*. Presumably septal pores in archaeocyaths, as with porous corallite walls in corals, enabled direct circulation of resources through the archaeocyath skeleton instead of only over the surface.

Figure 6.32 shows the percentage of archaeocyath genera in Regulars and Irregulars through the Early Cambrian that exhibited porous septa. Regulars show a trend towards occlusion of pores, whereas Irregulars maintain a high proportion of forms with septal porosity throughout their history. Interestingly, the trends shown in the proportion of Regular and Irregular genera with porous septa are quite independent of the trend of increasing modularity (compare with Fig. 6.15). The implication is that the trend towards increasing dominance of modular forms was not driven by the relative availability of forms with porous septa or by phylogenetic constraints.

Increasing integration and simplification of the skeleton can be correlated with an increase in soft-tissue flexibility. Even though high-integration, skeletally simple, stromatoporoid-like forms developed in Irregulars, low-integration, branching morphologies were by far the most successful. Irregulars show a strongly progressive trend of increasing modularity right up to the final extinction of the group, confirming the adaptive significance of integration in clonal organisms. Modular archaeocyaths also survived beyond the extinction of most solitary forms, suggesting a relative immunity to extinction because of a differential ecological response.

6.5 Summary

Notwithstanding the great diversity of invertebrates that have contributed to reef-building over the course of the Phanerozoic, there appear to be a limited number of functional organizations that have evolved repeatedly and independently with the appearance of each new group. This suggests that variations in life history according to growth form are profound.

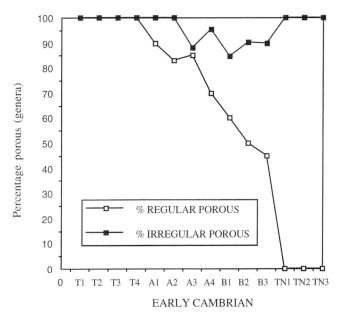

Fig. 6.32 Number of Regular and Irregular archaeocyath genera with porous septa per zone as a percentage of the total genera. Compare with the changes in the proportion of modular forms over the same time period (Fig. 6.15). T, Tommotian; A, Atdabanian; B, Botoman; TN, Toyonian. % = 138.49–8.7331 (time); r = 0.886. (From Wood *et al.* 1992a, by kind permission of SEPM Society for Sedimentary Geology.)

The processes controlling the distribution of ancient benthic organisms appear to have been remarkably similar to those observed today (Jackson 1983). Suitable substrate for colonization has always been a limited resource, and those organisms with either encrusting modular forms or densely aggregating growth have always been most successful at its acquisition. Metazoan reef communities have been differentiated into open-surface and cryptic communities since their first appearance; this strongly suggests that habitat selection for small refuges has been common since at least the Lower Cambrian. (Indeed, it seems likely that such a differentiation also took place in Proterozoic cyanobacterial and microbial reef communities).

However, statistically significant long-term trends (from 10 to several 100 Myr) are apparent in the history of many groups of reef-builders. An increase in the proportion of forms which show a modular organization is a widespread phenomenon, as is an increase in the degree of integration accompanied by soft-tissue flexibility. All these trends of increasing integration are polyphyletic, and suggest a general premium on improving physiological interaction between modules regardless of systematic placing (Jackson and McKinney 1991). However, many of these trends involve increases in variance, as is clear from the long-term persistence of solitary and low-integration morphologies. So in addition to the improved physiological interaction between modules, there has also been selection for increased specialization.

Other trends involve progressive shifts in the proportion of a given morphological form or trait, e.g. the development of multilayered encrusting bryozoans, or increasing rigidity in erect bilaminate bryozoans. Many of these trends are also polyphyletic, for example changes in the proportions of erect branching forms, their branch dimensions, and articulation, involve nearly all major bryozoan taxa.

But the reef landscape changed greatly during the course of Phanerozoic. The majority of Palaeozoic reef metazoans lived either rooted within or were attached to small skeletal debris and then grew over unconsolidated substrate. Multiserial, laminar forms were able rapidly to colonize such substrates, achieve large sizes (and possibly relatively long life spans) and had impressive powers of regeneration from partial mortality. But it is difficult to envisage such organisms living in the highly turbulent, surf zone which reef corals and coralline algae occupy today, as they lacked the means to gain secure and permanent attachment to a hard substrate. It is therefore not surprising that evidence for mutual encrustation by these organisms in reefs is lacking, and that the highly energetic areas of tropical marine environments may have been occupied by microbial and calcified cyanobacterial communities which were capable of forming structures of great rigidity and wave-resistance.

Before the Cenozoic, there was also an abundance of small, erect solitary and branching organizations occupying open, soft substrates. During the Palaeozoic in particular, tropical shallow level-bottoms often supported enormous populations of erect bryozoans and stromatoporoids: habitats which in the Recent show no apparent open-surface community, only a diverse burrowing one (Thayer 1983). Small cup-shaped, stalked and branching organisms do not use space to a maximum, which suggests that they would not have been good competitors where available substrate was at a premium. For example, archaeocyaths never occupied more than 50% of the total rock volume, in contrast to the dense populations achieved by many later, high-integration reef-builders. Indeed, many solitary calcified sponges (archaeocyaths and sphinctozoans) and rugose corals were often restricted to inter-biohermal areas or to cryptic niches, where they could form aggregations. There are therefore very few living analogies for the ecology of aggregating solitary forms in normal marine environments; most are now restricted to marginal settings.

Flexibility of growth form enables a variety of responses to changes in extrinsic environmental factors: fixed growth forms, such as rigid, branching growth, reduce the flexibility of response so reducing ecological options to often only one life strategy. It is clear that, especially after the Jurassic, clades with inflexible morphologies were outcompeted by flexible, encrusting forms on the open surface, and became reduced in diversity and often restricted to cryptic or off-shore niches (e.g. Jackson *et al.* 1971; Vermeij 1987; Jackson and McKinney 1991). During the Cretaceous, cheilostomes largely replaced cyclostomes, and narrowly branched or articulated erect bryozoans became more diverse and abundant than broadly branched or rigid forms, which became restricted to cryptic and deep-water habitats. Indeed, Mesozoic bryozoans occupied significantly higher latitudes than their tropical Palaeozoic counterparts (P.D. Taylor and Allison 1998). Encrusting species of bryozoans (with new budding styles), sponges, and coralline algae became more abundant and diverse than erect forms; and mobile free-living corals (see Fig. 6.23(h)) and bryozoans appeared. By the Late Cretaceous, diverse assemblages of large, erect, rapidly growing corals, gorgonians, and soft-bodied sponges had become common in reefs (Jackson and McKinney 1991).

Arguably, the most profound innovation in growth form and function was the evolution of an edge zone in scleractinian corals, which conferred the ability to secure permanent, encrusting attachment to extensive hard substrates. This can be hypothesized to have resulted in a shift from tropical, shallow marine reef communities being constructed upon relatively stable, but soft, substrates during the Palaeozoic, to those requiring hard substrates to initiate and perpetuate growth as in modern coral reefs which occupy high-energy environments. This shift also highlights a fundamental change in the community dynamics of reef growth. Many mid-Palaeozoic, soft-substrate reef communities show low diversities, often dominated by only a few species of reef-builders. This observation—together with the lack of common, direct, interspecific interactions (but possibly clonal fusion in groups such as fenestellid bryozoans)—suggests that the mechanisms responsible for the growth of many mid-Palaeozoic reef communities was determined more by rates of recruitment, larval aggregation, and habitat selection than by the complex postrecruitment competitive interactions and regeneration from fragments known to structure modern coral reefs (Section 6.1.6).

Secure attachment to a substrate also allowed the development of large, erect, branching morphologies. This may have enabled growth in highly energetic and disturbed regimes. The skeletons of scleractinian corals are highly porous, so meeting the mechanical requirements for skeletal strength: high growth and calcification rates are also promoted by photosymbiosis. Scleractinian corals also show a far higher proportion of highly integrated species than Palaeozoic corals, suggesting that their competitive powers and regeneration are superior.

Predictions that clonal organisms might show slower rates of speciation than sexual clades was tested by Coates and Jackson (1986). They found no statistically significant differences between aclonal and clonal species of 1381 species of scleractinian corals from the mid-Triassic to the Pleistocene. However, differences were highly significant when growth forms were considered separately, with erect multiserial species showing faster rates of evolution that all other forms (8.4 Myr compared to a mean of 9.45 Myr in other growth forms). This might be explained by the fact that erect growth forms are more specialized than others, and indeed we have seen that the rise in erect growth forms is essentially a latest Mesozoic and Tertiary phenomena.

What caused these profound changes? There is no evidence to suggest that changes in climate, sea level, or other aspects of the physicochemical environment have varied in a similar way to these major trends. Indeed, the only extrinsic events which substantially affected trends were those factors responsible for the major Late-Devonian and end Permian mass extinctions. Notably, trends appear to be independent of other extinction events and some are also remarkably linear. They can therefore best be explained by the progressive adaptation to routine physical processes and biological interactions, perhaps by ecologically based species-selection acting as a competitive process between species whereby higher extinction rates are sustained by forms with more 'primitive' characteristics (Jackson and McKinney 1991).

However, while physical processes have presumably remained relatively constant throughout the Phanerozoic, the nature and intensity of biological interactions have not. Many have suggested that the ability of organisms to bore, gouge, or scrape has increased dramatically since the mid-Jurassic, such that the armament which provided ample protection against Palaeozoic predators had proved inadequate by the Cretaceous (e.g. Vermeij 1983). If such a scenario is correct, then we would predict an increase in features adapted to withstand increased predation and herbivory. This hypothesis will be explored in the following chapter.

7 The rise of biological disturbance

The previous chapter illustrated how differences in the ability to withstand disturbance has been an intense selective force in the evolution of reef communities. But while there is no evidence that routine physical processes have changed significantly over the Phanerozoic, there are now considerable data to suggest that biologically induced disturbance has increased dramatically in shallow marine seas, especially since the Mesozoic (e.g. Papp *et al.* 1947; Vermeij 1977, 1987). This protracted event is known as the Mesozoic Marine Revolution (or MMR; Vermeij 1977) and involved the origin and diversification of many groups of *bioturbators* (organisms that disturb unlithified sediment), predators (those that consume either whole or parts of other organisms), and bioeroders (which penetrate or sculpt hard substrates to access prey or create dwelling places). The differential effects of the MMR are extremely difficult to isolate as many organisms cause disturbance in more than one way: some predators are also bioturbators; others are capable of significant bioerosion. The MMR thus involves coincident developments which might be predicted to have many common evolutionary consequences.

When prey possess traits that are proven to be effective against predation (and its subset herbivory), so promoting survival and reproduction, then predation can act as a powerful agent in evolution by encouraging the selection of antipredatory characteristics (Vermeij 1983). Not surprisingly, predation has been strongly implicated in the evolutionary selection of a vast array of morphological, physiological, and behavioural defence mechanisms. For example, the beginning of the Mesozoic saw a spectacular rise in the molluscivorous habit—which included the development of extravagant external ornamentation in gastropods, and the ability in bivalves to burrow, and to bore into, and cement to, hard substrates—all habits that have been suggested to be defensive adaptations evolved in response to increased predation pressure (S.M. Stanley 1977; Vermeij 1977; Harper 1991).

Avoidance of predation is of critical importance to any organism, but epifaunal, immobile organisms such as reef-builders are particularly vulnerable, and we have seen that modern reefs grow in particular environments where the avoidance of competition and disturbance are of prime importance. In particular, the inhabitants of shallow tropical coral reefs must be able to withstand considerable predation. The need for photosynthetic organisms to expose large areas of soft tissue to light increases the risk of both predation and fouling, and the control exerted by herbivores, particular fishes, in limiting the distribution and abundance of algae is probably crucial to the survival of coral reefs. As a result, many coral reef organisms are proposed to show a vast array of supposed antipredation mechanisms, yet details of their evolutionary origin and development are poorly known. But as many of the major groups of bioturbators and predators associated with modern coral reefs did not arise until the late Mesozoic/early Cenozoic this begs the question—what effect did their appearance have on reef ecosystems?

This chapter aims to explore the role of biological disturbance in the history of reef-building. As the basic organization of many reef-building organisms—being predominantly algae and lower invertebrates—is extremely simple, it might be possible to interpret any post-Palaeozoic changes in morphology or life habit as defensive. The central question then is to determine the importance of biological disturbance relative to other processes—such as physical disturbance, variable recruitment, and competition—and so evaluate the extent to which escalating biological disturbance is responsible for the morphological trends outlined in Chapter 6. Sometimes, supposed benefits may be in conflict, such that the outcome is a trade-off for many organisms. For example, the cost of investing in increased defence may be a reduction in growth rate or competitive ability.

Supposed beneficial traits (*aptations*) have been subdivided into two categories (Gould and Vrba 1982): *adaptations,* those which benefit a specific function or effect that has been enhanced by natural selection, and *exaptations*, whose benefits are secondary or incidental to the primary function to which they are adapted. Given that many traits are beneficial in a number of unrelated ways, which may also differ from those for which they evolved, most are likely to be exaptations. Correct identification of an adaptation must therefore be supported by observational and experimental proof that a feature is both beneficial in the respect proposed and also that there is a temporal coincidence between the appearance of the trait and the supposed predator (Harper 1994). In contrast, exaptations should confer notable selective advantage to so-endowed individuals on the appearance of the proposed selective agents.

These strict requirements are not easily satisfied (Harper 1994). Although it may be possible to demonstrate experimentally that a particular trait confers selective advantage against a given predator, it may be highly problematic to tie its evolutionary appearance to that particular threat. Predictions may be difficult to test, as many agents of mortality are not independent and indeed may occur simultaneously: in the marine environment, all potential threats (biological and physical) broadly decrease with increasing depth. Moreover, the ecological impacts which were important selective influences in the past may no longer operate today (Connell 1980). Identification of aptations is further complicated by the fact that predator–prey interactions do not often enter into tight coevolution because predators often prey upon several species, that is they are *generalists*. A given predator will not be universally similar in its effects, and certain prey taxa may be limited to or channelled towards particular defensive strategies by the morphological constraints of their own body plans (Harper and Skelton 1993). For example, while a toxic animal might evolve warning coloration in response to a threat from predators, a palatable species would become cryptic (Harvey and Greenwood 1978); a large sessile organism might evolve passive constructional defences, whereas a small, mobile organism may flee by developing the ability to burrow.

Crucial to the analysis of adaptation is to prove the existence of independent evolutionary events. This relies on an understanding of the phylogeny of a group, and a knowledge of which characteristics are primitive and which derived. If the same trait can be shown to have appeared independently in several evolving clades (that is it was derived repeatedly from the presumed more primitive state), then explanations based on an extrinsic cue, rather than shared ancestry, are likely to apply. Where similar traits have been acquired polyphyletically over a very short space of geological time is compelling evidence for an extrinsic selective force (Skelton 1991).

The focus of this chapter will be on the evolutionary changes in the processes of predation (and herbivory), and bioturbation, that characterize the modern coral reef ecosystem. The first section (Section 7.1) discusses the roles played by modern agents of biological disturbance, and Section 7.2 outlines their fossil record. Section 7.3 details supposed antipredator traits in major groups of reef-builders, which are then explored in relation to the rise of herbivores and predators. From this, we may hope to identify the relative effects of predation as a selective force responsible for many of the patterns and processes observed on reefs today (Section 7.4).

7.1 Biological disturbance on modern coral reefs

Coral reefs are host to a vast array of predators and herbivores with greatly varying mobilities, capabilities, and degrees of specialization (Fig. 7.1). The most abundant predators on reefs are the filter- and suspension-feeders, and planktivorous fish (the so-called 'wall of mouths'), which remove up to 60% of the biomass of zooplankton from the waters streaming across a reef (Glynn 1973). In particular, these predators may have a great impact upon the mortality of larvae, such that mechanisms which improve survivorship at this stage may represent important aptations. These include habitat or refuge selection, the behaviour and timing of settlement, and the release of toxins.

Coral reefs are, however, also characterized by diverse predators and herbivores, and borers, which prey upon or otherwise attack sessile organisms (Table 7.1). Of these, arguably the most diverse are reef fishes (see Box 7.1).

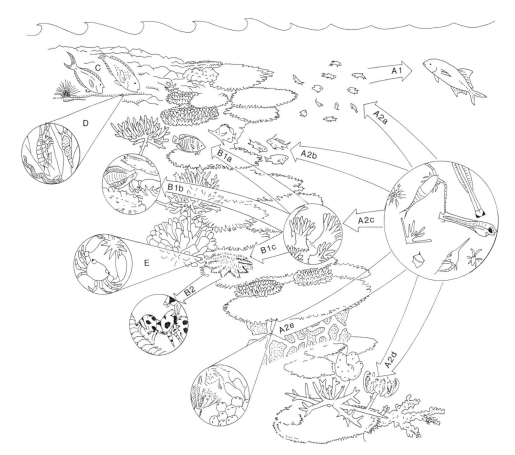

Fig. 7.1 Major predators and herbivores on modern coral reefs with inferred trophic pathways. Zooplankton (including larvae) enters the reef front and is preyed upon by consumers including planktivorous fishes (A2a—'wall of mouths'), benthic fishes (A2b), corals (A2c), and suspension feeders on open surfaces (A2d) and in reef cavities (A2e). Planktivores may be preyed upon by piscivores (A1) and corallivores (B1a–c). Herbivory (C) may involve indirect predation (D), and free-living symbionts that derive food from their coral hosts (E). Not to scale. (Redrawn from Glynn 1988.)

Table 7.1 Major groups of predators, herbivores, and bioeroders on modern coral reefs

Group	Ecology
Porifera	
Clionidae*	Borers
Mollusca	
Polyplacophora	Herbivores (scraping)
Gastropoda:	
—Archaeogastropods + mesogastropods (excluding limpets)	Herbivores (non-denuding and denuding)
—Patellacea	Herbivores (scraping)
—Nudibranchs	Corallivores
—Prosobranchs	Corallivores
Bivalvia	
—Lithophagidae	Borers
Echinodermata	
Echinodea:	
—Diadematoida*	Herbivores (excavating) and corallivores
—Arbacioida	Herbivores (excavating)
—Temnopleuroida	Herbivores (denuding)
—Echinoida	Herbivores (excavating)
Asteroidea	Corallivores
Arthropoda	
—Isopoda	Herbivores (non-denuding)
—Amphipoda	Herbivores (non-denuding)
—Decapoda	Corallivores
Annelida	
Polychaeta	
—Nereidae, Eucicidae, Dorvillidae	Herbivores (non-denuding) and corallivores
Pisces	
Perciformes:	
—Labridae	Carnivores
—Scaridae (79)*	Herbivores (excavating)
—Acanthuridae (76)*	Herbivores (denuding)
—Pomacentridae (~300)*	Herbivores (non-denuding and denuding) and corallivores
—Chaetodontidae	Herbivores (non-denuding) and corallivores
—Blennidae	Herbivores (denuding)
—Kyphosidae	Herbivores (denuding)
Tetraodontiformes:	
—Balistidae	Herbivores (non-denuding) and corallivores
—Monacanthidae	Herbivores (non-denuding) and corallivores

* Indicates most important groups. Species' numbers in brackets, after Choat 1991.

Box 7.1: **Coral reef fishes**

Reef fishes form a circumtropical assemblage with characteristic morphologies and ecologies. Many reef fish recruit directly onto reefs and remain within very specific habitats during their entire lives.

The most striking feature of coral reef fishes is their diversity: although the greatest diversity is developed in relatively few taxa, an estimated 3000 species of fishes live associated with coral reefs in the Indo-Pacific alone (Springer 1982), representing 18% of all living fishes. Most are advanced perciform teleosts: perciforms comprise 86% of the 20 most speciose families, and are overwhelmingly the most abundant individuals on reefs.

CORALLIVORES

Chaetodontidae
(butterfly fishes)

Balistidae
(triggerfishes)

Tetraodontidae
(puffers)

HERBIVORES

Pomacentridae
(damselfishes)

Scaridae
(parrotfishes)

Acanthuridae
(surgeonfishes)

Siganidae
(rabbitfishes)

Fig. Box 7.1 Main families of larger bodied modern reef fishes. (Redrawn from Nelson 1984.)

Over 100 families of fishes have coral reef representatives, but of these only a small subset exploit the sessile biota of a reef (Choat and Bellwood 1991):

(1) the chaetodontoids, comprising the families Chaetodontidae (butterflyfish) and Pomacanthidae (angelfish);

(2) the acanthuroids, comprising the Acanthuridae (surgeonfish), Siganidae (rabbitfish), and Zanclidae (moorish idols);

(3) the labrids, comprising the Scaridae (parrotfish); and

(4) the Pomacentridae (damselfish) and Labridae (wrasses).

Scarids and acanthuroids are largely restricted to coral reefs: most abundant are the labrids and pomacentrids.

Reef fishes are morphologically diverse, especially in terms of size and the variety of highly specialized feeding structures. For example, the smallest member of the Labridae has a standard length of only 30 mm, whereas the largest is nearly 2300 mm long (Choat and Bellwood 1991). Assemblages are known which comprise complexes of ecologically similar species co-occurring within localized areas, especially forms feeding on sessile benthos and small mobile invertebrates. Estimates suggest that between 27 and 56% of any reef fish community are benthic invertebrate predators, 7–26% are herbivores, and 4–20% are planktivores (G.P. Jones *et al.* 1991). Reef fish have very well-developed vision upon which they rely heavily not only for feeding but also for reproductive and ecological interactions. As a result, reef fishes invariably live in clear waters such as oligotrophic settings that allow complex behaviours to develop which are often associated with distinctive colour patterns.

Biological disturbance shows marked differences in distribution and intensity across a reef profile. Like physical disturbance, both predation and bioturbation decrease with depth, being generally greatest at shallow depths (from the lower intertidal zone to about 20 m), particularly on reef slopes with substrates of high topographic complexity (Hay 1984*a*). Herbivory is low above mean low water, usually reaches a peak at a depth of 1–5 m on forereefs, and then declines rapidly with depth (Steneck 1988). Planktivorous fish, though, may be numerous in very deep reefs. Regardless of depth, however, the effects of biological disturbance may be patchy and vary according to local environmental differences.

The factors responsible for the presence of so many predators on coral reefs are still unclear. It is likely that dietary plasticity, and the ability to exploit a wide range of prey, contribute to the success of many herbivores and predators associated with reefs. Niche diversification may also be a factor: predatory gastropods, for example, appear to respond to interspecific competition by niche contraction, thereby allowing the coexistence of more species.

7.1.1 Herbivores

The tropics are characterized by very diverse, but often small, seaweeds, exhibiting a great range of characteristics. These can be divided into three major functional categories:

● widespread and ubiquitous green *algal turfs* comprising often diverse communities of often microfilamentous species less than 10 mm in height. Numerous microscope invertebrates and protozoans live within the turfs, which also accumulate organic detritus. Algal turfs are among the most productive plant communities known, and have high growth rates and short life cycles that enable successful growth under conditions of frequent disturbance.

- *crustose algae* consisting of heavily calcified species which can withstand considerable physical and biological disturbance. This includes members of the Chlorophyta, Rhodophyta, and Phaeophyta.

- *macroalgae* (those greater than 10 mm in height), which are more rigid and anatomically complex than algal turf species. These are usually relatively uncommon components on coral reefs with restricted and patchy distributions.

Algae rapidly colonize bare substrates and exposed soft tissue on coral reefs, especially filamentous algal turfs. In the Caribbean, filamentous algae are estimated to produce an impressive $700 \, \text{g C m}^{-2} \, \text{year}^{-1}$ and macroalgae $1170 \, \text{g C m}^{-2} \, \text{year}^{-1}$, compared with $630 \, \text{g C m}^{-2} \, \text{year}^{-1}$ for corals (R.S.K. Barnes and Hughes 1988). Tropical intertidal and subtidal communities, however, are characterized by remarkably low-standing crops of algae, as these generally remain inconspicuous due to constant and intense grazing by herbivores. Some have estimated that between 50 and 100% of algal production is consumed by fish on shallow coral reefs (see discussion in Hay 1991).

A high diversity of herbivores live on coral reefs (for a summary see Steneck 1983), with varying degrees of mobility. These include *microherbivores* (tanaid isopods, amphipods, crabs, syllid polychaetes, and small gastropods) and *macroherbivores* (gastropods, echinoids, and fish). Two broad types of herbivorous feeding behaviour have been identified on coral reefs which influence prey populations and substrata in markedly different ways (Table 7.1):

- *browsers*, which consume plant tissues above substrates. These can have either a denuding or non-denuding affect. The most important denuders are some molluscs, some echinoids, acanthuroids, rabbitfish (siganids), and damselfish (pomacentrids).

- *grazers*, which crop very close to a substrate, so ingesting substantial portions of living plant tissue, associated small invertebrates and underlying skeleton, as well as algae. Many grazers are also capable of removing and ingesting calcareous material such as coralline algal skeleton and underlying substrate (see Fig. 7.2). These include *excavators* which are capable of deep excavation that removes large areas of substrate, and *scrapers* which have weaker jaw apparatuses, and take smaller bite sizes with limited substrate removal. The most important excavators and scrapers are limpets, chitons, some regular echinoids, and acanthuroids and scarids.

Despite this high diversity, relatively few groups of reef herbivores are quantitatively important. Of these scarids, acanthuroids, and urchins are commonly the most abundant (by mass) and have been observed to have the greatest impact on the distribution and abundance of benthic algae (Steneck 1988). But while invertebrate herbivores—mainly echinoids and gastropods—are often equally abundant at low and high latitudes (Sammarco *et al.* 1974), coral reefs are particularly characterized by the great abundance of grazing herbivorous fish, in marked contrast to shallow, benthic communities in temperate and high-nutrient tropical waters (see Box 7.2). However, in the Caribbean, both echinoids and fish are important, whereas on Indo-Pacific and Australian reefs, fish are the dominant herbivores (P.W. Sale 1980). Such geographic variation in distribution suggests the importance of historical biogeographic factors (Section 7.2.1).

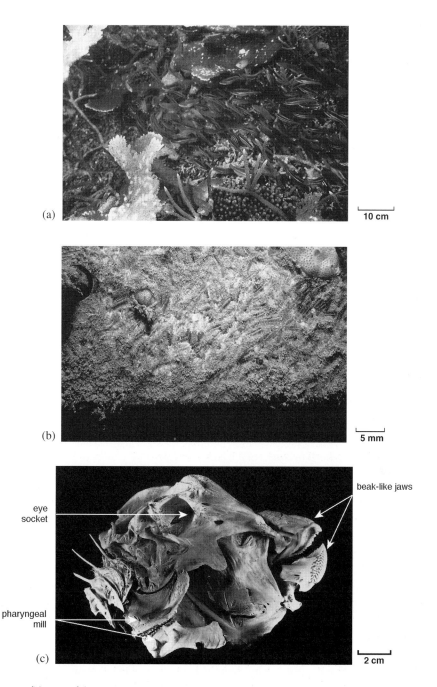

Fig. 7.2 Parrotfish—one of the major excavators on modern coral reefs. (a) Small parrotfish (*Scarus iserti*) feeding on algae. (Photograph: C. Birkeland.) (b) Heavily grazed artificial panal showing characteristic scraping traces of the Caribbean parrotfish *Scarus iserti*. Note how the tiny (~ 2 mm) coral (top left) was missed. (Photograph: C. Birkeland.) (c) Parrotfish remove large chunks of reef material using modified beak-like jaws, which is processed to fine sediment by a second set of highly modified jaws known as the pharyngeal mill. Powerful muscles pull the lower jaw upwards, crushing or shredding material between opposing teeth. (Photograph: D.R. Bellwood.)

Box 7.2: **Herbivorous reef fishes**

Herbivorous reef fishes are among the most abundant and widespread groups of vertebrate herbivores, and show characteristic structures and morphologies adapted for slow-swimming, precise movement and rapid, non-selective feeding (Fig. 7.2(a)). These include lateral compression and high body planes, an increase in the proportion of locomotory and feeding musculature as a proportion of body weight, and elaboration of the jaws and pharyngeal apparatus for powerful and efficient feeding mechanisms. This impression of uniformity, however, belies profound differences in food-processing mechanisms, including alimentary tracts and digestive regimes, which allow the targeting of particular plant food resources (Choat 1991). Many intrafamily differences exist, which makes the placing of herbivores in trophic groups problematic.

Some scarids have highly modified and innovative development of the teeth and jaw elements which allows them to excavate (Fig. 7.2(b)). Herbivory in the Scaridae is based on the development of the pharyngeal mill, its associated musculature, and a highly modified alimentary tract which allows non-selective feeding but yields rapid growth rates. The mill mechanically reduces calcareous material (Fig. 7.2(c)), and digestion is aided by a sacculate intestine with high pH. This allows scarids to utilize algal turfs which have high production rates but low-standing crops and low nitrogen value. It also explains why fish with grazing modes similar to acanthuroids, scarids, and siganids are unable to maintain populations on temperate reefs, as the turf assemblages there have insufficiently high productivities to obtain enough nitrogen to maintain their high metabolic rates. Algal turfs in temperate areas are dominated by invertebrate grazers with lower metabolic requirements. The long association of herbivorous fish with coral reefs and associated environments—notwithstanding ample opportunity to diversify into other environments—may be explained by the high productivity of turfs associated with coral reefs. This emphasizes the need for such fish to process large amounts of material: indeed, Hatcher (1981) has estimated that herbivorous fish on the Great Barrier Reef (GBR) consume up to ten times more carbon than is needed to meet their basic metabolic requirements. This alone is testament to the large impact of fish upon reef algae.

An exception are damselfishes (pomacentrids), which appear to be more selective, accessing microalgae of higher nutritional quality from algal turfs, or from detrital material within turfs. Pomacentrids show small body size and strict territoriality, and exert a strong local influence over algal populations (see Section 7.1.6).

Herbivorous fish feeding may also retain nutrients within the reef community via defecation (Polunin 1988). This has also been noted in planktivorous fish (Robertson 1982).

7.1.2 Carnivores

Benthic invertebrate predators include coral polyp feeders (*corallivores*), sessile invertebrate feeders, and those that consume mobile invertebrates. The latter group are far more numerous than other categories, with benthic crustaceans being the most substantial prey item for carnivorous reef fishes.

Corallivores include crustaceans (hermit crabs), polychaetes (amphinomids), gastropods (prosobranchs and nudibranchs), echinoids (diadematoids), starfish (Fig. 7.3), and numerous fishes (e.g. butterflyfish, damselfish, parrotfish, triggerfish (Balistidae), and pufferfish (Tetraodontidae).

5 cm

Fig. 7.3 Crown-of-thorns starfish (*Acanthaster plancii*) feeding on the branching coral *Acropora*. (Photograph: R. Steene.)

7.1.3 Bioeroders

Many organisms associated with coral reefs either indirectly or directly attack calcareous substrates by mechanical or chemical means (Fig. 7.4). Predators which feed directly upon sessile invertebrates or algae by etching, rasping, or scraping and so causing incidental skeletal or substrate damage are known as *epiliths*. Organisms which nestle into depressions, sometimes on living sessile organisms so deforming growth of the host, are known as *chasmoliths*, and those which live partly or wholly within substrates are known as *endoliths*. Endoliths include *etchers* (e.g. bacteria, fungi, and endolithic algae) and non-predatory *borers* (e.g. clionid sponges, bivalves, spionid polychaetes, and siphunculans). Some boring bivalves, sponges, and barnacles are capable of excavating into live corals, and borers also include mobile predators such as natacid and muricid gastropods which produce boreholes as a means of access to prey. Of these, the clionid sponges are among the most common and destructive endolithic borers on coral reefs globally, although polychaetes can be locally important, and lithophagid (mytilid) bivalves are common in the intertidal zone.

Living and dead corals on some reefs can carry massive infestations of endolithic organisms. These can severely weaken the skeletons and may lead to the eventual death of the coral. Although most bioeroders are small and well concealed, some have estimated that the biomass of endolithic organisms can equal, or even exceed, that of the surface biota on coral reefs (Grassle 1973). Surfaces that show intense bioerosion usually form only when rates of sedimentation are low.

The endolithic habit is relatively protected from most predators (although some tetraodontid and balistid fish will break off branches of living coral in order to feed on endolithic organisms). Consequently, many endoliths have reduced skeletal defences compared to their epifaunal ancestors and lack the armature, spines, and thick shells of open-surface dwellers.

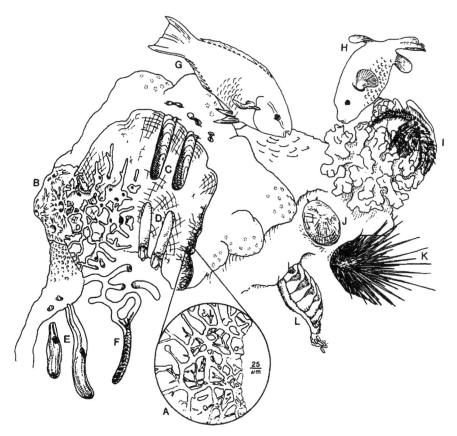

Fig. 7.4 Bioeroders commonly associated with modern coral reefs. Endoliths: A, Etchers—algae, fungi, bacteria; B–F, Borers—B, sponges (Clionidae); C, bivalves (*Lithophaga*); D, barnacles (*Lithotrya*); E, sipunculans (*Aspidosiphon*); F, polychaetes (Eunicidae). Epiliths: G, parrotfish (Scaridae); H, pufferfish (*Arothron*); I, hermit crab (*Aniculus*); J, limpet (*Acmaea*); K, urchin (*Diadema*); L, chiton (*Acanthopleura*). Not drawn to scale. (Redrawn from Glynn 1997*b*.)

7.1.4 Bioturbators

Many organisms contribute to the bioturbation of reef-associated sediments—including numerous bivalves, worms, crustaceans, and fishes—through sediment excavation, burrowing, or feeding activities. Representatives from 15 families of small teleosts and elasmobranchs (in particular rays), feed over carbonate sand or seagrass beds adjacent to reefs, but shelter within the reef itself at night (Suchanek and Colin 1986).

7.1.5 Importance of herbivores

Herbivores serve three main functions on coral reefs:

(1) they result in a higher, overall ecosystem primary-productivity;

(2) they facilitate the flow of energy to higher levels in the trophic web;

(3) intense herbivory leads to the predominance of well-defended sessile organisms, particularly those with calcareous skeletons.

Numerous field studies have demonstrated that herbivores limit the growth of marine plants on coral reefs. Reef herbivores are not evenly distributed on a reef, and differences in their size, abundance, and feeding mechanism can explain the distribution of algae with different life histories and morphologies across a reef profile. Many reef herbivores do not appear to be specialists, but select algae according to availability and abundance. Such behaviour allows potentially dominant species to be consumed such that they do not proliferate at the expense of subordinate species, which can survive only if recruitment and growth rates are higher than the rate of removal. Frequent cropping also provides new space for the invasion of opportunistic species.

The crucial relationship is between how much seaweed is eaten and the rate of production at any given site (Hay 1991). The high levels of excavatory grazing in the shallow parts of the reef slope, particularly by fish and urchins, leads to the apparent dominance of plant species which are resistant or tolerant to grazing. These forms are either relatively slow-growing but heavily skeletonized and long-lived, such as coralline algae, or those that possess repellent secondary metabolites or tough rubbery textures. But in such areas, the production rates of algal turf are very high, and even though these forms are highly susceptible to grazing their rapid growth rates allow regrowth from basal portions that have escaped herbivores due to the topographic complexity of their substrates. They are, therefore able to persist despite huge losses to grazers (S.M. Lewis 1986). Areas of intense herbivory are thus characterized by forms that are either very resistant—producing high-standing stocks which are conspicuous—or those which are highly tolerant and inconspicuous (i.e. with low-standing stocks), but nevertheless, highly productive. Both forms depend on herbivorous fish to prevent their exclusion by macroalgae.

In contrast, soft-sediment aprons at the base of reef slopes are colonized by a wide variety of fleshy macroalgae: most macroherbivores are absent from these sediment plains. That macroherbivores are responsible for this distribution has been confirmed by transplant experiments, which have shown that sand-plain species can have dramatically high success on the shallow reef slope if fish are excluded (Littler and Littler 1984). However, there is also some suggestion that in deep areas, the rate at which algal growth decreases in response to reduced illumination may in fact exceed the rate of herbivore consumption, so forming a refuge for macroalgae (e.g. Hay 1981, 1984a).

In the Caribbean, reef flats also often display abundant and diverse populations of macroalgae. Several studies have indicated that this habitat is relatively inaccessible to macroconsumers so forming a refuge for algal growth: reef-flat macroalgae translocated to reef-front habitats are rapidly consumed by herbivores (e.g. Hay 1981; Hay *et al.* 1983); and reef flats in the eastern Pacific, which are subject to significant tidal incursions allowing the invasion of large herbivorous fish, also have low macroalgae populations (Gaines and Lubchenco 1982). Further observations indicate that poorly defended plants are restricted to various other refuges, such as lagoons, reef crests, and cryptic niches (e.g. Hay 1981; Hay *et al.* 1983), to areas where herbivores have been removed, or to nutrient-enriched or turbid coastal waters. Indeed, macroalgae dominate in many tropical habitats where herbivory is reduced. Many of these may represent spatial refuges that are very close to areas of high herbivore pressure, so forming a mosaic of habitats that leads to high species diversity within small geographical areas.

However, the relationship between algal distribution and macroherbivore abundance may not always be straightforward, in that many herbivorous reef fish also inadvertently feed on the invertebrates and organic detritus that occupy algal turf. First, the exclusion of reef fish —either naturally or experimentally—may secondarily enhance herbivory by micrograzers such as amphipods. (Also, grazing by amphipods can induce an increase in the concentration of defensive secondary metabolites in brown algae; the algae can thus actively alter their susceptibly to herbivores through the lifetime of the genome (Cronin and Hay 1997).) Second, the low abundance of sessile invertebrates on open reef surfaces may, in part, be due to incidental damage caused by grazing herbivores (Day 1983). Juvenile corals are particularly vulnerable to such activity, until they are large enough to escape (see Figure 7.2(b); Birkeland 1977).

Differing food preferences, abundances, and activities of herbivores have also been used to explain both global and regional differences in the distribution of marine algae (Gaines and Lubchenco 1982). But whilst these described changes in algal assemblages can be plausibly attributed to concomitant changes in herbivore activity, they may also be associated with changes in physical conditions (both geographically and within individual reef environments) such as differences in turbidity, nutrient levels, temperature, and turbulence, whose individual effects are difficult to isolate.

Herbivory and the dominance of corals and coralline algae on reefs

Of particular importance for reef growth is the observation that both encrusting coralline algae and corals rely on grazers to prevent their overgrowth in shallow tropical waters. Many experiments that have artificially excluded large herbivores, particularly fish, from living reef surfaces have demonstrated that under such conditions species typical of heavily grazed communities (coralline algae and corals) are often excluded apparently as a result of competition—they become smothered with macroalgae and frequently die. In one study on a Caribbean back-reef, macroalgal cover increased from 2 to 30% in 10 weeks while control areas showed no change: when the herbivore exclusion cages were removed, all macroalgae were consumed in 48 h (S.M. Lewis 1985). This relationship has been confirmed on longer timescales, as the recent Caribbean-wide decline of the grazing echinoid *Diadema antillarum* has caused a tremendous proliferation of sheet, filamentous, and frondose algae resulting in subsequent coral death. Where human overexploitation of gastropods, fish, or urchins occurs, the normally limited quantities of frondose macroalgal stocks can also become extensive.

The presence of abundant coralline algae and corals in a shallow reef community is therefore indicative of moderate or intense levels of herbivory. In reef habitats <20 m deep, these two components may cover in excess of 80% of the substratum (Carpenter 1986).

Coralline algal reef formation

Algal ridges are shallow-water crests built almost entirely by crustose coralline algae. They represent a late successional stage in reef development forming a constructional cap over coral as the reef approaches sea level, and can vary from 1 to 8.5 m in thickness (Adey 1975).

The formation of algal ridges is thought to be related to grazing activity. Highly destructive parrotfish-grazing limits the abundance of coralline algae, such that algal ridges dominated by thick layers of algae with thick, robust morphologies (e.g. *Porolithon*) form as a result of severe wave shock (which excludes most corals and parrotfish, and minimizes sediment accumulation) and intense limpet grazing (Adey 1975; Littler and Doty 1975). The construction of algal ridges by delicate, branching coralline algae (*Neogoniolithum strictum*; Figs 7.5(a), (b)) is only possible in areas with low levels of herbivory and wave energy, but, unusually, this species can also withstand high sedimentation rates due to the presence of multiple cell fusions which may act as conduits for photosynthate translocation (Steneck *et al.* 1997).

(a)

1 cm

(b)

Fig. 7.5 (a) *Neogoniolithum strictum*. (b) An algal ridge formed by *Neogoniolithum strictum,* from Exuma Cays, Bahamas. This particular ridge is atypical because it is formed by branching coralline algae, and grows in areas of relatively low energy and rates of herbivory. (Photograph: R.S. Steneck.)

Free-living crustose algae, such as rhodoliths, are known from modern coral reefs in areas subject to moderate levels of wave action or currents but reduced herbivory.

Coralline algal reefs in Arctic settings, which produce remarkably high-standing stocks, also rely upon complex, biological feedback processes. These involve the attraction of specific mobile herbivores in order to reduce the otherwise intense fouling pressure by encrusting foraminifera, diatoms, filamentous algae, and the larvae of sessile invertebrates. Many of the herbivore larvae require living coralline algae to stimulate metamorphosis (Freiwald 1993).

7.1.6 The role of carnivores

Many coral reef carnivores have flexible diets according to the availability of prey. For example, when abundant, the main prey of the triggerfish *Balistes vetula* is *Diadema antillarum*, but following the mass mortality of this urchin *B. vetula* switched to a diet of crabs and chitons (Reinthal *et al.* 1984). This observation suggests that the ecological impact of carnivores may be influenced by changes in the relative abundance of potential prey.

Except for a few species, the direct effects of corallivores are not thought to affect coral populations as much as the indirect effects of grazers (although mortality from physical disturbance can be exacerbated by secondary predation (Knowlton *et al.* 1981)). Many corallivores produce free-space and influence community structure, but the extent of such effects remains largely unknown. Most corallivores have highly specialized diets, for example *Acanthaster* shows a marked preference for *Acropora* species (Fig. 7.3; Glynn 1990), and such non-random feeding directed towards plentiful species can actually increase coral species abundance. Like herbivore browsers, some corallivores are non-denuding whereas others (e.g. some polychaetes, gastropods, and starfish) entirely denude corals of their soft parts. However, few corallivores actively excavate the skeleton and their activities do not normally result in total colony death, only in partial mortality. In such cases, the capacity of a coral to heal or replace damaged areas of soft tissue becomes critical to its survival.

When corallivores reach abnormally high densities (*outbreak populations*)—which may be induced by catastrophic external disturbances such as low- or high-temperature events and hurricanes—their vast numbers combined with voracious appetites can lead rapidly to widespread coral mortality (Knowlton *et al.* 1981). The geographic extent of these outbreaks can be horrifyingly impressive; between 9 and 13% of the 14000 reefs comprising the Great Barrier Reef have experienced outbreaks of the crown-of-thorns starfish *Acanthaster* in the past 20 years, resulting in coral mortalities of up to 95% (e.g. Laxton 1974). Episodic outbreak predation does not necessarily produce irrevocable damage to coral communities, but it may play an important role in permanently shifting community composition if the preferred coral prey are numerous and the feeding intense. Numerous devastated coral communities have shown a remarkable capacity to replace newly generated free surfaces with similar, but not identical, organisms. Indeed, recruitment and growth has been more rapid than predicted.

7.1.7 Sediment production

Constant excavatory grazing and other bioeroding activity on substrate around live corals can lead to their eventual dislodgement and mortality, and dead skeletal material will be reduced further to rubble and sand by endoliths (Fig. 7.6).

2 cm

Fig. 7.6   The abundant skeletal debris associated with coral reefs is the result of intense physical abrasion and bioerosion. Great Barrier Reef. (Photograph: C. Oppenheimer.)

In the Caribbean, sea-urchin bioerosion may be more important than the activity of scarids, as most Atlantic parrotfish only scrape, rather than excavate, calcareous substrates. For example, approximately one-half of the gross carbonate production on a fringing reef in Barbados (10.7 kg m^{-2} year^{-1}; Stearn *et al.* 1977) was eroded and transformed into new sediment (5.3 kg m^{-2} year^{-1}; Scoffin *et al.* 1980) by grazing fish and a single species of sea urchin. However, Bruggemann *et al.* (1996) estimated that two species of scarid alone remove 7.62 ± 0.49 kg m^{-2} year^{-1} from the reef crest of some Caribbean reefs.

The most important bioeroders in the Indo-Pacific are excavating scarids, which are characteristic of reef crests and fronts. Scarids feed preferentially on living or dead convex surfaces, and pass substantial amounts of sediment through their alimentary tracts which is then redistributed into sediment aprons that form at the base of the reef. Bellwood (1995) has estimated that up to 5.6 kg m^{-2} year^{-1} may be removed by excavating scarids at Lizard island on the Great Barrier Reef. At this location, a single individual of the largest scarid, *Bolbometopon muricatum,* alone removes five tonnes of reef per year (Bellwood 1996a).

Scarids modify reef carbonate in three ways:

(1) by direct erosion;

(2) through decrease of particle size due to both erosion or sediment reworking; and

(3) by net transport of sediment away from its site of removal.

Such bioerosion may therefore be an important agent of structural change on reefs. The removal of material directly from shallow feeding areas, such as the reef crest where it might otherwise accumulate rapidly, to specific defecation sites often in deeper parts of the reef might significantly control the rate of prograding reef growth (Bellwood 1995). This, combined with the decrease in sediment size, increases the likelihood of hydrological transport so resulting in an active net movement of carbonate away from the reef (Bellwood 1996a).

7.1.8 Effects of bioturbation

Whereas predators greatly reduce populations of organisms which live on or near the sediment surface, they have little effect on organisms which live wholly within soft sediment (an *infaunal* mode of life). That disturbance might be a major factor in governing the distribution of infauna has been confirmed by fish exclusion experiments which resulted in significant increases in soft-sediment dwelling organisms, such as gastropods and molluscs (G.P. Jones *et al.* 1988).

Animals can disturb sediments in four ways:

1. Organisms which ingest sediment to extract food (*deposit-feeders*) often cause the resuspension of fine sediment into the overlying water. This may clog the filtering apparatus of surface-dwelling suspension and filter-feeding organisms, such that the presence of deposit-feeders in fine sediments often results in the reduction of heterotrophs.

2. Deposit-feeders may also move sediment from deep to shallow level, or vice versa. This causes sediment instability which may lead to the burial of organisms which rest on the sediment surface.

3. Many infaunal organisms displace and push aside considerable amounts of sediment in their search for food, or through rapid burial as a means of escape. This important process has been termed *bulldozing* (Thayer 1983), and may cause the sinking and burial of organisms which lie on the surface.

4. Active bioturbation increases the porosity and aeration of sediments, allowing oxygen to reach far deeper into sediments than in the absence of any infauna. This in turn promotes the growth of bacteria, further enhancing the nutrients available to infauna.

The growth of corals in back-reef environments are suggested to be inhibited by intense bioturbation (Aller and Dodge 1974): indeed few surface-dwelling organisms can withstand such disturbance. These include those that build tubes (polychaetes, tanaids, and amphipods) or have roots which penetrate sediment (seagrasses, mangroves), and those which preferentially attach to stable (which usually means large) substrates. Densely packed populations of organisms can also exclude mobile bioturbators and so stabilize and bind soft substrates. Some corals, such as *Porites*, grow in dense thickets with seagrass and algae which stabilize the sediment in which the coral lies partially buried (Thayer 1983).

7.1.9 Higher predation pressure

Higher order predators that feed on herbivores or corallivores can indirectly influence reef community structure. For example, large predators are thought to limit the foraging excursions of many herbivorous fish that seek shelter in coral reefs. Intensity of predation has been shown to be directly correlated with shelter quality, leading to the formation of haloes known as Randall zones around patch reefs in the Caribbean region. Fish and sea urchins grazing near shelter sites produce these heavily grazed bare zones, with algal growth flourishing only away from the reef itself.

If a causal relationship can be demonstrated between the predators of sea urchins and reef fish, then many modern reef communities may be markedly different to those of the recent past when large fish, green turtles, and sirenians were still abundant before exploitation by man (see discussion in Jackson 1997).

7.1.10 Indirect effects, territoriality, and diversity

Specialized predators can control the ecology of coral reefs in more indirect ways. Experimental manipulation of predator density has shown that the presence of predators will modify otherwise transitive competitive relationships, either by reducing the effects or totally preventing competition. The effect of predator removal from an assemblage of coralline algae was clearly demonstrated by Paine (1984), where diversity remained higher in assemblages growing in the presence of predators (Fig. 7.7). Grazing and predation are

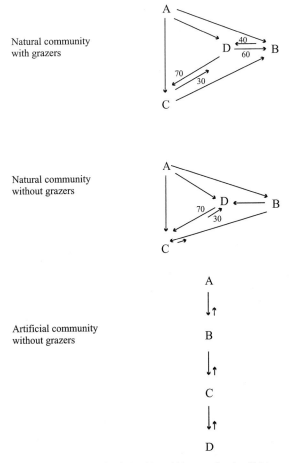

Fig. 7.7 Effect of predator removal on overgrowth relationships within natural and artificial communities of coralline algae. Arrows point from winner to loser, and numbers represent the percentage of total interactions where competitive dominance was unidirectional. A, *Pseudolithophyllum lichenare*; B, *Lithothamnium phymatodeum*; C, *Pseudolithophyllum whidbeyense*; D, *Lithophyllum impressum*. (Modified from Paine 1984.)

therefore important agents of diversification, as they retard the dominance of competitively dominant species (Benayahu and Loya 1977).

Many behavioural mutualistic associations also occur on coral reefs, such as ancillary feeding activities including 'cleaning stations', and those associated with defence. In particular, many reef fish—particularly damselfish—are highly territorial and aggressively repel intruders. These phenomena may have far-reaching consequences for maintaining reef diversity.

Damselfish defend algal mat territories of about 1 m^3, and these can cover up to 70% of flats adjacent to reef crests. By preferentially consuming macroalgae, removing fouling epibionts from corals, and excluding other herbivorous fish, different damselfish species can maintain territories with markedly different algal community compositions, higher productivities, and diversities than in surrounding, unprotected areas (Klumpp et al. 1987). For example, on the Great Barrier Reef the cryptofaunal density within the territories of damselfish can exceed 580 000 m^{-2}, up to four times that of adjacent areas (R.N. Hughes 1991). However, it is not yet clear whether this refuge is generated by reduced herbivory or an increased potential for growth and substrate acquisition. There is also some evidence that the abundance of coral recruits is increased within damselfish territories (Sammarco and Carleton 1982). Others have also argued that damselfish territories promote group-foraging in other fish such as acanthuroids and scarids, as aggregations of non-territorial herbivores.

7.1.11 Predation, nutrient level, and community differentiation

The trophic structure of modern, tropical epibenthic communities shows profound differences influenced by a complex interaction between predator characteristics and ambient nutrient levels (Fig. 7.8).

The most conspicuous primary producers on coral reefs are heavily calcified phototrophs (coralline algae) and *mixotrophs* (organisms which gain food both by photosynthesis and from organic matter, mainly scleractinian corals). These dominate generally where nutrient levels are low and disturbance is high (Littler and Littler 1984). Most scleractinian corals can only survive in any great abundance in low nutrient environments, as they are outcompeted by benthic fleshy algae or phytoplankton in higher nutrient regimes (Birkeland 1977, 1987, 1988; Hallock and Schlager 1986; Wood 1993). Unlike land plants, seaweed metabolites almost never contain nitrogen (with the exception of some cyanobacteria) are so they are very nitrogen-limited (Hay 1991). There is, however, some evidence to suggest that low nutrient levels may also interact with intense herbivory to restrict the success of macroalgae in the tropics. Nutrient levels may set the maximum size and biomass of algae but herbivores limit these levels to well below the constraints imposed by available nutrients.

The additional presence of many specialist herbivores in low nutrient regimes removes potentially fouling organisms, thus allowing scleractinian corals to become established and maintain dominance (Birkeland et al. 1985). Heterotrophic metazoans (suspension and filter-feeding organisms) also require hard substrate for attachment, but they are light-independent, often have finite growth, and are usually competitively inferior to corals. They are, therefore, far more abundant in cryptic niches: on the undersurfaces of corals, within the debris pile, and in submarine caves and grottoes.

In higher nutrient regimes, an increasing proportion of primary consumers (heterotrophs) is present in the open-surface community (Fig. 7.8). In mesotrophic conditions, the main

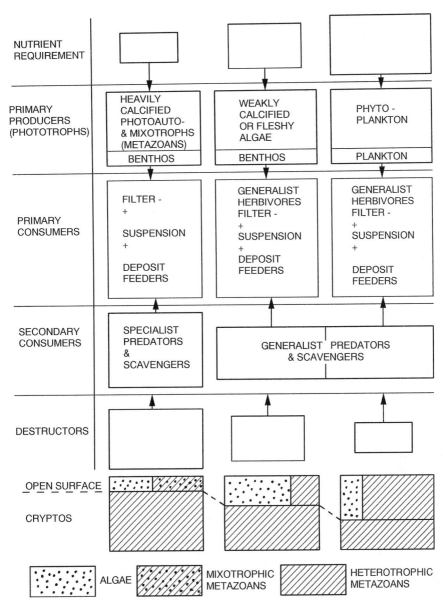

Fig. 7.8 The role of predation and nutrient input on the trophic structure of aggregating shallow marine tropical communities. Relative nutrient input and importance of destructor groups (predators and herbivores) is denoted by the size of the boxes. Also shown is the predicted biomass distribution and trophic type of open-surface and cryptic communities in different environmental settings. (From Wood 1995, by kind permission of SEPM Society for Sedimentary Geology.)

primary producers are fleshy or weakly calcified algae, such as the linear mounds of the alga *Halimeda*. Heterotrophs might still be dominant in cryptic areas however. In eutrophic areas, phytoplankton sequester nutrients and light such that abundant suspended food is available to primary consumers. Such areas are dominated by a sessile, open-surface heterotrophic community of sponges, bryozoans, and bivalves as well as some fleshy algae

(Wilkinson 1987; Birkeland 1988). Reefs built by monospecific stands of vermetid gastropods and sabellariid polychaetes might also grow under conditions which include high levels of nutrients.

Some limited experiments have shown that grazing pressure may also increase along a gradient of increased nutrient concentration. Grazing intensity was, on average, 25 times greater on a reef site in the eastern Pacific where upwelling occurs, compared to a reef site on the lower nutrient Caribbean coast (Birkeland 1987). This intensity is compounded by the fact that many grazers may forage as schools rather than singly or as pairs, and these consumers must also broaden their diets where population densities of predators and grazers become large in order to survive. Such intense grazing by generalists can set back succession by increasing the proportion of early successional biota. Under such conditions, even refuge in patchy distributions may not be effective for small sessile prey as they may be grazed indiscriminately.

Additionally, under higher nutrient conditions, the numbers of endoliths increases in both abundance and diversity. Highsmith (1980) has show than the percentage of massive corals bored by bivalves increases proportionately with phytoplankton productivity (Fig. 7.9). The reasons for this is unclear, but it may be that high nutrients favour survival of the larvae, or that, as heterotrophs, more food is available to the borers.

7.1.12 Zonation

Reef zonation forms in response to stillstand, that is when there has been little change in sea level over a considerable period of time. There has been negligible change in sea level over the past 6–7 kyr.

Classically, zonation has been explained by the differential response of organisms to physical controls, such as prevailing hydrodynamic regime (waves, currents, and tides) and

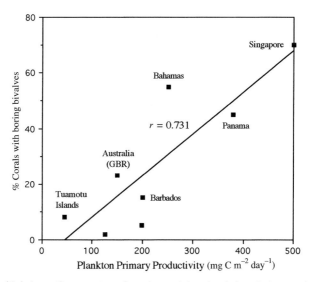

Fig. 7.9 The relationship between the percentage of massive corals bored and phytoplankton productivity (mg C m^{-2} day^{-1}). (Modified from Highsmith 1980, by kind permission of Elsevier Science.)

light, both determined in part by water depth. However, zones in Caribbean reefs are typically dominated by only one or a few species, even though most species of corals occur over a very wide range of depths even at single localities (Goreau and Wells 1967). While the types of dinoflagellate symbiont found with a single coral colony and species may vary with light intensity and therefore depth (see Section 8.1.3), organisms do have different competitive abilities under different environmental settings. In the Indo-Pacific, the zonation style is different, and the reefs may be classified into *Acropora* and non-*Acropora* dominated reefs which correlate with relative wave energy and postlarval settlement processes (Done 1982). It is now clear that zonation is achieved by interacting environmental and ecological controls, which may be highly complex. Here, the differential effects of predation and herbivory may be very important.

Many experimental manipulations of predators and prey have shown that the relative competitive abilities of organisms may be reversed in different environmental settings. Wellington (1982) provides an interesting example of how zonation of corals can be explained by the differential effects of territorial damselfish in the survival of different coral species (Fig. 7.10). On the fringing reefs of Panama, the tightly branching corals of *Pocillopora* are the main reef builders. These corals produce high topographic complexity which is required by the damselfish *Eupomacentrus acapulcoensis* for shelter sites. *Pocillopora* recruitment is high within and adjacent to damselfish territories as other herbivorous fish and corallivores are prevented from entry. The damselfish farm algal turf that grow on the *Pocillopora* tips which they graze (with a resultant loss of 2% total area of coral soft tissue), but they can also graze on the massive coral *Pavona gigantea*, but which might lose up to 59% of its soft tissue. So corals that form or occupy topographically complex surfaces

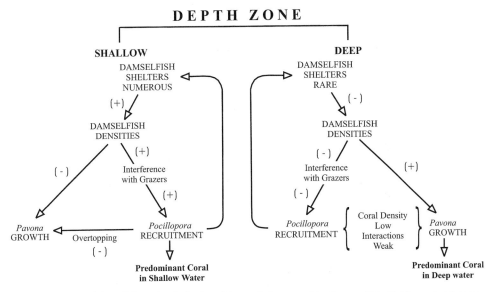

Fig. 7.10 The role of damselfish on the distribution of the corals *Pavona* and *Pocillopora* off the Pacific coast of Panama. Arrows indicate the direction of each effect; + and – indicate positive and negative effects, respectively. The direct negative effect of damselfish on *Pavona* in shallow waters is due to the removal of polyps, yet the direct positive effect in deep water is actually an indirect effect mediated by the scarcity of shelter for damselfishes. (Redrawn from Wellington 1982.)

(as well as those inhabiting cryptic niches) can preferentially survival predation, including outbreaks. Damselfish promote the growth of *Pocillopora* at shallow depths, where the closely branching organization appears to have antipredation characteristics.

7.1.13 Summary

Despite the low apparent standing stocks of algae on reefs, the extremely high growth rates of algal turfs make shallow coral reefs one of the most productive ecosystems known. Most of this production, however, is consumed by herbivores, particularly fish. The abundance of reef fishes is assumed to be of great importance on coral reefs, as evidenced by the dramatic effect of their decline on Jamaican reefs (T.P. Hughes 1994). Tropical marine hard substrata are usually sparsely vegetated, but a rich algal flora develops when herbivorous fish are excluded and/or nutrient input increases. Grazers not only allow the dominance of corals and coralline algae on coral reefs, they also contribute notably to carbonate sediment production and redistribution, to algal ridge formation, and to the maintenance of overall diversity. Like other predators, they can also ameliorate the effects of competition and may interact with physical controls to produce the characteristic zonation of modern coral reefs.

Many coral reefs organisms show highly complex, integrated sets of morphological, spatial, chemical, and temporal characteristics, many of which can be considered to be defensive. But the probability of any single defensive trait being effective is decreased as we have seen that much of the predation on coral reefs is non-specific: indeed, different predatory methods may even require the same defensive strategies. In studying supposed defensive aptations it is therefore necessary to consider the feeding methods used by predators. These categories are summarized in Table 7.2, and their evolutionary appearance and development are discussed in the following section.

7.2 The fossil record of biological disturbance

Many of the major groups of predators, bioturbators, and bioeroders on modern reefs have arisen only in relatively recent geological history. To analyse the effect of these new groups and their novel feeding methods requires the reconstruction of the feeding habits of extinct taxa through the Phanerozoic. The approach adopted here is to assume that shared anatomical and morphological features, especially of the feeding apparatus, might indicate shared feeding (including behavioural) characteristics (see Steneck 1983). Whilst the pitfalls and overgeneralizations of this methodology are well known (e.g. Choat 1991), this is a necessary approach given both the poor phylogenetic knowledge of potential prey and predators, and the preservational vagaries of the fossil record.

It should be noted from the outset, however, that our detailed knowledge of the first appearance, subsequent diversification, and rise to abundance of many of these groups is poor. Not only is the fossil record of soft-bodied biota particularly scant, but the known evolutionary history of other groups which bear hard parts with only poor preservation potential is likely to be very biased and dependent upon the finding of exceptionally well-preserved material. There will also be a taphonomic bias towards the preferential preservation of organisms with more robust and large skeletons. This is compounded by the fact that many of the feeding habits outlined in Table 7.2 are not easily recognizable in the fossil record, and more-

Table 7.2 General methods used by predators on sessile reef biota and their inferred first appearance in the fossil record

Predatory method	Predatory group and first appearance	Reference
Carnivory	?Polychaetes: Cambrian	Conway Morris 1977
Herbivory		
—Non-denuding	Archaeogastropods: Cambrian	Steneck and Watling 1982
	Polychaetes: Cambrian	Glaessner 1979
	Arthropods: Cambrian	Bergstrom 1979
—Denuding	?Orthothecimorph hyoliths: Lower Cambrian	Edhorn 1977; Kobluk 1985
	Mesogastropods: Devonian	Steneck and Watling 1982
	Teleost fish: Eocene	Tyler 1980
Excavation of live skeletal material (herbivory and carnivory)	Regular echinoids: Diadematoida: Late Triassic	Smith 1984
	Chitons: Chitoninae: Late Cretaceous	van Belle 1977
	Molluscs: Patellacea: Late Cretaceous	Lindberg and Dwyer 1983
	Teleost fish: Scaridae: Miocene	Bellwood and Shultz 1991
Live-boring (non- predatory)	Bivalves: Lithophagidae: Eocene	Savazzi 1982; Krumm and Jones 1993
	Porifera: Clionidae: ?	
	Barnacles: ?Eocene	D.S. Jones, pers. comm.
Substrate disturbance	Rays and skates: Devonian	Vermeij 1987
	Echinoderms: Holothuroids: Devonian	Thayer 1983
	Manatees: Eocene	
—Deep	Polychaetes: Triassic	Thayer 1983
	Gastropods: Late Triassic	Thayer 1983
	Irregular echinoids: Early Jurassic	Thayer 1983
	Decapods: Early Jurasssic	Thayer 1983

over many supposed predatory traces may be difficult to differentiate from other causes of skeletal damage. The range of morphologies of trace fossils formed by bioerosion is vast due to the diversity of organisms involved (Bromley 1992). But the sometimes distinctive, but often delicate, grazing or predation traces on reef-builders are evident only on the original upper surfaces which often suffer from some degree of postmortem destruction before final burial and fossilization. All these factors will make it extremely difficult to tie the evolution of a perceived defence to the appearance of a particular predatory group.

7.2.1 Predators and herbivores

Quantitatively important groups of herbivores and predators on modern coral reefs are the polychaete annelids, crustacean arthropods, chitons, gastropods, echinoids, and fish.

However, the first evidence of possible grazing activity is indicated by 'cropped' calcified cyanobacteria from the Lower Cambrian Bonavista Group of eastern Newfoundland (Edhorn 1977). This may have been due to the actions of the problematic Cambrian group, the hyoliths.

Annelids

We know that three common families of polychaete annelids contain herbivorous species, but even in vast numbers they are not capable of significantly denuding filamentous algal cover on hard substrates. Some are also carnivores, but none are known to excavate hard substrates. Although their fossil record is poor, polychaetes are known from the Cambrian (Glaessner 1979) and were diverse and probably abundant by the early Palaeozoic (Conway Morris 1979).

Arthropods

Herbivorous and carnivorous arthropods are known from the suborders Ostracoda and Malacostraca, but only three orders of malacostracans (isopods, amphipods, and decapods) are quantitatively important. These forms are unable to denude fleshy algae or excavate calcareous crusts. Although malacostracans appeared in the Lower Cambrian (Bergstrom 1979), their abundance is thought to have increased throughout the Phanerozoic (Sepkoski 1981). Arthopods capable of *durophagy* (shell-crushing) are thought to have been present by the Devonian (Signor and Brett (1984). Large decapod crustaceans (e.g. crabs) appeared in the Late Cretaceous, and may be non-denuding herbivores or carnivores (including corallivores).

Many behavioural mutualisms involving crustaceans also help to reduce predation and so affect community structure. For example, 13 living reef coral species are known to be defended by 18 species of crustacean guard (17 crabs and 1 shrimp; Glynn 1983) which have been shown to actively deter potential predators.

Molluscs

Molluscan herbivores include chitons and prosobranch gastropods, of which Steneck (1983) identified nine families. Most primitive gastropods bear chitinous teeth with finely denticulate margins, and feed on delicate filamentous and microscopic algae (Steneck and Watling 1982). Extinct groups such as the Bellerophontaceae and the Palaeozoic monoplacophorans are assumed to have had similar feeding capabilities. Mesogastropods possess stronger buccal musculature but retain a chitinous dentition, and are capable of denuding fleshy algae. Only chitons and limpets (Patellacea), which bear robust buccal muscles, unique dentitions with heavy mineral coatings of silicates and iron compounds (Steneck and Watling 1982) are thought to be capable of excavation (Steneck 1982).

Chitons are known since the Cambrian (Runnegar *et al.* 1979), but there is no indication from either trace or body fossils that they were an abundant component of marine communities during the Palaeozoic (Voight 1977). Families capable of excavation did not arise until the Cretaceous or Cenozoic (Vermeij 1987). Chitons excavate rock by secreting acid mucopolysaccharides and carbonic anhydrase from the foot and mantle edge, and then removing particles of substrate with the radula.

Archaeogastropods capable of grazing delicate algae are also known from the Cambrian; mesogastropods from the Devonian. Limpets appeared in the Middle Triassic and had become abundant by the Jurassic (Steneck 1982). The earliest limpet trace fossil is known from the Upper Jurassic (Voight 1977). This is also the earliest known mollusc grazing trace fossil.

Echinoderms

Of extant echinoid orders 78% are herbivorous and some (diademoids) are corallivores. Although Palaeozoic echinoids were probably herbivores, their weak and relatively rigid feeding apparatus suggests that they were not capable of effective macroalgae denudation or substrate excavation. The oldest echinoids (e.g. *Aulechinus*) possessed broad jaws which could probably only scoop material from a substrate; jaws capable of plucking and biting appeared in the Early Silurian (*Aptilechinus*), but the attachment of jaw muscles directly to the inner test wall only allowed limited horizontal movement so preventing the development of rock scraping. The Late Permian cidaroid, *Miocidaris*, was the first urchin in which muscles were attached to the perignathic girdle, which are a series of projections from the inner test wall. But in this genus, as well as all later cidaroids, only vertical movements remained possible, and so they are thought to have fed much like their Palaeozoic ancestors. Several new orders appeared in the Mesozoic, particularly in the Echinacea, which were capable of greater excavation: tooth structure changed from trough-shaped in Aulodonta lanterns, to more reinforced and keel-shaped in Stirodonta and Camerodonta lanterns. Muscle attachments in the Aristotle's lantern were more robust, and lantern supports enlarged allowing greater horizontal mobility and protrusion of the teeth (Kier 1974). The oldest, characteristic pentaradiate, grazing trace fossils (*Gnathichnus pentax)* are of Early Jurassic age (Bromley 1975), coincident with the development of strengthened teeth in a stirodont lantern. By the Late Cretaceous (Maastrichtian) the camerodont lantern appeared, and urchins attained modern morphologies, size, and feeding apparatuses. The earliest known occurrences of extensive areas browsed uniformly by pentaradiate apparatuses date shortly before the Maastrichtian (Bromley 1975). Few morphological changes occurred thereafter, but the abundance of urchins capable of excavation may have increased during the Cenozoic (Durham and Melville 1957).

Steneck (1983) analysed the excavating abilities of extant families and determined that there was a correlation between the excavating abilities of lantern type and the first appearance in the fossil record; there is a trend of increasing excavatory ability over time with 0% of families with cidaroid lanterns capable of excavating hard substrata, 20% Aulodonta lanterns, 25% with Stirodonta lanterns, and 60% with Camerodonta lanterns. On modern coral reefs, most herbivorous echinoids possess camerodont lanterns.

The ability of starfish to feed extraorally was probably acquired in the Early Mesozoic (Gale 1987).

Reef fishes

Fish capable of durophagy appeared in the Devonian (Signor and Brett 1984). These include placoderms and chondrichthyans. However, there is no evidence that such forms fed on lower invertebrate reef-builders.

Perciform fish probably arose from a beryciform-like acanthopterygian stock in the Late Cretaceous (Carroll 1987). From this time until the early Tertiary, the perciforms underwent a period of extremely rapid anatomical radiation. Most modern reef fish derive from long-established groups, with most families appearing in the Eocene. So within 20 Myr of the first appearance of perciform fish, they achieved morphological complexity and diversities that are almost indistinguishable from that of living fishes, followed by a long period of relative stasis. There is little evidence of fish with the structural and morphological characteristics of herbivores in earlier radiations of actinopterygian fishes.

Much of what is known of fossil reef fish comes from the Eocene deposits (50 Ma) of Monte Bolca in northern Italy (Box 7.3). These exceptional faunas demonstrate that not only had most modern reef fish families appeared, but a morphological specialization equivalent to that of modern species was present by this time. The complex pharyngeal apparatus of labrids was present, and major labrid clades were already differentiated. Balistids first appeared in the Oligocene. The first record of a confirmed scarid—which bears a remarkably close resemblance to modern species—is known from the middle Miocene, only 14 Ma (Bellwood and Schulz 1991).

Box 7.3: **Monte Bolca: The first reef fish fauna**

The Eocene deposits (50 Ma) of Monte Bolca in northern Italy contain a remarkable fossil fish assemblage. This region was part of the northern margin of the Tethys Sea, and supported coral reef development until at least the mid-Miocene (Rosen 1988). The fish faunas are very diverse, with 227 species, 177 genera, 80 families in 17 orders recorded (Blot 1980). Many of the families represented are characteristic of modern coral reefs.

The assemblage appears to have died in a series of mass mortality events with excellent preservation of complete individuals, sometimes with pigmentation patterns and

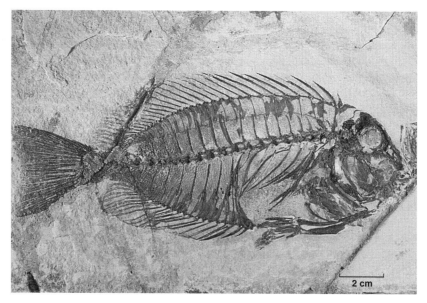

Fig. Box 7.3 A herbivorous acanthurid from the Monte Bolca reef fish biota (50 Ma). (Photograph: L. Sorbini.)

squamation. With no apparent ecological selectivity, a faithful reconstruction of the fauna has been possible (Bellwood 1996*b*).

Individual fossil species from Monte Bolca are readily identifiable and bear a striking resemblance to modern forms. For example, *Eozanclus* from Monte Bolca is virtually indistinguishable form the Recent *Zanclus*. In fact, to date 22 genera can be considered close to forms now living, including herbivores (Sorbini 1983). The Monte Bolca assemblage, however, shows some differences to modern reef fish faunas. First, it retains distinct links with Mesozoic communities. Beryciform and generalized perciform fish are well represented, which are now minor components on modern reefs. There are three species of pycnodontids known from Monte Bolca, a group which were dominant in shallow marine waters throughout most of the Mesozoic (Carroll 1987) but appears to have died out in the Eocene. Second, in modern reefs the herbivore community is dominated by acanthurids and scarids, which today have comparable species' diversities. Although acanthurids are well represented at Monte Bolca, no scarids are known.

Limited information on the evolutionary history of reef fishes has been available from the fossil record as fossil fish remains are often fragmentary and frequently misidentified (cf. Bellwood and Schulz 1991). However, recent cladistic phylogenetic analyses have revealed new data concerning biogeography and habitat associations.

First, differential extinction appears to have taken place in the Atlantic in the mid–late Cenozoic: large excavating scarids, herbivorous siganids, and planktivorous caesionids are conspicuously absent from Atlantic reefs. Family-level comparisons between Atlantic and Indo-Pacific sites also show more haemulid and sparid species in the Atlantic but fewer species in characteristic reef families (Labridae, Pomacentridae, Apogonidae and Chaetodontidae). There is also an absence of several widespread Indo-Pacific taxa (Caesionidae, Lethrinidae, Nemipteridae and Siganidae; Bellwood 1997). Broadly coincident extinctions in other reef biota from the Caribbean have been reported, including many corals (Budd *et al.* 1994).

Second, although reef fishes evolved in shallow tropical seas, throughout their 50-Myr history they appear to have maintained strong links with non-reef habitats. The range of all extant coral reef fish families extend beyond coral reefs. Moreover, in at least two major clades, living on associated habitats appears to be the primitive condition followed later by a move into coral reef habitats: the highly modified jaws and pharyngeal apparatus of scarids probably evolved in taxa which fed on seagrasses or seagrass epiphytes (organisms living attached to seagrass fronds) and only later (~5 Myr) were they used to feed on coral reefs or rocky reefs (Bellwood 1994), and primitive acanthuroids live associated with soft sediments. This raises the possibility that early reef fish assemblages may have been composites of clades whose origins were from a range of non-reef habitats, some with possible temperate links (Bellwood 1997). These past associations are supported by the coastal non-reef habitats described from the Monte Bolca region (Sorbini 1983). The history of reef fishes may, therefore, depend on the connectivity of habitats associated with reefs (Bellwood 1997).

7.2.2 Endoliths

Encasement within a hard substrate provides ample protection from many predators, and the ability to bore has evolved many times since the end of the Neoproterozoic (see Warme

(1975) and Vermeij (1987) for compilations; summarized with modifications in Fig. 7.11). Not all organisms readily fossilize within their boreholes, such that while some boring traces are very characteristic, the identification of others is not always possible.

The first traces of bioerosion are delicate borings of algal or fungal origin from the Late Archaean or Palaeoproterozoic. The first boreholes of unknown, but probable metazoan origin, are from the late Neoproterozoic problematic sessile organism, *Cloudina* (Bengtson and Zhao 1992). It is difficult to determine whether these were predatory, but they certainly did not constitute a major cause of death (Harper and Skelton 1993). Cylindrical borings

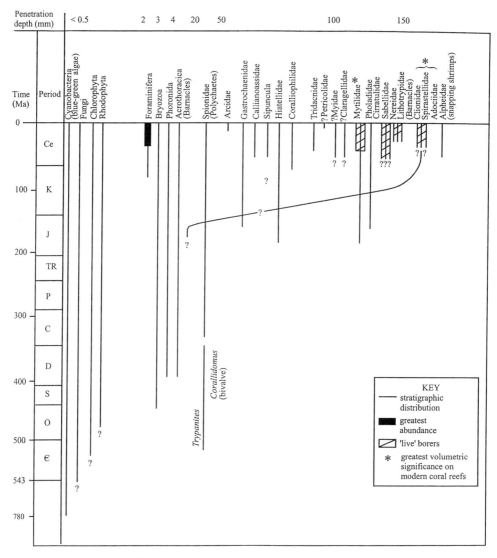

Fig. 7.11 Stratigraphic ranges of endolithic organisms arranged according to the first appearance of taxa with bioeroding representatives, and the depth of penetration. (Drawn from data modified from Vermeij 1987, Table 5.1 therein.)

with a single aperture comprise the form known as *Trypanites*. *Trypanites* appears in the Lower Cambrian (associated with hard grounds), and remains abundant—but rarely constituting serious levels of erosion—throughout the Phanerozoic. The simplicity of these borings makes them impossible to assign to a particular maker—polychaetes, sipunculans, as well as echinoids and crustaceans can all form such traces.

Non-predatory borers had diversified into at least four genera by the Middle Ordovician, but were of limited abundance, except locally, during most of the Palaeozoic, although 11 ichnogenera are known by the Late Devonian (Kobluk *et al.* 1978). All endolithic groups originating in the middle Palaeozoic or earlier did not penetrate to depths greater than 20–30 mm below the surface of hard substrata (Fig. 7.11).

A major radiation of endoliths occurred from the Triassic onwards, and deep borings (those greater than 50 mm or more) are known only from the Mesozoic and Cenozoic (Fig. 7.12). Records of predatory gastropod-like boreholes are known in Triassic bivalves (Fürsich and Jablonski 1984), but boring gastropods may not have been common until the Cretaceous when the first unequivocal natacid and muricid boreholes are recorded from the Albian of England (J.D. Taylor *et al.* 1983), although it has been thought likely that they did not become a significant threat until much later (Harper and Skelton 1993). However, recent findings of exceptionally well-preserved aragonitic bivalves with abundant gastropod-like boreholes from a variety of Lower Jurassic localities in Britain suggests that boring may have been a common phenomenon some 70 Myr earlier than previously thought (Harper *et al.* 1998).

Of the major clades of boring bivalves, the lithophagids (Mytilidae), gastrochaenids, pholads, and hiatellids all have an early Mesozoic origin (Harper and Skelton 1993). From the Jurassic until today, bivalve borings have increased in diversity and abundance. The first documented reefs to contain a diverse and abundant assemblage of endoliths are bivalve–algal reefs from the Portlandian (Upper Jurassic) of Portland, southern England (see Case study 3.12). Here, 11 taxa of macro- and microborers (those smaller than the milli-metre scale) have been described, and up to 50% of the reef framework has been estimated

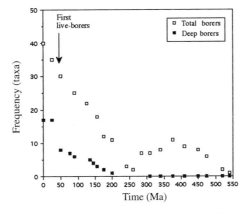

Fig. 7.12 The history of bioerosion though the Phanerozoic, showing the number of bioeroding ichnogenera and the number of deep bioeroders, based on the first appearance of such forms within families. Deep bioeroders are those which penetrate more than 50 mm into the substrate. (Drawn from data modified from Vermeij 1987, Table 5.1 therein.)

to have been removed by bioerosion (Fürsich *et al.* 1994). This is in marked contrast to the relative paucity of borings found in Triassic and most older Jurassic reefs, although coral heads can locally be conspicuously bored by bivalves (e.g. *Gastrochaena* and *Lithophaga*). Notwithstanding the facts that the excellent preservation of borings at this locality is favoured by prodigious early cement formation (Fürsich *et al.* 1994)—which probably records the diversity of small, delicate borers more faithfully—and that many of the taxa recorded occur in older sequences (e.g. *Gastrochaenolites, Talpina,* and *Rogerella)*, this profusion of endolithic activity probably records the genuine evolutionary proliferation in the diversity and abundance of the boring habit which occurred sometime in the mid–late Jurassic. Of particular note is the presence of abundant borings by clinoid sponges, which is only rarely known in older rocks; indeed these examples may in fact be the result of borings by the haplosclerid sponge *Aka*—which is known from the Triassic (Reitner and Keupp 1991)—or even other phyla (Fürsich *et al.* 1994).

These endoliths described only attack postmortem—that is abandoned skeletal parts of living organisms, or clasts of skeletal material. However, some groups are capable of overcoming the defences of an organism and boring through living tissue to seek refuge within the skeleton. Only in such forms might we predict any defensive response on the part of the ' host' prey, although any bioerosive activity will not only alter the substrate conditions but also the patterns of sediment production and distribution within a reef complex.

Live boring into corals occurs in several groups (Fig. 7.13). Little is known as to the first appearance of this capability in sponges, but the first live-boring bivalves in corals are known from the Eocene—from a solitary fungid (Savazzi 1982) and a colonial coral (Krumm and Jones 1993), as are the first live-boring barnacles (D.S. Jones, personal communication). Regular urchins commonly feed on boring sponges, and Morton (1990) has suggested that live-boring in bivalves may have evolved in particular to the threat of predatory gastropods, and that the bivalve gains protection by being encircled by the nematocysts of the coral. The style of modern endolithic bioerosion can be traced back to Oligo-Miocene reefs (Pleydell and Jones 1988).

The evolutionary origin of many endolithic groups is obscure, but the ancestry of most molluscan borers is well known. The ability to bore has evolved at least nine times in the Bivalvia, with most appearing from the late Triassic onwards (Fig. 7.11). The lithophagids and tridacnids were derived from byssate ancestors (Harper and Skelton 1993). Coralliophilid gastropods (Ordovician) probably descended from surface rock-dwelling forms (Vermeij 1987). With common ancestry in epifaunal forms, it is tempting to suggest that the rise in the boring habit might have evolved in response to the MMR.

7.2.3 Bioturbators

The stratigraphic distribution of taxa containing substantial numbers of bioturbating predators and suspension feeders show that their rise is a post-Palaeozoic phenomenon (Thayer 1983; summarized in Fig. 7.14). Post-Palaeozoic burrowing taxa feed deeper, rework sediment faster, and have shorter turnover times than can be inferred for Palaeozoic taxa. Sediment reworking as a result of deposit feeding is also more rapid in sandy, warm waters than in cold waters: shallow-water, subtropical and tropical communities are therefore most vulnerable and they are often sites of intense activity. In particular, with the appearance of

Fig. 7.13 Live-boring bivalve *Lithophaga* sp. in the massive coral *Porites*, from the upper Miocene of Majorca.

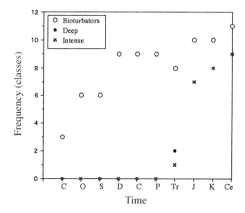

Fig. 7.14 The history of bioturbation though the Phanerozoic, showing the number of bioturbating classes, the number of classes with deeply bioturbating members, and the numbers of classes with intense bioturbating members, based on the first appearance of such forms within classes. Deep bioturbators are those which penetrate more than 100 mm into the sediment; intensive bioturbators record the sediment at a per capita rate of 100 mm³ per day, or higher. (Drawn from data derived from Thayer 1983.)

holothuroids (sea cucumbers)—probably in the Devonian—deposit-feeding in sands may have become important for the first time.

From the Late Carboniferous to the Jurassic, the number of bioturbating families increased from 13 to 32 (Thayer 1983). From the beginning of the Triassic, predators also became the major sediment bioturbators. A large-scale episode of infaunalization occurred in the Late Triassic–early Jurassic, including those forms capable of processing large quantities of sediment per individual (such as deposit-feeding organisms capable of 'bulldozing'): deep burrowing lugworms (polychaetes), appeared in the Triassic, irregular sea urchins—with their substantial capability for sediment reworking—appeared in the early Jurassic. Asteroids, which can excavate deeply and rework huge amounts of sediment, appeared in the early Jurassic, and bulldozing infaunal gastropods were present by the Late Triassic and radiated further in the Cretaceous and the Cenozoic. The deepest known burrowers from modern seas—the ghost shrimps (thalassinidean decapods)—are known from the early Jurassic, and the earliest burrowing crabs are of Late Jurassic or early Cretaceous age.

Ray-like fish appeared in the Devonian, but the true rays are only known from the late Jurassic onwards. Other burrowing teleost fishes appeared in the latest Mesozoic–early Cenozoic. Sea cows, once common on seagrass beds associated with reefs, have an origin in the Eocene.

7.2.4 Summary: the rise of biological disturbance

Table 7.2 indicates the first known appearances of the different predatory methods in the fossil record. From this it is evident that grazers and carnivores throughout the Palaeozoic and early Mesozoic were relatively small individuals, with limited foraging ranges, and were incapable of excavating calcareous substrates. Although there appears to have been a radiation during the Devonian of durophagous, mobile predators (Signor and Brett 1984), it seems that these forms relied upon manipulation to crush, or ingest, whole shells (Harper and Skelton 1993). By the early Mesozoic, sessile organisms had to contend with an increasing battery of more advanced feeding methods, in particular those involving excavation, as well as sediment disruption due to deep bioturbating activity (Fig. 7.14). Bioerosion increased in intensity from the mid–late Jurassic, but, as far as can be ascertained, this was exclusively postmortem. Biological disturbance reached new heights of intensity from the latest Cretaceous–early Tertiary when deep-grazing limpets, urchins, and especially the highly mobile reef fishes appeared, and with them the increasing ability to excavate substantially hard substrata over a greater area (Fig. 7.15). A concurrent radiation of endoliths occurred from the Triassic onwards, and deep borings are known only from the Mesozoic and Cenozoic (Figs 7.11 and 7.12). Clionid sponges—one of the major bioeroders on modern coral reefs—had become abundant by the latest Jurassic. The first live-borers are known from the Eocene, as are fishes similar to modern reef faunas (50 Ma). The complex pharyngeal apparatus of labrids was present at this time, and major labrid clades were already differentiated. Balistids first appeared in the Oligocene, and the oldest scarid fossil currently known is from the Miocene (14 Ma). It seems likely that sometime during the Oligocene–Miocene, reef bioerosion gained a modern caste.

How did post-Palaeozoic reef communities respond to these new threats? From our knowledge of the role played by organisms with various predatory methods (and by endoliths) on modern reefs, we might be able to make a series of predictions concerning

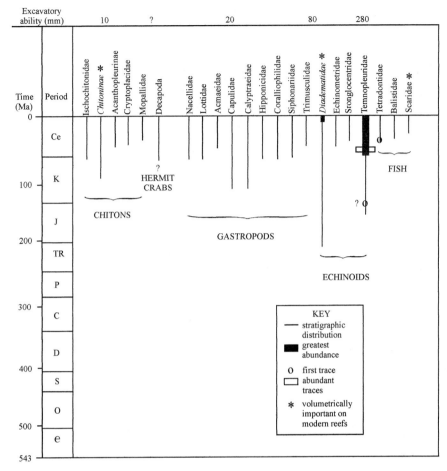

Fig. 7.15 Stratigraphic ranges of chasmolithic and epilithic organisms arranged according to the first appearance of families with excavating representatives, and depth of penetration. (Drawn from data modified from Vermeij 1987, Table 5.2 therein.)

changes in reef-community ecology based on their rise to abundance in the fossil record (Table 7.3). In the sections following we will test these predictions, and see how the evolutionary history of reef-associated herbivores and predators has been intimately bound with the origin and assembly of the modern coral reef ecosystem.

7.3 Evolution of antipredatory traits

Although most grazers and many predators are non-selective, the foraging activities and diets of consumers are influenced by a number of factors. In the evolution of prey taxa, predation failure is an important process: high failure rates in species which display variability in defences are likely to result in the selection of antipredatory characteristics (Vermeij 1982). Some taxa, however, may be restricted from evolving suitable defensive adaptions by

Table 7.3 Predicted changes in reef community ecology based on the rise to abundance of new predatory methods (and endoliths) as evidenced in the fossil record and inferred timing

Event	Prediction	Timing
The rise of macroherbivores	A shift to more conspicuous, well-defended macroalgae on reefs	?Late Mesozoic
The rise of excavatory grazers and predators	A shift to organisms with deterrent traits and those that tolerate partial mortality	Late Mesozoic to Miocene
	A shift form macroalgal-dominated to coral and coralline algal-dominated reefs	Late Mesozoic–Eocene
	Increase in the diversity of the cryptos and other spatial refugia	Jurassic onwards
	Increase in multiserial scleractinian corals	Throughout the history of the group
	Increase in multiserial, branching corals	Cretaceous onwards
	Algal ridge formation by branching coralline algae	Cretaceous
	Thick coralline algal crusts	Eocene–Miocene
Increase in abundance of endoliths	An increase in skeletal sediment production	Late Jurassic
	Formation of sediment aprons	Eocene–Miocene

the constraints of their own body plans. In particular, sessile organisms have a highly restricted range of antipredatory options at their disposal as these must be based upon passive constructional defences. Moreover, susceptibility to partial mortality and reliance upon herbivory/predation to remove competitors or foulers usually entails loss of the prey's own tissues. This means that particular anatomies are required that allow resumption of normal growth as quickly as possible, or indeed even create some advantage from this adversity.

Antipredatory traits in immobile, epifauna can be divided into three sets:

(1) *avoidance*—those that reduce initial accessibility to potential predators;

(2) *deterrence*—structural traits that discourage predation by increasing the cost of successful predation;

(3) *tolerance*—traits that attempt to reduce or minimize, or even capitalize upon, the damaging effects of partial mortality.

Avoidance

● Inhibitive or evasive life habits that reduce accessibility or availability to predators by adoption of restricted spatial or temporal distributions

Spatial refuges include infaunal, endolithic, and cryptic habitats where predator access is difficult, as well as deep-sea or intertidal refuges which are beyond the ranges of many predators. Nocturnal, ephemeral, or seasonal occurrence represent restricted temporal distributions.

The composition of exposed coral reef organisms at night is very different from that during daylight hours, and this is thought to be due in part to the avoidance of daytime predators. Numerous crabs and echinoderms congregate under overhangs or within rubble during the day but only become active at night to forage. Zooplankton, which live in cryptic areas during the day, swim freely at night, as do many benthic invertebrates that migrate into the water column at night (Alldredge and King 1977). A significant proportion of reef fish, particularly herbivores (Acanthuridae, Labridae, Scaridae, and Pomacentridae) also 'sleep' or undergo torpor after daylight hours. For example, estimates suggest that the population size of nocturnal swimming coral reef fish is only about 10% that of the diurnal population at One Tree Island, GBR (Talbot and Goldman 1972), and that their feeding habits are more restricted than diurnal fish, with a greater consumption of immobile organisms occurring during daylight hours (Hobson 1968).

We have already seen that some protection from both biological and physical disturbance can be gained by adopting closely aggregating growth (Section 6.1.5).

Deterrence

● The acquisition of structural traits which are effective at discouraging predation by increasing the cost of successful predation

Deterrent structural characteristics include permanent and secure attachment to a stable substrate, heavily armoured skeletons, tough rubbery textures, the secretion of mineralized sclerites or tough fibrous components, and unpalatable, noxious, or toxic secondary metabolites which inhibit ingestion or digestion.

Many reef organisms, especially those living on open surfaces, bear heavily calcified skeletons. Most modern, immobile reef biota are permanently attached to hard substrates. While permanent attachment to a stable substrate may reduce predation from those predators which rely upon prey manipulation, it provides those which do not with prey that are virtual sitting targets (Harper and Skelton 1993).

Some morphologies are better resistant to predatory breakage than others. For example, colonies with closely spaced branches can make predator access difficult by forming hidden, protected areas. Flattened terminations of branches also offers greater resistance to breakage, and this trait is found in erect species of bryozoans, gorgonian corals, and stylasterine corals.

The presence of noxious or toxic chemicals may be accompanied by the acquisition of warning or cryptic coloration and form. Such defence relies on sight recognition by the potential predator. Many reef organisms bear toxins, particularly forms which are found on open surfaces (Bakus 1981), and, therefore, this has been proposed to be the result of a selective predator–prey interaction (e.g. Bakus 1974). Indeed, there are more toxic sponges, cnidarians, arthropods, holothurians, and fishes in the tropics than at higher latitudes. Some have also argued that reef-dwelling, tropical macroalgae generally bear stronger chemical defences (mainly terpenoids) than temperate seaweeds (Bolser and Hay 1997). Although it is likely that some of the most common, exposed invertebrates on reefs are subject to little or no fish predation because they use defences such as toxicity to prevent even partial mortality, this has not yet been convincingly demonstrated. Indeed, toxins do not confer total immunity to predation: many algae assumed to be protected by antipredatory compounds are still readily consumed by some species. Moreover, compounds with similar structures can differ dramatically in their effects on herbivore feeding, and those which might deter one herbivore may have no effect, or even stimulate feeding, on another (Hay 1991).

● Acquisition of reduced nutritional quality

The potentially deterrent effects of low nutritional characteristics have received little experimental attention, but this strategy would operate best only if alternate food sources of better quality were rarely available (Hay 1991).

● Acquisition of behavioural characteristics which increase the energetic cost of successful predation

Numerous behavioural aptations have been suggested to occur on coral reefs in response to predation pressure. These include rapid or erratic movement, active escape, assault, migration of demersal zooplankton, and even synchronized spawning (Babcock *et al.* 1986). Many behavioural mutualistic associations occur in reef organisms, for example the shrimps (*Stenopus*) and fish (*Amphiprion*) associated with sea anemones, and the numerous crustaceans (e.g. *Trapezia* and *Alpheus*) which live amongst the branches of the large coral *Pocillopora*, whose aggressive territorial behaviour has been shown to deter some predators.

● Internalization of soft tissue, or exposure of large areas of relatively thin tissue

Some sessile organisms enclose their vulnerable soft parts within heavily armoured skeletons. Many sessile reef organisms possess a modular habit which incidentally reduces soft tissue to a relatively thin veneer over a larger basal skeleton. This not only decreases accessibility and the ease of prey manipulation by predators, but also minimizes the tissue biomass available to predators whilst maximizing the cost of collection. For example, in a typical domal colony of *Porites*, only about 0.5% of the colony's radius is occupied by soft tissue (Rosen 1986). In branching and platy colony forms, the relative proportion of skeleton is much lower.

Tolerance

● Enhanced ability to regenerate soft tissue and/or skeleton

Strategies which rely upon herbivores/predators to remove competition often entail loss of the preys' own tissues. Algal turfs grow very rapidly and so can regenerate from basal portions that have escaped herbivory. Well-defended coralline algae can also tolerate intense herbivory due to their ability to rapidly regenerate removed material (see below).

In clonal/modular organisms, partial predation may remove either individual or a few modules, or large areas may be cleared of living tissue, sometimes together with the excavation of underlying skeleton. Newly cleared areas of skeletal substratum may be rapidly colonized by fouling or endolithic organisms presenting further threat to the colony. Clonal/modular organisms tend to have far greater powers of regeneration than solitary forms (Section 6.1.4).

Differential regenerative abilities may be a major factor in determining the resistance of many clonal organisms to partial predation, thus affecting their distribution and abundance. But the ability to regenerate lesions varies widely between different species of clonal organisms, and colony size, lesion size, and position also affect the regeneration pattern, making the outcome difficult to predict.

● Adoption of large/small body size

Larger colonies not only win more competitive overgrowth interactions and are more fecund than smaller colonies, but also are more likely to survive partial mortality by regenerating injuries more rapidly than smaller colonies. Large body size also allows organisms to achieve a refuge in size (Section 6.1.4).

In contrast, adoption of a small body size might allow access to cryptic or otherwise protected refugia.

7.3.1 Evolution of antipredator characteristics in reef builders

The fossil record is silent on many of the antipredatory defences concerning behaviour and physiology: only skeletal anatomy and morphology, spatial distribution, skeletal attack or breakage, and regeneration might be detected—or inferred—in the fossil record of reef organisms.

The origin and diversification of such fossilizable traits are considered below. First, experimental and field evidence for the defensive value of anatomical characters in major reef-building groups is examined, followed by an analysis of their status as either aptations or exaptations using the fossil record.

Reef algae

Many groups of algae are prominent in the history of reef-building (Fig. 7.16). These can be assigned to the following broad functional groups:

(1) stromatolite, thrombolite, and microbialite-forming microalgae (mainly presumed cyanobacteria), and calcified cyanobacteria;

(2) articulated calcareous macroalgae;

(3) crustose algae (including phylloid algae, solenopores, and corallines).

Soft-bodied macroalgae have a poor fossil record.

Functional anatomy and differential susceptibility to herbivory
Microalgae and calcified cyanobacteria   While coral reef algal turfs clearly thrive on intense herbivory, it has long been suggested that bioturbation and grazing prevent the growth of microalgal mats that form stromatolites. It is this vulnerability that is supposed to restrict the distribution of modern stromatolites to environments, such as those of stressed salinities, where these biological disturbances are thought to be minimal or non-existent (Garrett 1970; Awramik 1971).

Algal mats are unable to grow on soft substrates due to burial and mat disruption, and to the increased local rates of sedimentation (Pratt 1982). In the Modern, they are therefore limited mostly to hard substrates (Dravis 1982). Recently, however, the assumption that grazing is inimical to stromatolite growth has been challenged, on two fronts.

First, non-denuding grazers (such as isopods and amphipods, molluscs (gastropods and bivalves), and fish) are associated with stromatolites growing under natural conditions (McNamara 1992). Second, experimental evidence shows that stromatolites can actually grow effectively under high grazing pressure (L. Moore, unpublished data). Cyanobacteria possess many secondary metabolites which may inhibit herbivory, but this has received only limited experimental attention to date (e.g. Wylie and Paul 1988).

Calcified cyanobacteria consist of delicate filamentous, globose, or branching morphologies (see Box 3.2). No close modern analogues are known, and no information is available on

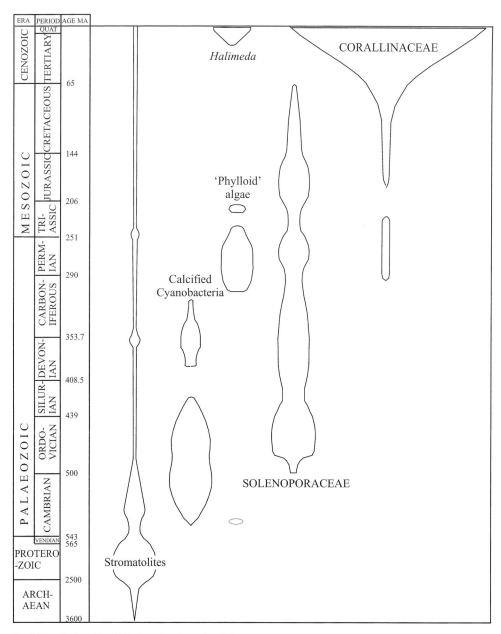

Fig. 7.16 Stratigraphic distribution of reef-associated algae.

growth or regeneration rates. Apart from a calcified habit (which may, however, be acquired postmortem), there are no apparent structural defences to combat herbivory. However, prostrate, finely branching, growth forms are known to be relatively resistant to grazers (as found in the brown alga *Padina*), and it is possible that the closely branching habit of some calcified cyanobacteria might have excluded some microherbivores. Moreover, many adopted a cryptic habit throughout the Palaeozoic, although this is more likely to be a result of competition for limited hard substrates.

Articulated calcareous algae   In the Modern, this group is represented by forms such as *Halimeda* and *Amphiroa*. *Halimeda* (a calcified green alga from the Udoteaceae) possesses a combination of a structurally resistant (calcified) thalli and secondary plant compounds (diterpenoids) which have been proved effective against some herbivores (Steneck 1988). Indeed, *Halimeda* is one of the few abundant algae in well-grazed areas of coral reefs. One remarkable, potentially antiherbivore defence shown by *Halimeda* is that this alga has somehow decoupled photosynthesis from growth and produces new growth exclusively at night when herbivorous fish are inactive (Hay *et al.* 1988). These new young portions bear a high concentration of chemical defences. Within 2 days after new growth, however, calcification increases and the onset of this structural defence is coincident with the reduction of chemical defences. The concentration of chemical defences in older, more heavily calcified plant tissues are < 10% of the youngest portions (Hay *et al.* 1988).

Crustose algae
● Corallines

In addition to the general ability of clonal, modular organisms to regenerate from partial mortality, coralline algae are able to withstand the most intense herbivore onslaught by virtue of four distinct morphological structures (Steneck 1983); Fig. 7.17):

(1) a heavily calcified thallus that is resistant to attack,;

(2) a thallus differentiated into an outer protective layer (epithallus) which overlies the more delicate meristem, and a basal layer (hypothallus);

(3) intercellular conduits (fusion cells and secondary pits) for translocating photosynthates;

(4) armoured reproductive structures (conceptacles).

These traits have been shown experimentally to serve an antipredation function.

   Often in coralline algae, only the top 100–200 μm is photosynthetic, so the outer cell layers support the rest of the plant by translocation of photosynthates (Steneck 1983). When a deep grazing herbivore bites into algal crust, the removed tissue can rapidly regenerate into a meristem together with the necessary photosynthetic apparatus. So in coralline algae, primary substrate is never exposed and therefore protected from the invasion of fouling microalgae. The epithallus has been shown to protect the meristem from deeply excavating herbivores (Steneck 1982), which also protects the region that produces both reproductive structures and photosynthetic tissues (Steneck 1983). The thickness of the underlying hypothallus relative to the thickness of the entire crust, however, is positively correlated with the rate at which an individual can grow laterally.

   Fusion cells, together with primary and secondary pits, allow the translocation of photosynthates throughout the thallus, and so are important for both the growth and survival of the hypothallus. This is supported by the fact that:

(1) the growing edge of corallines is white because the apical meristem and hypothallus tissue is composed of non-photosynthetic cells;

(2) living hypothallial cells are commonly buried under a thick opaque layer of perithallus (the middle layer of tissue between the epi- and hypothallus, Fig. 7.17), where most photosynthesis occurs.

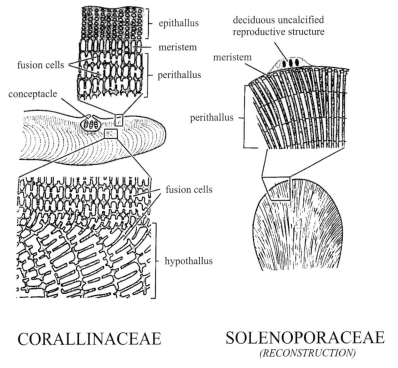

CORALLINACEAE   SOLENOPORACEAE
                              (RECONSTRUCTION)

Fig. 7.17 The anatomy of solenopore (reconstructed) and coralline algae. (Modified from Steneck 1983.)

Crusts grazed by parrotfishes show that wounds which penetrate beyond the zone of photosynthesis are healed by regeneration of the perithallial cells aided by effective translocation from fusion cells. Coralline algae respond to very deep wounds by forming new hypothalli laterally over dead tissue.

Conceptacles are enclosed cavities within the perithallus that contain reproductive structures. These structures have been demonstrated to protect reproductive anatomy from intensive grazing (Steneck 1982, 1983). Conceptacles are, however, no match for intense excavation by parrotfishes, perhaps explaining why they are only found on non-tropical thickened crusts.

The several hundreds of species of living coralline algae can be divided into three convergent morphologies—thin, branched, or thick—which can be considered as broad adaptive strategies (Steneck 1985). Which growth form dominates in any habitat can be considered as a trade-off between the cost of investing in increased defence, and a concomitant reduction in growth rate or competitive ability.

These growth forms can be either free-living or firmly adherent to a hard substrate (encrusting). In shallow, tropical waters, their distribution corresponds to the intensity of disturbance in their environment (Fig. 7.18). With increased disturbance, the proportion of rapidly growing thin corallines decreases and the proportion of slower growing thick forms increases. The rapid growth and recruitment rates of leafy crusts enables them to overgrow other crusts in the absence of disturbance. But the intensity of herbivory in the shallow tropics restricts such forms to refuges of low disturbance such as intertidal, cryptic, or deep-

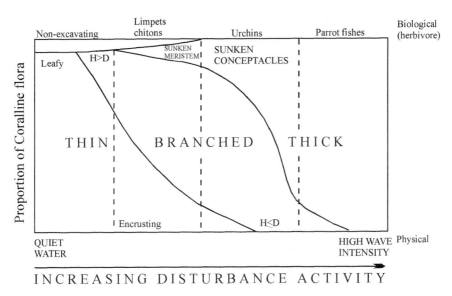

Fig. 7.18 Adaptive strategies of coralline algae with respect to disturbance from biological (upper) and physical (lower) sources. H: algal height, D: algal diameter. (Modified from Steneck 1985, by kind permission of Springer-Verlag.)

water niches. For example, herbivorous fish graze the top peaks more than the valleys such that thick crust (*Porolithon pachyderma*) dominates the peaks but thin crusts dominate in the valleys. A further example of the significance of growth form is given by the brown alga *Padina,* which forms prostrate growth forms (with thin branches) in areas of intense herbivory, but within 96 h of excluding herbivorous fish it will adopt a rapidly growing, upright blade-like form that, while being very susceptible to fish grazing, is more competitive than prostrate forms and may even overgrow or kill corals (S.M. Lewis 1986).

Branched morphologies not only increase the area of photosynthetic tissue, but are also effective in defence from the grazing by limpets or urchins, as closely spaced protuberances make adherence difficult such that only the branch tips are accessible for grazing. However, branched morphologies are easily consumed by mobile parrotfishes, such that only thick corallines can survive their attack. On the GBR for example, branched coralline algae are found only above and below the zone of intense parrotfish grazing (Steneck 1982). This might explain the preponderance of branching growth forms in temperate and Arctic coralline algal reefs, which do not suffer intense excavatory herbivory.

Both encrusting coralline algae and corals rely on grazers to prevent smothering. That the survival and growth of recently settled coral and crustose coralline algae is a consequence of antipredatory traits rather than competitive abilities is confirmed by the observation that if these slow-growing algae were present in areas where algal turfs and macroalgal were able to flourish, they would rapidly be overgrown and die (e.g. S.M. Lewis 1986).

● Solenopores

The extinct solenoporacean algae differ in their physiological integration and reproduction from corallines in that they were characterized by a meristem and probably bore raised reproductive structures (Fig. 7.17). They bore no epithallus, fusion cells, and for much of their history they also lacked a hypothallus. The lack of epithallus suggests that the meristem and underlying perithallus were unprotected from excavating herbivory, and the absence of fusion cells predicts that the solenopores also had poor regenerative abilities from excavatory herbivory. The is confirmed by the common occurrence of wounding in post Mid-Jurassic solenopores (Steneck 1983).

The lack of a well-developed hypothallus in solenopores suggests that, like corallines with reduced hypothalli, their ability to encrust hard substrates may have been limited and their growth rate slow (e.g. Wray 1977). As a result, solenopores probably showed relatively slow rates of lateral growth over unconsolidated substrates, and this together with the raised, uncalcified reproductive structures would have made them highly susceptible to herbivory. Modern coralline algae which most closely resemble solenopores have massive crusts with reduced (single-layered) hypothalli (e.g. *Hydrolithon* sp.). Such crusts grow slowly, possess few antiherbivore defences, and thrive only in back-reef and deep-water habitats where biological disturbance is low. These environments are similar to those in which solenopores once grew (Wray 1972).

Distribution and abundance

Microalgae and calcified cyanobacteria   Stromatolites appeared 3.5 Ga, and show greatest abundance and diversity from 2.5 to about 1.2 Ga. After the late Palaeoproterozoic, stromatolites suffered three periods of decline, in terms of both diversity and abundance: at around 1 Ga, at the base of the Cambrian (~545 Ma), and a final decline after the Early Ordovician (some 490 Ma). Lithified microbial/algal/stromatolitic crusts in conjunction with skeletal forms continued, however, to be important components of reef ecology, especially in the Late Palaeozoic to Early Mesozoic.

The decline of Proterozoic stromatolites is commonly cited as being due to the rise of grazers such as gastropods and monoplacophorans (Awramik 1971) and burrowing organisms which disrupt lamina formation (Walter and Heys 1985). But in addition to the lack of experimental evidence for the deterrent properties of non-denuding herbivores, the temporal gap between the decline (1000–545 Ma) and the first known occurrence of metazoan body fossils (~600 Ma) questions the validity of this hypothesis.

What unites modern stromatolites—whether they grow in normal or abnormal salinities—is that they all develop in environments rich in calcium carbonate combined with low nutrient levels that allow cyanobacteria, but not other faster growing, but nitrogen-limited, algae to thrive (Box 3.1). High carbonate levels not only provide a source of carbon, but also enable stromatolite construction. Here it would seem that competitive exclusion, rather than herbivory, is an important determinant of stromatolite distribution. This is confirmed by the observation that in areas of stromatolite growth now suffering nutrient pollution—such as the increasing levels of phosphate in the groundwaters that feed Lake Clifton in western Australia—the prolific growth of other algae is smothering cyanobacterial communities and impairing stromatolite formation (McNamara 1992). But while filamentous green algal turfs are capable of rapid growth and regeneration, it appears that the slow-growing, stromatolite-forming cyanobacteria recover only very slowly from disturbance. Substrate competition by the evolution of faster growing, higher algae (Pratt 1982) in non-nutrient-limited settings, and the decrease in the carbonate saturation of surface sea water through the Proterozoic (Grotzinger 1990) may have been responsible for the decline of stromatolites. The first seaweeds appeared approximately 1100 Ma and the first eukaryotes 1800 Ma.

Calcified cyanobacteria were commonly associated with reefs until the end of the Palaeozoic. The calcified cyanobacterium *Renalcis* survived the Late-Devonian (Frasnian–Famennian) mass extinction into the Lower Carboniferous (Tournaisian) (Chuvashov and Riding 1984), while all calcareous reef algae went extinct in the mid–late Famennian. Although there is some possible evidence of grazing activity as indicated by 'cropped' calcified cyanobacteria from the Lower Cambrian Bonavista Group of eastern Newfoundland (Edhorn 1977) which may have been due to the activities of orthothecimorph hyoliths, and from the Toyonian Shady Dolomite of Virginia (Kobluk 1985), there is no evidence that the decline and extinction of calcified cyanobacteria is due to increased herbivory.

Articulated calcareous algae   *Halimeda* appeared in the late Miocene (~6 Myr), and extensive banks and bioherms which grew at subtidal depths are known from this time (Martín *et al.* 1997).

Crustose algae   The systematic placing of many phylloid algae is problematic: many may be members of the Chlorophyta and Rhodophyta. From well-preserved examples, some Late Carboniferous algae such as *Eugonophyllum* are known to have udoteacean affinities (Kirkland *et al.* 1993), but they had phylloid ('leaf-like') rather than articulated organizations (see Box 3.9). However, *Eugonophyllum* possesses sunken, calcified oval chambers which might represent cryptic reproductive structures, that are similar to those found in some dasycladacean algae (order Siphonales).

Corallines.   The anatomical features unique to modern corallines appeared suddenly with evolution of the oldest abundant coralline algae, *Archaeolithophyllum,* known from the Late Carboniferous. *Archaeolithophyllum* formed a thin, leafy ('phylloid') crust which had a differentiated thallus and strongly raised conceptacles (Wray 1964).

Archaeolithophyllum grew in shallow agitated reef environments, forming thick algal banks of unattached laminar crusts or cup-shaped forms that could grow rapidly laterally or upward. They proliferated on unconsolidated substrates—a growth habit that is unknown today except in some peysonnelids. Although *Archaeolithophyllum* bore raised conceptacles, several Mesozoic corallines appear to have had sunken conceptacles.

Coralline algae are unknown from the Triassic, but from the Jurassic their diversity increased, and they underwent a spectacular radiation beginning in the Late Cretaceous which continued through the Cenozoic (Fig. 7.19(a)). Unlike the largely free-living habit of thin leaf-like crusts or erect growth forms of Palaeozoic representatives, post-Palaeozoic corallines are most commonly found encrusting hard substrates, often in association with reefs. The first known coralline algal ridges are from the Eocene of north-east Spain, formed by foliaceous frameworks of *Mesophyllum* and the subsidiary *Lithophyllum* which show some development of branches (Bosence 1983).

Other trends inferred to be adaptive continued through the Cenozoic (Steneck 1983). Changes in proportion of different morphologies correspond with the trend of escalating herbivory: thin and branched (herbivore-susceptible) forms declined, and correspondingly thick (herbivore-resistant) morphologies increased (Fig. 7.19(b)). Delicately branched corallines reduced in importance in the tropics after the Eocene (J.H. Johnson 1961).

Multilayered epithalli first appeared in the late Cenozoic, which has been demonstrated to offer protection from deep-grazing chitons and limpets (Steneck 1977). The thick, coralline algae that dominate zones of intense grazing today did not become abundant on reefs until the Pleistocene (J.H. Johnson 1961).

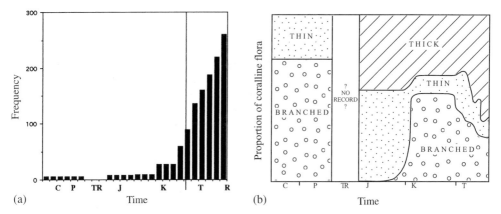

Fig. 7.19 (a) Diversity of coralline algal species through time. (b) Proportion of coralline species with branched, thin, or thick crusts (*n* = 527 fossil species). C: Carboniferous, P: Permian, TR: Triassic, J: Jurassic, K: Cretaceous, T: Tertiary, R: Recent. (Modified from Steneck 1986.)

There is evidence for the coevolution of prey and herbivore in the tropical genus *Porolithon* and the cold-temperate *Clathromorphum*. These genera have lost the ability to slough epithallial cells (without which reproduction cannot occur) and require deep-grazing limpets to perform that function (Steneck 1983).

Solenopores.   Solenopores first appeared in the Cambrian and rapidly increased thereafter in diversity and abundance. Two periods of maximal abundance were the Ordovician and particularly the Jurassic (Wray 1964). The group declined after the Jurassic, became rare in the early Tertiary and went extinct in the Miocene (Elliott 1973). Most solenopores were massive or nodular and lived in quiet waters such as back-reefs, or in deeper waters as unattached nodules.

The prediction that solenopores were poorly defended from excavatory herbivory has been confirmed by a survey of micrographs of fossil solenopores from 72 monographs (Steneck 1983). The proportion of wounded species increased after the mid-Jurassic, and none of the Cenozoic representatives were without wounds. By contrast, coralline algae from the same assemblages showed an insignificant proportion of wounds, demonstrating that corallines were either better defended from excavation or healed so effectively as to leave no trace of disruption.

Summary

The temporal distribution of reef-associated algae through the Phanerozoic broadly follows a trend of increasing ability to withstand disturbance (Fig. 7.16). The decline of the stromatolites towards the end of the Proterozoic may only be indirectly due to the rise of herbivores; it was more likely the result of competition with newly evolved groups of faster growing algae, and a possible decline of supersaturation levels with respect to calcium carbonate (Section 4.4.4).

Poorly defended calcified microalgae declined at the end of the Palaeozoic, but for reasons that appear to be unconnected with any escalation in herbivory. Many Palaeozoic tropical shallow reef-associated algae (including algal mats) were free-lying forms which were able to cover soft substrates rapidly, whereas most post-Palaeozoic reef floras encrusted hard substrates. This shift in dominant growth morphology to thick, encrusting forms can plausibly be laid at the feet of increasing herbivory, even though the trade-off was slower rates of growth. Solenopores sharply declined in the late Mesozoic–early Cenozoic with the onset of excavatory grazing, and this was coincident with a reciprocal rise in the abundance and diversity of coralline algae. Herbivore-susceptible, delicately branched corallines reduced in importance in the tropics after the Eocene (J.H. Johnson 1961), and the first algal ridges are known from the Eocene—both coincident with the evolution of herbivorous fish, although the first scarid is not known until the Miocene, and indeed may not have become important in reef communities until even later (Bellwood 1994). Escalating herbivory is also conjectured to have resulted in the progressive disappearance of fleshy macroalgae from shallow marine environments through the Mesozoic (Steneck 1983).

The well-developed hypothallus of corallines initially allowed the rapid lateral growth required for life on a soft substrate. The hypothallus, together with the presence of fusion cells, also allowed the Corallinaceae to encrust (and so reduce/ease predator manipulation), to acquire branching morphologies (and so increase photosynthetic powers and growth rates), to produce conceptacles, and to regenerate from deep wounds—all features which probably enabled corallines to radiate as herbivory intensified through the mid–late Mesozoic and Cenozoic. Yet more than 100 Myr elapsed between the first appearance of coralline algae and the subsequent radiation of the group (Steneck 1983), strongly suggesting that these distinctive features of coralline algae are all exaptations. One exception might be multilayered epithallial growth, which might be an adaptation which arose in response to intense grazing by particular species of limpets in the Pleistocene.

Palaeozoic free-living sessile invertebrates

Functional anatomy and differential susceptibility to predation

Organisms without secure attachment to a stable substrate are susceptible to the adverse affects of bioturbation. Invertebrates that are inverted and buried will die unless they are able to re-establish contact with the water column to continue feeding and respiring. Newly settled juveniles will be most susceptible to disturbance, and so initial settlement and growth will be considerably hindered thereby reducing successful recruitment. In Recent shallow-shelf seas, many immobile epifauna are excluded from most soft substrates, which are dominated by mobile deposit-feeders. While immobile (but unattached) corals are common today they are restricted mainly to areas protected from high biological and physical disturbance. Most modern suspension-feeders require a hard substrate, even if these are only isolated patches within areas of unstable, soft substrate ('benthic islands', see Fig. 6.5(b)), or in dense aggregations.

Many mid–late Palaeozoic reefs were dominated by large, sheet-like invertebrates (stromatoporoid sponges, tabulate and rugose corals, and trepostome and cystoporate bryozoans) that were initially attached to small, ephemeral skeletal debris and then spread over the surrounding sediment, often covering a substantial area. These forms were unspecialized and immobile, and only bryozoans were able to regulate their position by current action. Continued thickening of the colonies often led to the development of hemispherical morphologies, and some examples could reach several metres in diameter. Some forms could attach to more stable hard substrates, but in most there is little evidence for any active recruitment onto extensive hard substrates. Small, branching forms (stromatoporoids and bryozoans) lacking extensive attachment sites were also common, and they were presumably partially rooted in soft sediment.

This substrate preference is somewhat complicated by the effects of physical disturbance. Areas of strong currents and wave impact will remove much loose and fine sediment, yielding a preponderance of more stable hard substrata—the distribution of substrata is thus in part dependent upon the ambient hydrodynamic regime. So while the frequent attachment of much Palaeozoic reef immobile benthos to ephemeral substrates might reflect an ability to gain secure attachment to extensive substrates and so preclude growth in energetic regimes, it might also reflect the limited availability of more stable substrates.

However, tabulate and stromatoporoid corals could achieve some stability in large size, and so some resistance from both physical disturbance and bioturbation. Some also produced lateral 'outriggers' that hovered over the sediment (Fig. 7.20); others had laminar morphologies that formed hollow domes. These novel growth habits may have been a bid to escape smothering by episodic sedimentation, or to avoid the destabilizing effects of the infauna therein. In the absence of excavatory predation in the Palaeozoic, such thin plates would not have been susceptible to biological destruction. Many of these immobile, modular Palaeozoic fauna also showed considerable powers of regeneration from partial mortality.

It difficult to determine the relative sediment-stabilizing or binding ability of Palaeozoic immobile reef invertebrates, but that they grew predominantly on muddy or sandy soft sediment is beyond dispute. Many were clearly gregarious.

Although stromatoporoids and many Palaeozoic corals show the ability to reorient growth after partial overturning, toppling, or burial by sediment incursions, none had any mechanisms for self-righting (with the possible exception of some hadrophyllid rugosans (Gill and Coates 1977)), and none appear to have acquired the free-living, mobile habit.

Distribution and abundance

Unattached epifauna, such as reclining brachiopods, maclauritacean and euomphalacean gastropods, as well as unattached stabilizers of sediment, such as crinoids, laminar bryozoans, corals, and stromatoporoids diversified through the Phanerozoic, especially in the early and middle Palaeozoic (Fig. 7.21). However, analysis of the stratigraphic record of immobile versus mobile groups during the Late Palaeozoic shows some striking statistically significant differences in diversity (Thayer 1983). Immobile organisms show a marked decrease in the Late Devonian, but the mobile groups capable of burrowing (such as malacostracans and holothurians) increased (Fig. 7.21). Likewise, the diversity of molluscs and brachiopods predicted to be immune to bioturbation (e.g. free-burrowers and cementers) increased, but the proportion of susceptible, immobile groups (pedunculate, endobyssate, and free-lying) declined. However, this decline is also coincident with the Late Devonian mass extinction event, making the differential effects of physicochemical environmental change, and increasing biological disturbance of substrates difficult to disentangle.

The Late Permian extinction event removed most of the remaining major sessile metazoan components from Palaeozoic reef communities: epifaunal immobile suspension-feeders have been rare since that time (McKerrow 1978), and the re-establishment of an abundant

Fig. 7.20 Some Upper Devonian stromatoporoids produce lateral, plate-like 'outriggers' (S) that hovered over the sediment. The undersurface of this example has been colonized by the calcified cyanobacterium *Renalcis,* which is encased within early marine cement. The shelter cavity has been subsequently filled by layered internal sediment and cement. Windjana Gorge, Canning Basin, western Australia.

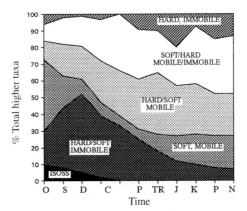

Fig. 7.21 The proportion of epifauna in functional groupings through the Phanerozoic. O: Ordovician, S: Silurian, D: Devonian, C: Carboniferous, P: Permian, TR: Triassic, J: Jurassic, K: Cretaceous, P: Palaeogene, N: Neogene. (Modified from Thayer 1983, by kind permission of Plenum Press.)

Mesozoic fauna included relatively few free-lying forms—most post-Palaeozoic sediment-stabilizers are plants such as seagrasses. In most cases, however, the decline began in the late Palaeozoic, but in the case of the stromatoporoids as early as the Late Devonian. All these groups either became extinct or restricted to hard substrates or marginal habitats where many deep bioturbators are excluded.

Summary

The late Palaeozoic decline of immobile epifauna coincides with the rise of major bulldozing taxa, which passed through the end of the Permian extinction unscathed (Fig. 7.14). But while it seems possible that the rise of bioturbation is responsible for the decline of the Palaeozoic free-lying epifauna, this hypothesis must remain conjectural until tested experimentally.

Thayer (1983) has postulated that the Modern deep sea may suffer the same degree of bioturbation as early Palaeozoic shelves, and indeed the deep sea appears to harbour an immobile soft-substrate fauna of shallow marine Palaeozoic caste. Stalked crinoids, articulate brachiopods, hexactinellid sponges, and free-living immobile bryozoans are all concentrated, often in considerable abundance, in the deep sea. It has long been suggested that archaic Cambrian and Palaeozoic faunas migrated to deep-sea environments from shallow shelves. But alternatively, rather than being controlled by levels of bioturbation, this migration from near-shore to deeper waters might equally reflect the competitive displacement of these groups by the subsequent appearance of better substrate competitors.

The decline in erect bryozoans since the Jurassic may, however, be related to the rise of bioturbation, and this is discussed in the following section.

Bryozoans

Functional anatomy and differential susceptibility to predation

Erect growth forms in bryozoans are more susceptible to predation than encrusting morphologies. For example, the sea urchin, *Centrostephanus coronatus* feeds preferentially on erect bryozoans (and other sessile erect organisms) on shallow, hard substrates off the coast of southern California. Erect growth forms are therefore rare except in areas where the urchin has been experimentally removed, or is naturally absent (Vance 1979). Likewise, the Atlantic bryozoans *Steginoporella* spp. and *Stylopoma spongities* grow as encrusters on coral reefs, but as foliaceous erect colonies offshore where they are few crushing predators (J.B.C. Jackson, unpublished data). (Such a flexible morphological response to differing biological disturbance is also notable in the sponge *Cliona celata*; this has a wholly endolithic habit in the presence of heavy urchin grazing, but in their absence grows as an encrustation and eventually forms small, erect branches (Guida 1976).) These data suggest that predation might be more important that substrate stability for the occurrence of erect bryozoans, at least in shallow waters. Adoption of a free-living, mobile mode of life—with powers of self-righting—would be favoured in soft-sediment environments of high physical or biological disturbance.

Species of encrusting bryozoans differ in the way in which new zooids are added to grow across a substrate, and also in their potential for vertical budding away from the substrate (Section 6.3.3). Zooidal budding (in particular, multizooidal (giant) budding) dominates bryozoan communities on modern stable substrates, and is associated with rapid rates of lateral growth, large colony size, and colony thickness. These forms also have greater flexibility at colony margins and have increased regenerative capacity. Thick, multilayered colonies, formed by frontal budding of new zooids one on top of another, are better defended against overgrowth than single sheets, as well as from predators and grazers (McKinney and Jackson 1989). Their capacity to repair injured tissue is also markedly greater than other budding types. For example, *Gemelliporidra belikana* apparently survives heavily grazed areas of coral reefs by the early onset of frontal budding (before the colony has reached 10 zooids) which imparts increased relief to this already heavily calcified species (M. Gleason and J.B.C. Jackson, unpublished data).

All abundant bryozoans of open, modern reef surfaces are heavily calcified encrusters. The heavily calcified zooids of *Reptadeonella costulata*, with obvious surface fortifications and thick lateral and frontal walls, makes them 15 times harder to puncture, and nearly twice as hard to crush as the zooids of *Steginoporella* sp. which lack any secondary calcification (Best and Winston 1984). The presence of well-developed cryptocysts (zooids with a shelf-like interior wall) has been shown to limit feeding by small carnivores such as isopods that consume individual zooids (Buss and Iverson 1981).

Abundance and distribution

Palaeozoic faunas are overwhelmingly dominated by erect species both in terms of abundance and diversity, but from the Jurassic–Cretaceous these forms became increasingly restricted to mid-shelf and deeper waters. Since the late Mesozoic, they have been largely confined to cryptic niches, and have narrow-branched, flexible, and articulated forms, rather than the robust and rigid morphologies shown by those which dominated shallow seas before the Cretaceous. From the Jurassic onwards, bryozoan faunas became increasingly dominated by encrusting forms, especially in the Neogene (Fig. 6.26), such that the modern depth distributions of bryozoans show that the upper depth limit of rigidly erect species is very different from Palaeozoic distributions (McKinney and Jackson 1989). Multilayered growth (in cheilostomes) did not become common until the Cenozoic.

Several bryozoan groups show a trend of increased calcification of zooids, such as that associated with the rise of fenestrates (Ordovician–Permian), and the sequential change in the relative abundance of anascans, cribrimorphs, and ascophorans (McKinney and Jackson 1989).

The four 'orders' of post-Palaeozoic bryozoans form a series of increasing skeletal defence, interzooidal integration, and zooidal polymorphism that follows their sequence of stratigraphic appearance (Jackson and McKinney 1991). In particular, there is a marked increase in the proportion of cheilostome relative to cyclostomes since the Early Cretaceous (P.D. Taylor and Larwood 1988; Lidgard *et al.* 1993). Zooidal budding appeared in the early, Late Cretaceous (~100 Ma), and the proportion of species which have independently acquired this type of budding show a marked increase from the latest Cretaceous (~70 Ma) to the Recent (Lidgard 1986). Frontal budding likewise shows an increase from its first appearance in the Early Palaeogene (~60 Ma) (Fig. 6.27(b)).

The proportion of free-living, immobile bryozoans also decreased during the Palaeozoic (Fig. 6.26). Free-living bryozoans, however, gained the ability to be mobile on sediments in the Late Cretaceous to Eocene with the appearance of several specialized groups (including the Lunulitidae, Conescharellinidae, Setosellinidae, Mamilloporidae, Orbituliporidae). From that time the proportion of the mobile representatives increased dramatically and has remained high to present times.

Summary

Of these surveyed trends in bryozoan morphology, the increase in the proportion of fauna with encrusting morphologies, continuous zooidal and frontal budding, and rapid growth and good powers of regeneration have markedly increased since the Mesozoic, especially from the Late Cretaceous–Eocene. These trends all coincide with the rise of biological disturbance. Moreover, the free-living, mobile habit also appeared at this time.

Scleractinian corals

Functional anatomy and differential susceptibility to predation

Scleractinian corals show a range of traits which have been suggested to serve an antipredatory role. Those that might be detected in the fossil record include:

- the ability for secure and permanent attachment to a hard substrate. Some clades also acquired a free-living mobility.

- the potential to reach a very large size. Competitive ability, regenerative abilities, and sexual reproductive capacity of colonial animals are all strongly correlated to individual colony size (Jackson 1985).

- the possession of predominantly multiserial massive or erect colony, i.e. the overwhelmingly dominant growth forms of living photosymbiotic forms (Coates and Jackson 1987). Multiserial modularity, in addition to promoting architectural diversity and flexibility, also allows compartmentalization of damage and enables some colonies to regenerate from fragments (Jackson and Hughes 1985).

- rapid growth and calcification rates. These may be conferred by photosymbiosis (but see Section 8.3.2).

- invasion of spatial refuges, such as intertidal and deep-water habitats.

Permanent attachment and the free-living habit

Possession of an edge zone allows scleractinian corals to gain permanent attachment to a stable substrate. Not surprisingly, scleractinian corals dominate reef framework environments, especially those in high-energy settings, but they may be scarce in adjacent unconsolidated sediments. However, while secure attachment to a hard substrate allows some scleractinians to grow successfully in such environments, this may be a result of the preponderance of wave-swept, extensive hard substrata for colonization. Permanent attachment allows, however, for the development of very large branching morphologies.

Attachment is ample defence from bioturbation, and considerably reduces the ease of manipulation by potential predators. But as none of the major predators upon corals need to manipulate their prey, this factor is unlikely to have any direct antipredatory value. Indeed, there is no experimental data to show that attached corals are any more susceptible to predation than unattached forms.

Mobile, free-living, corals such as fungiids are able to creep, right themselves, and are capable of extracting themselves from burial. Some corals (e.g. *Heteropsammia*) grow around a symbiotic siphunculid, which 'tows' its coral host. Although many are solitary, all living mobile corals are photosymbiotic and clonal, suggesting that either rapid growth and/or regenerative powers may be important prerequisites for the acquisition of this condition. Again, while a free-living habit might provide defence from bioturbation, it is difficult to imagine that such forms have achieved sufficient mobility to flee from potential predators.

Large size

In most marine invertebrates, larger colonies are less susceptible to the agents of mortality than smaller ones (Section 6.1.4). Juvenile corals are particularly vulnerable to predation until they can reach an escape in size—that is become sufficiently conspicuous to be avoided (Birkeland 1977). There is some evidence to suggest that corals direct their energies initially to colony growth before the onset of reproductive maturity, that is they delay reproduction so as to increase the chance of growth to a larger, safer size. Also, older colonies may preferentially allocate resources to the regeneration of damaged tissues rather than for further growth (Bak 1983). In all reef-building species examined in one study, there was a colony threshold size below which no mature gonads were found (Hall and Hughes 1996). Interestingly, older colonies will continue gametogenesis even if they become fragmented below the threshold size required for younger colonies to become sexually mature (Kojis and Quinn 1985). It is therefore likely that the relative investment in growth declines with colony size, and often age.

However, we have seen that large size offers multiple benefits for sessile reef invertebrates, and in itself is not an antipredatory trait.

Growth form and regenerative abilities

Growth form is an important determinant of survival. Preliminary evidence suggests that in primary (solitary) polyps there is no regeneration when the central part of the polyp (the mouth-stomadaeum) has been destroyed by predation (Bak and Engel 1979). Corals that form or occupy topographically complex surfaces (as well as those inhabiting cryptic niches) can preferentially survive predation, as well as outbreaks. Damselfish have been demonstrated to promote the growth of *Pocillopora* at shallow depths (Section 7.1.12), suggesting that the closely branching colony form in this coral might represent an antipredatory trait.

Generally, predators prefer erect branching or platy corals to encrusting forms (Glynn *et al.* 1972; Rylaarsdam 1983). But as most partial mortality is caused by many processes which occur close to the substrate, 'escape in height' by the acquisition of a branching habit will increase the chances of survival dramatically (Meesters *et al.* 1997), and so represents a significant defensive feature. Most significantly, branching corals also show tremendous powers of regeneration: *Acropora palmata* has one of the highest rates recorded (Bak 1983). Indeed, unlike massive, platy or encrusting forms, damage to branching corals often leads to an immediate increase in growth rate so causing an increase in size rather than simply repairing damaged tissue (Fig. 7.22(a)).

Populations of staghorn corals (*Acropora cervicornis*) frequently form dense, monospecific stands on shallow Caribbean reefs. The fragile organization of this species results in easy breakage as a result of both high wave activity and bioerosion, especially by boring sponges which infest the colony bases. However, such corals are able to reanchor fragments and rapidly regenerate and grow, often fusing with other colonies, at rates up to 150 mm year[-1], but there is little evidence of sexual recruitment (Tunnicliffe 1981). Such branching corals have turned adversity into considerable advantage, and appear to flourish because, and not in spite, of breakage.

When the living tissue of a coral is damaged, the surrounding polyps respond by generating new tissue. Initially, tissue regeneration is very fast, but then slows down exponentially (Bak 1983). The size–frequency distributions of naturally occurring lesions have been found to be very skewed to the right in the three reef-building corals *Acropora palmata, Porites astreoids,* and *Diploria strigosa,* showing that most partial mortality is both small in size and well within the regeneration abilities of these species. The presence of permanent lesions on other coral colonies, however, may indicate that regeneration capacity has been limited.

Different coral species show widely varying abilities of regeneration, which are species-specific and related both to lesion size and to environmental factors. Large colonies are able to regenerate lost tissues faster than smaller representatives of the same species (Connell 1973; Loya 1976; Bak *et al.* 1977), and the same species may show different regenerative abilities when growing at varying depths (Bak and Steward-Van Es 1980). In the Modern, partial mortality is almost absent in very small colonies or individuals. Since a lesion is often as large as the total living surface this results in total mortality, and, as a result, naturally occurring damaged colonies less than 2.5 cm in diameter are rarely found (Meesters *et al.* 1997). Brooding is associated with small maximum colony size, and this mode of reproduction has been suggested to have evolved to overcome the high rates of mortality associated with small size (Meesters *et al.* 1996). Also, different types of lesions may influence the regeneration rate. For example, some species (e.g. *Montastraea annularis*) are able to regenerate faster when the skeleton as well as soft tissue is damaged (Fig. 7.22(b)). Such observations suggest that different predators may have differential responses on coral prey. *Montastraea annularis* will recover more slowly from denuding predators than, for example, *Agaricia agaricites*, but will not suffer so greatly from excavating predation (Fig. 7.22(c)).

Rapid rates of growth and calcification

Rapid rates of growth can give clear competitive superiority to sessile organisms, and calcification too offers a structural defence from predation. The autotrophic capabilities of most reef-building corals also removes the need to expose tentacles during day, when most predators are active. This raises interesting questions as to the possibility that non-zooxanthellate reef-building corals, such as *Tubastraea,* may bear toxic metabolites which counteract predation. The fact that corals—like *Halimeda*—show skeletal extension during the night when herbivorous fish are inactive (see Section 8.2.3) might also be inferred to be an antipredatory trait. This has yet to be proved experimentally and the search for, and understanding of, the secondary metabolites present in corals is still in its infancy.

Spatial refuges

The intertidal habit has reduced predation pressure, but presents physiological difficulties as organisms must overcome the problems of thermal stress, ultraviolet irradiation, and desiccation. Some corals can withstand the intertidal habitat by a variety of methods including the presence of secondary metabolites which act as sunscreens and retraction of polyps (see the summary in Brown 1997*a*).

Many relatively poorly defended living corals such as those with small, solitary, or low-integration branching organizations with slow rates of regeneration may be found within protected refuges of the reef framework—under overhangs or within crypts.

Deep waters have been successfully invaded by some scleractinian corals. Azooxanthellate forms can thrive at depths of up to 300 m; the zooxanthellate species *Leptoseris fragilis* can photosynthesize at depths of up 145 m, due to the presence of additional photosymbiotic pigments and specialized skeletal structures that aid heterotrophic feeding in waters of highly reduced illumination (Schlichter 1991; see Section 8.3.1—Specialized skeletal structures).

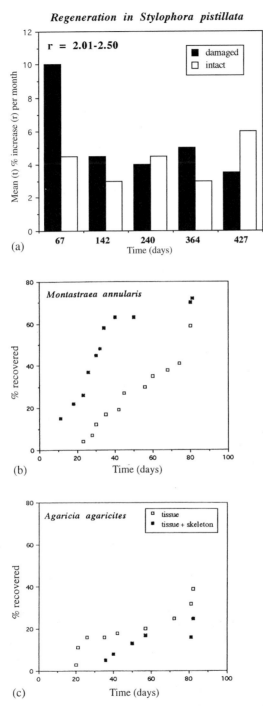

Fig. 7.22 Regeneration rates in modern corals. (a) *Stylophora pistillata* (replotted from Loya 1976, by kind permission of Macmillan Magazines Limited). (b) *Montastraea annularis* (replotted from Bak *et al.* 1977). (c) *Agaricia agaricites* (replotted from Bak *et al.* 1977).

Distribution and abundance

All except a few genera of the most primitive scleractinian corals (Wells 1956) possess an edge zone and so were capable of encrusting a hard substrate. This raises the interesting question of when did permanent attachment become important? When this transition occurred, however, is difficult to ascertain. Attachment sites are problematic to detect in outcrops, and many Jurassic and Cretaceous reef corals appeared to have lived on unconsolidated substrates in areas of high sedimentation rates, although some possessed elevated platy morphologies (see Fig. 3.31(b)). Large, branching Upper Triassic reef corals are thought not to have been free-standing, but encased and supported by sediment (Satterley 1994).

Active, free-living microsolenid and caryophyllid corals first appeared in the Jurassic, and Fungiidae and Flabellidae during the Cretaceous (Wells 1956; Gill and Coates 1977). The apparently mobile cyclotid corals from the Upper Cretaceous are remarkably convergent to the fungiids in the modern Australian fauna (Gill and Coates 1977). The first 'towed' corals are also known from the Upper Cretaceous.

When did colony size increase? Large size offers many advantages to reef-builders, but there is only limited evidence available to assess how colony size has changed through the history of the Scleractinia, although we might predict that the average size of shallow-water forms may have increased through the history of the group. Very large colonies (up to 10 m in height)—albeit with fragile branches—were certainly present by the Upper Triassic (e.g. '*Thecosmillia*'). Anecdotal evidence suggests that large colony size did not become common until the Cenozoic. In an attempt to analyse the causes of accelerated turnover in Caribbean Pliocene–Pleistocene (4–1 Ma) reef corals, K.G. Johnson *et al.* (1995) analysed susceptibility to extinction in Miocene–Recent faunas within different ecological groupings based on colony size, colony form, corallite size, and reproductive characteristics. Only colony size showed significant differences in evolutionary rate. Extinction rates were higher for species with small, massive colonies, which tend to live in small, short-lived populations with highly fluctuating recruitment rates and limited dispersal abilities. This differential extinction resulted in an increase in the proportion of species with larger colonies, which tend to have larger population sizes, longer generation times, and more constant rates of population increase.

Scleractinian corals also show a steady and marked increase in the proportion of high-integration forms since the Mid-Triassic, which appears to be uninterrupted by the end of the Cretaceous extinction event (Fig. 6.24(a)). In contrast, the percentage of scleractinian erect species (mainly low-integration, phaceloid–dendroid growth forms) decreased until the Turonian, but increased markedly—particularly in multiserial forms with inferred rates of rapid regeneration—after that time (Fig. 6.24(b)).

Although there is abundant evidence of damage and subsequent regeneration in scleractinian corals, regeneration from predation has proved difficult to differentiate from that due to other sources of partial mortality. One exception is the identification of distinctive gall-like structures or 'chimneys' found in Pleistocene and modern faunas, which are known to form in response to the grazing of the three-spot damselfish (*Eupomacentrus planifrons*) on *Acropora cervicornis* (Kaufman 1981). Similar cavities are formed by other corals regenerating over lesions colonized by algae (Bak *et al.* 1977). The highly territorial, three-spot damselfish grazes upon algal turf by killing all live coral within its territory and excluding other herbivores. However, remnants of *A. cervicornis* can regenerate rapidly from such behaviour.

Mean corallite diameter does not appear to have remained the same through the history of the Scleractinia. In a survey of 1600 species, Coates and Jackson (1985) showed that whilst corallite diameter has been remarkably consistent in clonal forms, there was a marked drop in the mean diameter of solitary/aclonal scleractinians in the Eocene (Fig. 7.23).

Scleractinian corals occupied the abyssal plains, cool high-latitude shelves, deep-water coral banks and tropical–subtropical carbonate banks, as well as platforms and reefs by the end of the Early Jurassic (Toarcian) (Roniewicz and Morycowa 1993). Low-diversity associations of platy corals with skeletal structures similar to the living *Leptoseris fragilis* are known from a variety of Jurassic (Oxfordian), Early Cretaceous (Barremian–Early Aptian, and Albian), and Late Cretaceous localities (see Insalaco 1996; Rosen 1998).

Summary

Many of the skeletal characteristics of modern reef-building corals have antipredatory qualities, but most of these offer multiple benefits, and are at best exaptations. Many of these traits have been present in some representatives of the Scleractinia from the early origins of the group, but have proliferated as they subsequently proved useful, including for withstanding partial predation.

Against the steady increase in the proportion of high-integration forms capable of regeneration from partial mortality and probable acquisition of large colony size, is the dramatic and spectacular rise of multiserial, branching forms from the Late Cretaceous onwards. This was coincident with the appearance of new groups of predatory excavators. However, only with increased phylogenetic knowledge will it become clear if this increase is the result of a limited number of differentially proliferating clades, or whether it represents a truly polyphyletic response to changing predation pressure. The fact that coincident with this proliferation, was the appearance of a mobile free-living habit in both corals and bryozoans is certainly suggestive of the action of an external factor—although not necessarily the same one. The substantial, shallow epicontinental seas of the Cretaceous were characterized by shifting soft substrates, and such environments might also be conducive to the evolution of forms with the ability to right and disinter themselves. Interestingly, although branching

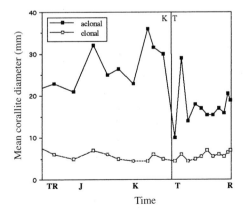

Fig. 7.23 Changes in mean scleractinian corallite diameter in aclonal and clonal species. TR: Triassic, J: Jurassic, K: Cretaceous, T: Tertiary, R: Recent. (Replotted from Coates and Jackson 1985.)

acroporoids appeared in the Eocene, they did not become dominate reefs until the Early Pleistocene. The rise of this group—with its particularly remarkable powers of regeneration from fragmentation and rapid growth—would then seem to be independent of any known changes in predation style.

The decrease in mean corallite size in solitary/aclonal corals in the Eocene might suggest the preferential removal of large, open surface-dwelling forms with poor powers of regeneration, with the retention of small, possibly cryptic forms protected from intense predation. This coincides with the rise of reef fishes, but such inferences require experimental testing. Again, this proposal needs to be tested with detailed distributional data of such growth forms.

Also noteworthy is the appearance of most families of modern scleractinian corals during the Eocene that spread throughout Tethys. Groups appearing at this time, such as the poritids, their relatives the actinids, and the favids (which had survived the Cretaceous extinction), dominate most coral reef communities throughout much of the Cenozoic (McCall *et al.* 1994). By the end of the Eocene, all modern coral families had appeared.

7.4 Summary

Biological interactions have structured reef communities from their inception, and we have seen that an analysis of functional morphology can successfully predict the outcome of such interactions. But while predators, bioturbators, and bioeroders have been associated with reefs since the late Proterozoic, there can be little doubt that a dramatic escalation occurred from the mid-Jurassic to the Eocene in the diversity and abundance of organisms with innovative and destructive predatory methods—including those with the ability to bore, scrape, and excavate living skeletal benthos. The important regulatory role played by such organisms on modern coral reefs—particularly gastropods, echinoids with camerodont lanterns, and herbivorous fishes—makes it difficult to believe that there was not a profound response manifest in the defensive methods employed by their reef-building prey—and so in the community ecology of the reef.

We have seen that there has been a proliferation of traits with proven antipredatory benefits since the Mesozoic, and that some sessile reef organisms (e.g. corals and coralline algae) even appear not only to thrive but actually require conditions of heavy grazing for their survival in shallow tropical seas. The rise in herbivory may well have allowed the dominance of the highly tolerant coralline algae and corals in modern reef habitats, particularly after the appearance of reef fishes in the Eocene. Moreover, many of the functional organizations which proved intolerant of excavatory attack became largely absent from shallow marine tropical reef biotas from the late Palaeozoic to the early Mesozoic, e.g. the soft-sediment dwelling, finely branching benthos of rigid bryozoans and calcified sponges. But many of the devices employed to tolerate or resist predation—such as sunken conceptacles (coralline algae), multiserial organizations, and the rapid translocation of metabolites (corals)—aid recovery from all agents of partial mortality. So while the Palaeozoic environment was characterized by relatively little biological disturbance, it appears that the physical agents of partial mortality which have always been present in shallow marine tropical environments—such as episodic sedimentation—had already selected for the traits which subsequently became beneficial for combating escalating, and excavatory, predation. Most of the antipredatory traits surveyed here then are exaptations, as they offer multiple benefits, with these traits appearing, and in some cases proliferating, independent of escalating predation. Indeed, in many cases it is difficult to differentiate the effects of competition from predation, particularly in the knowledge that predation may act to ameliorate competition in many interactions.

Traits acquired polyphyletically over a very short space of geological time is compelling evidence for the operation of an extrinsic selective force (Skelton 1991). For example, the evolution of mobility in bryozoans coincides with that of scleractinian corals, with both appearing in the Jurassic–Cretaceous. While there is a notable increase in the proportion of forms with high-integration modularity in scleractinian corals throughout their history, this is particularly marked from the Eocene onwards. In addition, while these corals display the full range of morphological forms and corallite size by the Late Triassic, certain morphological types did not become dominate until far later in their history. Thick crusts in coralline algae become more dominant, and branching forms become noticeably less conspicuous on reefs from the Eocene. And while the incidence of erect growth in relatively small and fragile organisms such as bryozoans plummeted in the Mesozoic, this rose markedly in scleractinian corals in the Late Cretaceous (Turonian). Branching coral taxa such as *Acropora, Porites,* and *Pocillopora* appeared during the Eocene, and the rapid appearance and radiation of reef fish coincides with a major expansion and reorganization of the coral reef ecosystem. However, much more data is required to assess the incidence of different forms of partial mortality and the ability to regenerate damaged skeletons. But the coincidence between forms which are both tolerant of excavatory predation and have the ability to withstand substantial physical destruction (abrasion and wave-shearing), raises the intriguing possibility that increased tolerance to biological destruction may, in turn, have enabled some reef building corals and coralline algae to invade more physically disturbed habitats. It is also possible that as many algal and aggregating heterotrophic communities thrive in areas of relatively high nutrient input, the appearance of specialist predators and herbivores from the late Mesozoic onwards favoured the diversification of photosymbiotic corals in relatively low nutrient habitats where macroalgal growth is not prolific. Coral reefs could not have existed in their present forms until corals either occupied low nutrient settings or until herbivores removed their algal competitors. Only under such conditions, are modern reef organisms rendered better competitors than fast-growing macroalgae.

As much of the predation on reefs is generalist, diffuse coevolution can be inferred to be a common phenomenon. Some coevolution while important—such as the action of herbivores upon coral growth—is only indirect. In such situations, the interaction is essentially one way, as prey evolve mechanisms to reduce predation but the selective pressure of the predators to evolve a counter-measure is limited. The selection pressure on the prey is thus far higher than on the predator to increase capture efficiency. With the exception of some apparently tightly evolved interactions in seaweeds, tight coevolution may be important only over small geographic ranges. There are, however, some examples of genuine adaptations, such as the coevolution between the tropical *Porolithon* and the cold-temperate *Clathromorphum* coralline algal genera and molluscs—which have lost the ability to slough epithallial cells without which reproduction cannot occur—and where deep-grazing limpets are required to perform that function (Steneck 1983).

As an intriguing aside, since fish have very well-developed vision upon which they rely heavily not only for feeding but also for reproductive and ecological interactions, we might speculate that shallow marine coral reefs became more colourful and dazzling places since the Eocene.

Almost nothing is known as to changes in the style of skeletal sediment production and distribution within reefs after the appearance of abundant bioerosion from the Upper Jurassic, especially after the appearance of fish in the Eocene to Miocene. We would predict that substantial aprons of sediment may not have been present on pre-Eocene reefs, and likewise in the absence of the grain-size reduction activities of acanthurids, the modern style of coral reef lagoon may also not have appeared until that time. Also barely explored are sedimentological consequences of differences in the geographical distribution of bioeroders—which is especially marked in fish populations due to differential extinction in the Atlantic.

All marine hard-substrate communities were also profoundly affected by the Mesozoic and later diversification of deeply excavating bioeroders and endoliths. This resulted in greater microhabitat complexity which presumably allowed the further diversification of cryptic organisms. First, the increased topographic complexity and shelter afforded by bioeroders may have allowed the diversification of small predators such as stomatopods and alpheid shrimps (Moran and Reaka 1988), as well as offering a refuge for other immobile organisms such as sponges, bryozoans, solitary corals, and brachiopods (Jackson *et al.* 1971). Indeed, many modern, non-cryptic hard substrates are covered in thick coralline algal crusts, whilst most other encrusting organisms (except corals, gorgonians, and sponges) are restricted to crypts. Second, greater topographic complexity may have facilitated the settlement of coral larval planulae where they were protected from grazers and other predators, particularly on surfaces where faster growing algae might smother or shade these new recruits (Birkeland and Randall 1982). The evolution of light-dependency in scleractinian corals in conjunction with excavatory predation may have restricted many previously open-surface heterotrophs to cryptic niches, but these trends have only been preliminarily documented. While bryozoans predominated on open reef surfaces during the Palaeozoic, today they are more common in crypts (Kobluk 1988), and it has been suggested that this shift occurred during the Cenozoic (Cuffey 1974). Available evidence from calcified sponge groups, however, indicates that these modern crypt dwellers are not displaced former open-surface dwellers, but represent the remnants of communities which have always occupied cryptic niches. These patterns require phylogenetic analysis to differentiate between patterns of the selective extinction of some clades and the proliferation of others, and those that are truly polyphyletic.

There is a pressing need to understand the ecological (or genetic) conditions that favour a sustained arms race rather than some other coevolutionary outcome such as extinction, the development of polymorphisms, or a change in the type of interaction (e.g. a parasitism becoming a mutualism). Very little quantitative data is available to assess the relative importance of different sources of partial and whole mortality in ancient communities, and as yet we have only a poor understanding of how the nature of constraints and the conflicting demands of different defensive strategies have resulted in the ecological characteristics of the coral reefs. This is one area where the fossil record can contribute in a fundamental way to our knowledge of the functioning of modern ecosystems.

8 Photosymbiosis: access to a new metabolic capability

All reefs are sites of increased carbonate production, but few are significant beyond a local scale. Of these, coral reefs are of global importance, scattered over some $584–746 \times 10^3$ square kilometres in modern tropical seas (Kleypas 1997).

Coral reefs and their associated platforms are responsible for about half of the Earth's present production of calcium carbonate (global carbonate production is 1.00 Gt year^{-1}), and this is formed faster than it can be dispersed. Such reefs are estimated to fix between 5–20 g m^{-2} day^{-1} of organic carbon (e.g. Odum and Odum 1955; J.B. Lewis 1977, 1982), producing a colossal 1–5 kg C m^{-2} year^{-1} (Hatcher 1990), 100 times greater than open tropical waters (20–50 g C m^{-2} year^{-1}). This corresponds to carbonate production rates of 10 kg CaCO$_3$ m^{-2} year^{-1}. Most remarkably, the vast majority of this carbonate is derived organically—many coral reef organisms bear substantial calcareous skeletons.

Much of this rapid calcification—so vital to coral reef growth and maintenance—is closely linked to light through complex photosynthetic processes. Light is one of the most abundant resources in the shallow marine tropics, especially in low-nutrient environments which provide more potential habitats for photosynthesizing organisms than where trophic resources are freely available and water clarity reduced. Apart from coralline algae, the most plentiful calcareous organisms on coral reefs are protists (foraminifera) and invertebrates (most notably scleractinian corals) which have entered into *photosymbioses* with a variety of photosynthesizing microorganisms, mainly single-celled algae. These associations constitute particular non-parasitic associations known as *endosymbioses*, which involve two species of unequal size where the whole body of the smaller (the symbiont) is housed entirely within the larger (the host). Photosynthetic products (*photosynthates*) released by the symbiont to the host often result in increased rates of growth and hence increased skeletal calcification when compared to organisms without photosymbionts (non-symbiotic forms), a process known as *light-enhanced calcification*. Photosymbioses underpin the functioning of modern coral reefs, as not only are they the main source of primary production, they are also responsible for construction of the physical structure itself.

Symbioses are abundant in nature, and may play a central role in many ecological processes. There is now little evidence to support the long-held notion that many symbioses are *mutualistic*, that is where both symbiont and host can be shown to benefit from the long-term association (Douglas and Smith 1989). Microbial symbionts often possess a capability lacking in their host, frequently one related to nutrition. So although symbiosis is a costly acquisition, involving mechanisms for the control of symbiont cell division, and regulation of the symbiont population and distribution, its great importance lies in the fact that it allows a host to gain a novel metabolic capability (Douglas 1994). New taxa can arise indirectly from the acquisition of a new metabolic capability, and so symbioses are often regarded as important triggers for the abrupt appearance of new clades, especially when this new capability can enable the host to exploit a new food resource or invade a previously inhospitable environment. Indeed, it has been suggested that expansion of the metabolic repertoire of eukaryotes by symbioses laid the foundation for many radiations, and eukaryotes themselves are thought to be symbioses (Margulis 1970).

Modern coral reefs are facing multiple anthropogenic threats, including a progressive rise in sea surface temperature, as well as eutrophication and increased sedimentation (Birkeland 1997). High temperatures and increased UV radiation are widely believed to operate synergistically in the collapse of the algal symbiosis in corals and other invertebrates, causing the mass expulsion or *in-situ* degradation of the algae resulting in *bleaching* (Glynn 1993). Extreme bleaching events can lead to widespread coral mortality, sometimes with a resultant community shift from a benthos dominated by corals to one dominated by macroalgae (T.P. Hughes 1994). These ecological observations suggest that the loss of symbioses might have a profound impact on reef communities on evolutionary timescales.

The clear correlation between photosymbiosis and the success of living scleractinian corals in reef construction has led to the widespread supposition that the ability of metazoans to build reefs is contingent upon possession of photosymbionts. Many have argued, therefore, that extinct groups of reef-building metazoans also possessed this metabolic capability (e.g. Cowen 1988; Talent 1988). In addition, the loss and subsequent slow re-establishment of photosymbiotic relationships has been implicated in the long recovery time (up to 10 Myr) of reef communities after mass extinction events (e.g. Fagerstrom 1987; G.D. Stanley and Swart 1995). Such assumptions have not, however, been tested adequately.

This chapter explores five topics concerning photosymbiosis and the history of reef-building:

1. Is photosymbiosis necessary for reef-building?

2. Is photosymbiosis necessary for reef-building in scleractinian corals?

3. Can photosymbiosis be implicated in the varied contribution of scleractinian corals to reef-building over geological time?

4. How can we explain the apparent anomaly of the vulnerability of living scleractinian coral photosymbioses to anthropogenic environmental changes with their long-term stability on geological timescales?

5. Can the widespread loss of photosymbiosis be implicated in the recovery time of reef communities after mass extinction events?

Consideration of these questions requires an understanding of the evolutionary biology of photosymbioses in modern reef organisms. This is explored in the following section.

8.1 The formation of photosymbioses

A survey of photosymbioses indicates that while few major eukaryote groups become hosts, they do so with a very diverse range of algae, namely chlorophytes (green algae), rhodophytes (red algae), cyanophytes (blue-green algae), and diatoms (Table 8.1). Of these, the most common photosymbionts in the marine environment are gymnodinioid ('naked'; Order Gymnodiniales) dinoflagellates, known as *zooxanthellae* when occurring in symbioses.

Table 8.1 Symbiotic associations of protists and invertebrate hosts with photosynthetic algae and bacteria associated with modern reefs (modified from Douglas 1994, by permission of Oxford University Press.)

Group	Photosymbionts	Light-enhanced calcification?
Protists		
Foraminifera	Diatoms, dinoflagellates, chlorophytes, rhodophytes	Yes
Radiolaria	Dinoflagellates	[siliceous]
Acantharia	?	n/a
Invertebrates		
Porifera (sponges)	Cyanophytes, dinoflagellates	n/a
Cnidaria		
—Anthozoans		
Corals	Dinoflagellates	Yes
Anemones	Dinoflagellates	n/a
—Scyphozoans	Dinoflagellates	n/a
—Hydrozoans	Chlorophytes	n/a
Ascidiacea (sea squirts)	Cyanophytes	n/a
Turbellaria (flatworms)	Dinoflagellates, chlorophytes, and diatoms	n/a
Mollusca	Dinoflagellates	Yes

Zooxanthellae are surrounded by single or multiple membranes of host origin known as the *symbiosome* (Fig. 8.1(a)). The only widespread zooxanthella is *Symbiodinium* (Trench 1997), which resides in an impressive array of reef invertebrates, including hydrozoans, tunicates, and bivalved molluscs (such as the 'giant' clam *Tridacna*; Fig. 8.1(b)), as well as corals.

Many microbial symbionts cannot be cultured in the laboratory and so are difficult to accommodate within traditional classification schemes. Indeed, a full appreciation of the diversity of photosymbionts was not possible until the development of molecular techniques. Long believed to be represented largely by one species—*Symbiodinium microadriaticum*—numerous functionally distinct and distantly related taxa, or types of zooxanthellae are now known on the basis of considerable biochemical, morphological, and molecular evidence (e.g. Rowan and Powers 1991; Trench 1997). These analyses reveal that the algae in scleractinian corals have probably evolved from a single adaptive radiation, which alone includes as much diversity as an entire family of free-living, non-symbiotic dinoflagellates.

8.1.1 Origin and radiation

A common feature of most endosymbioses, including photosymbioses, is that the number of host genera or species may be at least tenfold greater than the number of symbiont genera or species (Law and Lewis 1983). Notwithstanding the fact that the diversity of endosymbionts has been underestimated substantially by traditional taxonomic methods, the greater diversity of hosts may in fact be real, and may reflect the evolutionary processes of diversification of intact associations (Douglas 1995).

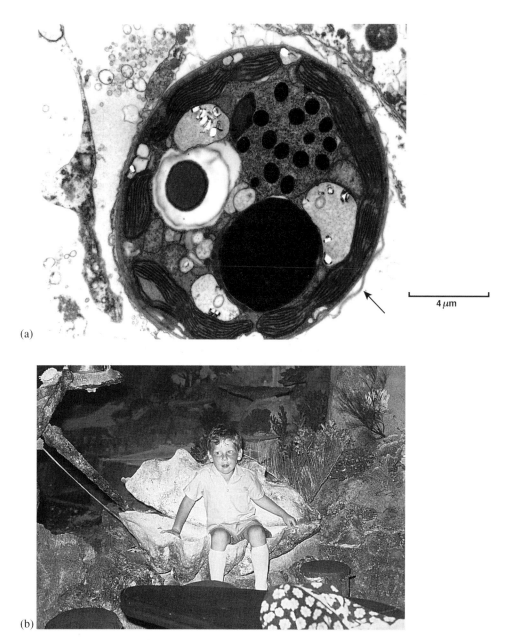

(a)

(b)

Fig. 8.1 (a) Cell of dinoflagellate alga *Symbiodinium* in the sea anemone *Anemonia viridis*. The algal cell is separated from the host by symbiosome membranes (arrowed). (Micrograph: A.E. Douglas.) (b) Possession of photosymbionts can confer considerable size to hosts: the photosymbiotic 'giant' clam, *Tridacna*. (Photograph: G. Galletly.)

Symbiosis, in general, is accepted to have triggered adaptive radiations in many host taxa (Douglas 1994), but endosymbionts would not necessarily show a parallel diversification with their hosts. Symbiotic microorganisms appear to have evolved relatively rarely, and have been acquired on many occasions by taxonomically diverse hosts. This suggests that the ability of hosts to form symbioses may not be the most important determinant of the incidence of symbiosis. Rather, the origin of symbioses may be determined by the availability of an appropriate symbiont, together with its effectiveness in promoting host 'success', that is increased survival, growth, and reproduction.

Some photosymbioses may evolve gradually, with the future host parasitizing or feeding on the potential symbiont and then later developing strategies and modifications for the two life histories to become more fully integrated. Symbionts that enhance host fitness can become so integrated with their hosts that they frequently lose those characteristics required solely for a free-living existence, so that their survival becomes wholly dependent upon the association (Douglas and Smith 1989). The integration of symbionts into the host environment has involved dramatic changes in key biochemical and cellular processes of the microorganisms, such that those that enter into endosymbioses form a coherent group that is ecologically distinct from those which do not (Douglas 1994).

8.1.2 Modes of transmission

Symbionts are transmitted to the next host generation by a number of means which are ultimately linked to the reproductive traits of the host. First, symbionts can be transferred directly from host to offspring, a process known as *vertical transmission*. Vertical transmission is the norm for asexual reproduction as daughter clones will automatically receive a complement of algae. Parental algal cells may also be placed in egg cytoplasm immediately prior to fertilization and subsequent release. This occurs in many 'brooding', but few 'broadcasting', corals (Trench 1987).

In other hosts, such as broadcasting corals, scyphozoans, and tridacnid bivalves, symbionts must be acquired anew by each generation from the open environment after metamorphosis to the adult form. This is known as *horizontal transmission*. Some host organisms (e.g. *Tridacna*) are restricted to this method, as vertical transmission is made impossible due to structural barriers in the host which bar access of the symbiont to the host gametes. Hosts capable of asexual and sexual reproduction are able to employ both vertical and horizontal transmission of symbionts.

Re-establishment of symbioses at each generation offers the potential for colonization by symbionts that are genetically distinct from those of the parent, and might allow juveniles to select locally optimal symbionts. Also, this mode of transmission implies that symbionts are widely available for acquisition from the environment. Dinoflagellates are mostly free-living and motile making them available for ingestion by a prospective host. But, as corals will not ingest free algae it is probable that they and other cnidarian hosts acquire symbionts whilst feeding on herbivorous zooplankters which contain viable algal prey in their guts (Douglas 1994). However, the concentration of *S. microadriaticum* in sea water is likely to be low under normal conditions; this organism is not thought to lead a permanent free-living existence, and may only be available for reinfection when newly released from former hosts.

Vertically transmitted symbioses evolve by the diversification of permanently associated clades, such that the distribution of symbionts is restricted and the phylogeny of the symbionts will mirror that of the host. Moreover, the abundance and distribution of the symbionts will be indistinguishable from that of their host, and the success of the symbiont will be determined by its effectiveness in promoting host survival or increased fitness. But if hosts are able to select their symbiotic partners (for example from the environment), then no congruence between host and symbiont phylogenetic relationships would be expected. Also, the abundance and distribution of the symbiont will be determined not only by its effectiveness, but also by its ability to colonize different hosts.

Until recently, most symbioses were thought to be ancient, long-lived associations which persisted to the extinction of either host or symbiont (e.g. Wilkinson 1984). However, considerable new molecular data now supports the hypothesis of successive recombination for many symbioses. Most eukaryotic groups include closely related symbiont-bearing and symbiont-free representatives; in many host clades symbiotic associations have evolved several times and some have even secondarily lost symbionts, e.g. the photosymbionts found in lichens (Douglas 1992). Some host taxa are able to form stable associations with more than one taxon of symbiont, for example the foraminifera *Amphistegina* may bear either *Chlorella* (a chlorophyte), or diatoms, or the more usual *S. microadriaticum*.

In particular, variation in the 18S rRNA gene sequence of *Symbiodinium* has distinguished three distantly related symbiont clades, A, B, and C, in cnidarians which show no congruence with the phylogeny of its hosts (Rowan and Powers 1991; Fig. 8.2). This is surprising as *Symbiodinium* is transmitted asexually in virtually all hosts and by sexual reproduction in only half of the host species (Douglas 1995). It has been suggested that the tremendous plasticity of reproductive traits found in the Cnidaria may explain this lack of congruence between host and symbiont clades (Fautin 1991).

8.1.3 Specificity and selection of symbionts

One important aspect of symbioses is the taxonomic range of partners with which an organism can form a symbiosis. This is known as *specificity*, and is a consequence of both the degree of specialization achieved by an organism for a particular partner, and the capacity to select between alternative potential partners (Douglas 1994). To analyse specificity requires rigorous taxonomic understanding of both host and symbiont; a full appreciation of the diversity of photosymbionts was therefore not possible until the development of molecular techniques, which are now enabling investigation of genetic diversity of symbionts among hosts within a single taxon.

Although, at present, there are insufficient data to deduce the precise way in which symbioses evolve, what is clear is that many symbioses have complex evolutionary histories. Host specificity varies between associations, with some being more tightly bound than others. Many hosts which solely employ vertical transmission are dependent upon the association for their survival: if the association is broken, the host will die.

Contrary to the widely accepted belief that corals harbour only one symbiont, work on the important Caribbean reef coral species *Montastraea annularis* and *M. faveolata*, and *Acropora palmata* and *A. cervicornis*, has shown that individual colonies of these species growing at different depths (Fig. 8.3(a)) can bear different types of zooxanthellae (Rowan and Knowlton 1995; Baker *et al.* 1997). Moreover, *Montastraea* can be polymorphic, that is a single colony can bear more than one zooxanthella type. A and B types are common in shallow-water corals (high-irradiance habitats), whereas C predominates in deeper corals (low-irradiance habitats). Mixed populations are common at intermediate depths, where A or B may dominate unshaded colony or column tops, and C the shaded colony sides (Fig. 8.3(b)). These symbionts occupy distinct, but overlapping, areas of the colony. This suggests that symbionts exist as complex communities that can track gradients in environmental radiance within a colony.

Algae from

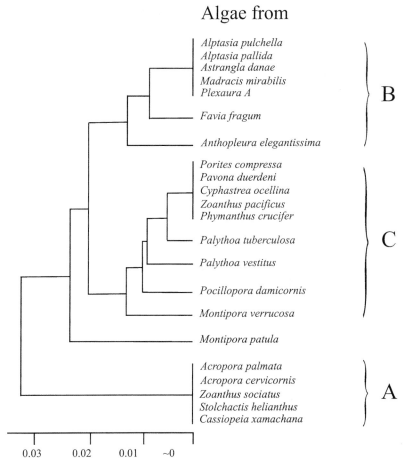

Alptasia pulchella
Alptasia pallida
Astrangla danae
Madracis mirabilis
Plexaura A

Favia fragum

Anthopleura elegantissima

B

Porites compressa
Pavona duerdeni
Cyphastrea ocellina
Zoanthus pacificus
Phymanthus crucifer

Palythoa tuberculosa

Palythoa vestitus

Pocillopora damicornis

Montipora verrucosa

Montipora patula

C

Acropora palmata
Acropora cervicornis
Zoanthus sociatus
Stolchactis helianthus
Cassiopeia xamachana

A

0.03   0.02   0.01   ~0

Fig. 8.2 The phylogeny of the dinoflagellate algae *Symbiodinium* isolated from various cnidarian hosts based on 18S rRNA sequences. Genetic distances (estimated from the average fraction of nucleotides substituted) are shown. The algae can be divided into three major groups (A, B, and C) which show a marked discordance with host taxonomy. (Redrawn from Rowan and Powers 1991, reprinted with kind permission from the American Association for the Advancement of Science.)

It is not yet clear how common such genetic variation is in photosymbionts whose hosts occupy environments characterized by different solar irradiances. Preliminary work suggests that partitioning of *Symbiodinium* with depth and irradiance may not be a general phenomenon in Cnidaria as it is not present in some sea anemones (Billinghurst *et al.* 1997; Bythell *et al.* 1998).

We are only just beginning to appreciate the complexity of many symbiotic relationships. Future development of molecular taxonomic methods will certainly revolutionize our understanding of specificity, such as the extent of intracolony and intraspecific geographical variability within different taxa and how this might be determined by local symbiont availability. Such potential biogeographical constraints on symbiont diversity might have profound evolutionary implications. The selection pressure on a host to be resistant to ineffective symbionts may be an important factor that limits the host range of symbionts. The

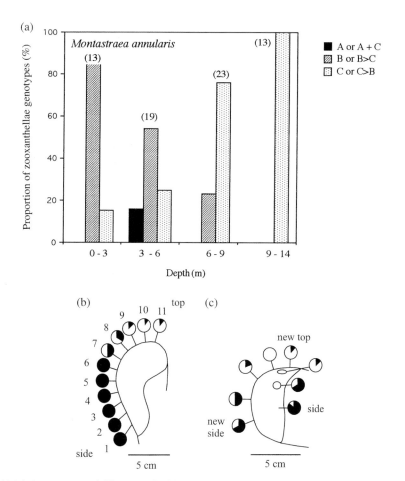

Fig. 8.3 (a) Relative occurrence of different *Symbiodinium* groups (A, B, and C) in single samples from different colonies of the coral *Montastraea annularis*, found growing at different depths. The three types show significant differences in depth distribution. The number of coral samples analysed in each of the four depth ranges are given in parentheses. (Data replotted from Rowan and Knowlton 1995, copyright (1995) National Academy of Sciences, USA.) (b) A single column of *M. annularis*, showing the sampled distribution of *Symbiodinium* type C (black) and *Symbiodinium* type B (white). (c) New zonation in a column of *M. annularis* 6 months after being transplanted to 90° from the vertical, demonstrating that the algae were able to re-establish predicted patterns of symbiont zonation. ((b) and (c) redrawn from Rowan *et al.* 1997, by kind permission of Macmillan Magazines Limited.)

host range of *Symbiodinium* in reef communities may be particularly important as the competitive interactions between reef corals are so intense and because—exceptionally among intracellular symbionts—*Symbiodinium* is routinely expelled from cnidarians (Douglas 1995).

8.2 What does photosymbiosis confer upon reef-building metazoans?

The following section considers five potential advantages that photosymbiosis might confer upon reef-building metazoans. These are: as an additional source of nutrition; the ability to occupy low-nutrient habitats; the ability to calcify and grow rapidly; the potential to achieve large size; and the ability to adapt to the highly dynamic shallow marine tropical environment. The section concludes with a discussion as to whether photosymbiosis is necessary for reef-building in scleractinian corals.

8.2.1 An additional source of nutrition

In Cnidaria, zooxanthellae are intracellular within gastrodermal cells (the lining of the central gut cavity of each polyp). In *Tridacna*, they reside extracellularly within the terminal branches of the diverticulum of the stomach (Norton *et al*. 1992). Both these physiological settings require that all nutrients gained by the zooxanthellae be acquired from the host and not directly from the surrounding sea water. In corals, *Symbiodinium* can exceed concentrations of 1–2 million cells per square centimetre, although the algae are neither distributed uniformly within a single colony nor between different colonies of the same species, especially when they grow at different depths.

Symbiodinium provides its hosts with photosynthetic carbon (Muscatine 1990). In corals, this energy source is in addition to that gained from zooplankter capture by tentacles, bacteria trapped in mucus, and dissolved organic molecules taken up from sea water across the body wall.

Zooxanthellae release photosynthates in the form of glycerol and triglyceride (Douglas 1994) which are used in lipid and protein synthesis. Calculated proportions of photosynthate transfer to a host vary: for example, between 50 and 95% of total production has been recorded in corals (Muscatine 1990), and 62–84% in *Tridacna maxima* (Trench *et al*. 1981). Such similarities in the nutritional interactions between coral and mollusc host with zooxanthellae might indicate that neither the identity of the host nor the structural relationship (extracellular or intracellular) has any major effect on the quantity of photosynthate translocation.

Symbionts can photosynthesize at substantial rates, and may release virtually all photosynthates to host tissues which can be sufficient to fuel the entire respiratory carbon requirements of the host. In such cases, this can then effectively make the coral independent of the food sources available in the ambient environment. It is likely, however, that the importance of photosynthates to the energy budget of a host is determined by environmental factors such as solar radiation: shallow-water cnidarians growing in full sunlight fuel a greater proportion of their energy demands by respiration of photosynthates than those growing in the shade or deep-water species (Muscatine 1990).

8.2.2 Occupation of low-nutrient environments

Symbiodinium populations in marine cnidarians increase very slowly, with growth rates rarely exceeding 0.1 day^{-1} (equivalent to a doubling time of 10 days) (Douglas 1994). But the growth rates of their hosts is even lower. For example, the Red Sea coral *Stylophora pistillata* in its natural habitat increases by 0.001–0.005 day^{-1}—up to 100 times slower than its algal symbionts (Hoegh-Guldberg *et al*. 1987). Therefore action must be taken to maintain a relatively constant biomass of symbiont and host. This is achieved through two mechanisms: expulsion or lysis of excess symbiont cells and, more importantly, regulation of the proliferation rate of symbionts by the host.

Hosts are thought to control the rate of symbiont proliferation by controlling access to inorganic nutrients (phosphorus and nitrogen). Experimental observation shows that growth and division of algal cells occurs soon after the host has fed, suggesting that algal cells may be deficient in nitrogen, phosphorous, and carbon. That symbiont growth is nutrient-limited is also supported by other experimental evidence. Elevated concentrations of ammonia can markedly increase algal populations, causing host tissues to acquire the chocolate-brown colour of *Symbiodinium* cells (Table 8.2). In *Tridacna gigas*, nutrient-limitation of zooxanthellae is a function of the availability of ammonia to symbionts (i.e. it is not regulated by the host), but irrespective of the nutrient levels in sea water, the zooxanthellae do not have access to phosphorus, probably as a result of host control (Belda *et al.* 1993*b*).

These data suggest that the proportion of photosynthate transferred to the host declines in nutrified conditions. This may, however, simply imply a general breakdown in the regulatory mechanisms of the host. But the release of nearly all photosynthetically derived carbon to the host under normal growth conditions suggests that this mechanism also aids control of algal populations. Symbiotic algae appear to release photosynthate because they are deficient in nitrogen (Douglas 1994), and so are only beneficial to a host in low-nutrient environments.

A necessary condition for the release of photosynthate to the host appears to be an imbalance between the carbon and nitrogen metabolism of the photosymbionts (Rees and Ellard 1989). Because symbionts have limited access to nitrogen, they have an excess of carbon compounds derived from photosynthesis which may be released to the host. This 'metabolic imbalance hypothesis' is supported by the facts that:

(1) photosynthates released do not contain nitrogen;

(2) the growth of algal symbionts is probably nitrogen-limited;

(3) the proportion of photosynthate translocated to the host declines with elevated nutrient levels; (McCloskey, unpublished data in Douglas 1994).

Such data does not support the widely held view that nitrogen recycling occurs in the coral symbiosis. *Tridacna*, however, appears to recycle phosphorous very efficiently, as there is very little leakage of metabolically derived compounds compared with those released from large non-symbiotic bivalves (Belda and Yellowlees 1995).

An increase in algal symbiont population may result in the algae sequestering inorganic carbon which would otherwise be released to the host (Marubini and Spencer Davis 1995).

Table 8.2 The effect of ammonia enrichment of sea water[*] on zooxanthellae density in the corals *Stylophora pistillata* and *Seriatopora hystrix*, and the giant clam *Tridacna gigas* (data from Hoegh-Guldberg and Smith 1987 modified from Douglas 1994 (with kind permission of Oxford University Press), and Belda *et al.* 1993*b*)

| Enrichment of sea water | Density of algal cells (10^6 × no. cells per mg^{-1} protein) | | |
	Stylophora pistillata	*Seriatopora hystrix*	*Tridacna gigas*
None	0.55	2.11	0.8–1.0
Ammonia	1.49	2.78	2.5

[*]Corals were incubated in sea water supplemented with 10–40 µM ammonia for 19 days, with corals in unenriched sea water as control. Giant clams were grown in sea water supplemented with 10 µM NH_4Cl for 3 months, with clams in unenriched sea water as control.

This can lead to the disruption of normal skeleton production. *T. gigas* grown artificially under elevated nutrient conditions showed serious perturbations in shell formation, including a general weakening of the shell and reduced calcification rates (Belda *et al.* 1993a). A similar loss of skeletal strength and calcification rate has also reported in corals as a result of exposure to phosphate (C. Rasmussen 1988). Indeed, a reduced rate of overall coral calcification was recorded on the reef at Kaneohe Bay, Hawaii, with only a minute increase in the input of ammonium (Kinsey 1991).

While the foregoing discussion outlines some experimental evidence to suggest that symbiotic algae are nutrient-limited, whether it is this metabolic requirement that restricts many coral reefs to low-nutrient environments, or the fact that elevated nutrient levels enhance the growth of macroalgae that compete with corals for light and space, is not clear (see Section 7.1.7). These alternatives cannot be explored until there is a global compilation of the distribution of photosymbiotic corals, and coral reefs, in relation to ambient nutrient levels. Moreover, this also provokes a confusion between those environmental factors conducive to photosymbiosis and those to reef growth. Indeed, we have already established that these two phenomena do not necessarily have the same environmental requirements (see Chapter 4).

8.2.3 Increased rates of calcification

There is no correlation between symbiosis and the ability to secrete either calcite or aragonite. Of the variety of living organisms that have entered into photosymbioses (see Table 8.1), only scleractinian corals, bivalved molluscs, and foraminiferans secrete calcareous skeletons. Some radiolarians with siliceous skeletons bear symbionts, and many cnidarians and ascidians lack a skeleton altogether. Symbiosis promotes calcification where it is already part of the host biology, but cannot instigate it.

Photosynthesis by symbiotic algae consumes CO_2. Marine calcification precipitates carbonate ions, forcing the re-equilibration of the HCO_3^--dominated marine inorganic carbon system which creates a source of CO_2. As a result, it has long been assumed that photosymbiosis is causally linked to high rates of calcification in hosts compared with non-symbiotic relatives.

Growth rate can be measured by either skeletal extension (the process by which an organism increases in length over time) or calcification (which includes both extension and the thickening of skeletal elements i.e. mass/area/time). While photosymbiotic organisms bear calcification mechanisms which operate independently of photosynthesis, in numerous experiments the presence of algae growing in warm and highly illuminated conditions has been shown markedly to enhance this process (light-enhanced calcification). Calcification in scleractinian corals on sunny days may be twice as high as on cloudy days, and although calcification occurs continuously, rates under light conditions are typically 2–3 times greater (D.J. Barnes and Chalker 1990; Fig. 8.4). Likewise, the rate of incorporation of ^{45}Ca-labelled calcium into the tests of foraminifera can be increased up to 50-fold by increased illumination. Also, marked reductions in the extension rate for both branching and massive corals were noted during a 12-month study of controlled growth under artificially shaded conditions (Wellington 1982).

Coral and coral reef growth is clearly light-limited, and appears to be a function of photosynthetically available radiation (wavelengths 400–700 nm) and its attentuation with depth. While the growth rates of different coral species are highly variable, all measured photosymbiotic coral extension rates decline logarithmically with depth (Fig. 8.5). As a result,

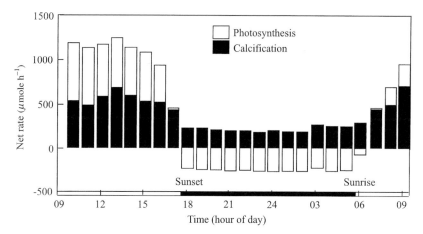

Fig. 8.4 Net photosynthesis and calcification rates shown by a 20-cm diameter colony of the Pacific staghorn coral *Acropora formosa*, growing at a depth of 5 m, Bowl Reef, GBR, Australia. Rates are for 50-min periods within an incubation period of 24 h. Photosynthesis and calcification were determined from changes in oxygen concentration and pH in the seawater surrounding the coral. (Replotted from D.J. Barnes and Chalker 1990, with kind permission from Elsevier Sciencce.)

although individual corals present a wide range of calcification versus depth profiles, it can be assumed that calcification rate is a linear function of photosynthesis (Kleypas 1997).

Calcification mechanisms in zooxanthellate corals

Photosynthesis is clearly intimately associated with rapid calcification, but details of this metabolism are far from understood, although a number of models currently exist to explain the relationship (for a summary see D.J. Barnes and Chalker 1990). The most fundamental requirement of calcification is the creation of an isolated environment in which crystal precipitation can occur, and the most rapid calcifiers, such as corals, are those which most effectively achieve this. This isolation strongly implies that rapid calcification is not simply fortuitous, as the organism must facilitate nucleation and crystal growth by the introduction, exclusion, or removal of organic substances—as well as continued modification of this internal system with respect to surrounding sea water. The system must maintain precipitation, with photosynthesis providing the energy.

Skeletal extension in non-symbiotic and photosymbiotic corals occurs mainly at night (Vago *et al.* 1997). This confirms that extension is not dependent upon the concomitant daytime photosynthesis of zooxanthellae, and that skeletal extension and calcification rates are not necessarily coupled processes. All corals may possess the same mechanism of extension, but under light conditions zooxanthellae may help to optimize the harvest of calcium carbonate from supersaturated waters in the tropics (Buddemeier 1997).

Indeed, we have already seen that there is a correlation between the distribution of photosymbiotic corals and supersaturation levels with respect to calcium carbonate in tropical sea waters (Section 4.4.1). While neither calcification nor photosymbiosis is latitudinally limited, the distribution of photosymbiotic calcification in corals corresponds closely to the distribution of at least 2–3 times supersaturation with respect to aragonite (Buddemeier and Fautin 1996*b*; Fig. 4.2). Aragonite supersaturation may be as important an environmental control as temperature on the distribution of symbiotically induced calcification.

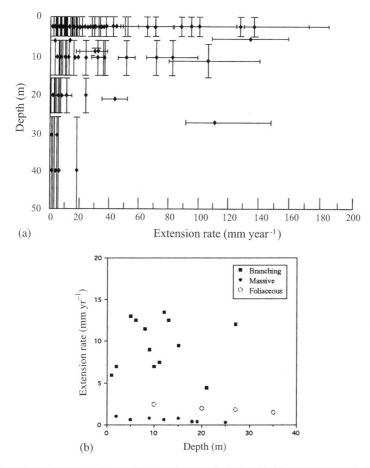

Fig. 8.5 (a) Annual extension rate in photosymbiotic corals versus depth. (Modified from Kleypas 1997.) (b) Annual extension rate of some photosymbiotic corals (by colony growth form) versus depth. (Data from E.H. Gladfelter *et al.* 1978; Hubbard and Scaturo 1985; Tunnicliffe 1983.)

Although calcification rates vary widely among zooxanthellate corals, photosymbiosis may increase calcification by providing photosynthates either as an energy source or as specific substrates for the organic matrix of biomineralization. Also, the uptake of phosphate-waste poisons the calcifying environment, perhaps explaining why in darkness corals with zooxanthellae calcify faster than those without. Calcification in corals certainly appears to depend upon a specific, energy-requiring mechanism for translocating calcium and photosynthates to areas of skeletal growth (Pearse and Muscatine 1971). Although the transfer of a photosynthate does not imply that the photosynthate itself is necessary for calcification, the fact that the growing tips of branches contain fewer zooxanthellae than older parts is evidence that translocation does indeed take place. Also in support of translocation is the integrated regeneration response shown by coral colonies to damage (Loya 1976). One group of models suggest that photosynthesis provides organic material as an energy source or a component vital to skeletal growth. This evokes the tendency of sea water to precipitate calcium carbonate which can be enhanced by a matrix which orders ions so as to cause nucleation. A second set of models suggest that photosynthesis modifies the internal calcification environment by removing inorganic substances, such as phosphate, which would otherwise inhibit crystal growth. These models are not mutually exclusive, and the most widely accepted hypotheses call upon photosynthesis to shift the sea-water carbonate equilibrium towards precipitation (see D.J. Barnes and Chalker 1990).

The widespread assumption that symbiotic corals generally calcify more rapidly than asymbiotic forms has recently been challenged by A.T. Marshall (1996a). He found that the calcification rate in an azooxanthellate species (*Tubastraea faulkneri*) actually appeared to be higher than that found in a zooxanthellate species (*Galaxea fascicularis*). He also proposed that the mechanisms of calcification differed between these two species, with the zooxan-thellate coral being effectively 'dark-repressed' instead of 'light-enhanced'. Calcification may enhance photosynthesis, as calcification provides protons for the conversion of HCO_3^- into CO_2 which is used for photosynthesis. This raises the possibility that light-induced calcification may simply be a by-product of algal photosynthesis (McConnaughey 1989). Zooxanthellae may therefore control calcification in order to enhance CO_2 availability in the light, and conserve metabolites of the host in the dark when photosynthesis does not occur. This model is consistent with the observation that dark calcification in *Galaxea* and *Acropora* occurs particularly in the regions where zooxanthellae are absent.

While the methodology employed by A.T. Marshall (1996a) has been subject to some crit-icism (e.g. Goreau *et al.* 1996), his findings are not necessarily inconsistent with existing data. They simply suggest that there must be a range of calcification rates for azooxanthellate and azooxanthellate corals, and while zooxanthellate corals usually show higher calcification rates, these ranges can overlap (A.T. Marshall 1996b). Indeed, Marshall suggests that the ability of corals to form reefs might not be a direct consequence of light-enhanced calcification, but may be due to other characteristics such as their autotrophy (an organism that uses CO_2 as its main, or sole, source of carbon), tremendous plasticity of growth form, and predator-resistant traits.

Calcification mechanism in coralline algae

Calcareous algae can have very rapid rates of calcification—*Halimeda*, for example, shows rates similar to those of corals (Bak 1976).

The calcareous habit has been adopted by very few algae (less than 4% of the 23 000 known species: Dawson 1966), most notably in coralline algae. Closely related algae may or may not bear skeletons, and environmental conditions can induce or inhibit calcification. Indeed, skeletal precipitation by the coralline algae occurs within a pre-existing organic area. As calcification increasingly isolates living cells from their environment, nutrients are transported inwards from the exterior parts of the organism, possibly combined with an outward movement of photosynthate.

8.2.4 Large size

Light-enhanced calcification can allow for the rapid growth of large, often robust, skeletons. For example, living tridacnid bivalves can reach up to 1.5 m in width (Fig. 8.1(b)) and benthic foraminifera up to 150 mm in diameter—representing an order of magnitude greater size than their azooxanthellate counterparts. Likewise, single zooxanthellate coral colonies (which moreover may be massive, domal forms) may reach many cubic metres—azooxanthellate colonies rarely reach more than a few cubic decimetres (represented by loose branching forms). But large size is not always conferred by possession of algal symbionts, and not all exceptionally large organisms may bear symbionts, e.g. deep-sea and high-latitude photo-symbiotic foraminifera are no larger than most non-symbiotic forms. Large size correlates with many measures of success and fecundity (see Section 6.1.4) and may offer increased resistance to high wave energies, but it appears to be most advantageous to organisms when food supplies are scarce, juvenile mortality rates are high, and adult mortality is predictably low (Hallock 1988)—precisely the conditions which persist in the modern coral reef environment.

8.2.5 Adaptation to environmental change

We have already seen that the coral reef environment is both highly dynamic and hetero-genous; corals show extensive intraspecific genetic variability as well as *acclimatization* (the ability to adapt to temperature change during the organism's lifetime) that may permit rapid adaptation to local environmental conditions (Section 4.4.1). In particular, the coral–zooxanthellae partnership shows a remarkable range of adaptations to varying flux and quality of solar radiation, and much of this capacity is due to the considerable ability of zooxanthellae to adjust rapidly to different radiant environments (see summary in Brown 1997*a*), as well as to the variation in behaviour and growth form of the coral host. The mor-phological plasticity of coral growth form itself may provide a mechanism that maximizes the supply of carbon to zooxanthellae under different water-flow regimes. Shaded or unshaded parts of a colony may not only bear different types of zooxanthellae (Fig. 8.3(b)), but zooxanthellae will often also exhibit different degrees of photoadaptation: those occupy-ing shaded areas (known as shade-adaption) are often larger and contain more light-harvest-ing, pigment–protein complexes. There may be little exchange of algal symbionts between these different areas; this suggests that corals have overcome the potentially high costs asso-ciated with the maintenance of mixed symbiont populations within a single host, by achiev-ing some degree of regional autonomy within a colony which might arise from differential growth rates. Physiological adaptations to environmental variability include fluxes in photo-synthetic pigments and protective carotenoids in the zooxanthellae, as well as the production of stress proteins and possibly amino acids that may act as sunscreens for both zooxanthellae and the coral host (Brown 1997*a*). So much of the remarkable tolerance of corals to fluctuat-ing temperature, water flow, and irradiance regimes may be both a direct, and indirect, result of the coral–zooxanthellae symbiosis. However, limited evidence suggests that the thermal tolerance of the zooxanthellae may be less than that of the animal host. This suggests that while zooxanthellae may enable corals to live in highly fluctuating shallow, tropical envi-ronments, the thresholds of those fluctuations may be set more by the algae than by the host.

The symbioses between corals and algae are stable, with the notable exception of coral bleaching. Bleaching is a poorly understood response to environmental stress, particularly the synergistic effects of high irradiance and temperature (Fitt and Warner 1995). It has been suggested that bleaching might be an adaptive strategy that allows corals to recombine with a different algal type during bleaching episodes (Buddemeier and Fautin 1993). That recombination with different algal types might potentially occur is supported by the obser-vation that symbiont polymorphism in *Montastraea annularis* can explain both the depth-distribution of bleaching (which predominates at intermediate depths) and within-colony patterns, as symbiont type C associated with low irradiance has been shown to be eliminated preferentially from the brightest parts of its distribution, presumably due to reaching some limit of physiological tolerance (Rowan *et al.* 1997).

The potential for the exchange of symbionts between or within different parts of a host has been demonstrated experimentally. When naturally occurring *S. microadriaticum* symbionts of the clam *Tridacna squamosa* were removed, and individuals artificially reinfected with a different type extracted from a coral, no change in growth or calcification rate was detected (Fitt and Trench 1981). Moreover, artificially toppled columns of *M. annularis*, which experienced instant and profound changes in irradiance gradients, were able to re-establish predicted patterns of symbiont zonation within only 6 months (Rowan *et al.* 1997; Fig. 8.3(c)). This is clear evidence that the patterns of photosymbiont distribution are highly dynamic in at least some corals, and can minutely acclimatize or adapt to different light (and temperature) conditions. As noted earlier, however, it is not yet clear if the partitioning of *Symbiodinium* with depth and irradiance is a general phenomenon in Cnidaria.

Established symbioses with zooxanthellae may be able to respond to rapid or profound environmental change by switching partners to a more favourable combination (Buddemeier and Fautin 1993). Indeed, it is advantageous for a host to maintain a relatively broad specificity if the most effective symbiont is not available, or if the most effective symbiont changes with the environment. Such continuous sampling of available algae would promote long-term stability in the face of environmental stress and change by enhancing the chances of survival of both host and alga. However, the long-term consequences of such recombination would depend on how rates of coral growth and reproduction are affected (Rowan *et al.* 1997).

Unequivocal data to support the hypothesis that bleaching is adaptive is, as yet, not forthcoming: the retention of up to a third of the original zooxanthellate population during bleaching suggests, alternatively, that it may serve as a strategy to limit damage during stressful periods (Brown 1997*a*). There are also consistent taxonomic differences in vulnerability to stress-related bleaching. In particular, species with high rates of recruitment and rapid growth (that is weedy species such as *Acropora cervicornis*) appear to be more susceptible to bleaching and subsequent mortality than slower growing corals found in the same environment, perhaps as a result of the predominance of symbiont type C in such species.

However, if scleractinian corals are able to respond to environmental change—especially rapid change—by switching symbionts, then this might explain the apparent anomaly of the vulnerability of living scleractinian coral photosymbioses to anthropogenic environmental changes with their long-term stability on geological timescales.

8.2.6 Is photosymbiosis necessary for reef-building in scleractinian corals?

Although there is no doubt that living corals show a strong correlation between reef construction and photosymbiosis, not all reef-building corals are photosymbiotic; and, moreover, not all photosymbiotic corals form reefs. For example, subtropical areas, such as the eastern Pacific, while supporting zooxanthellate coral faunas, show only patchy and ephemeral reef formation due to wide temperature fluctuations (due to El Niño events), nutrient upwelling, and high productivity that favours the growth of suspension-feeders and bioeroders. Similarly, the development of reefs along the Brazilian and west African coasts is minimal due to high coastal run-off and turbidity, even though these areas support zoo-

Table 8.3 Comparison of the environmental distribution of modern zooxanthellate and azooxanthellate scleractinian corals (modified from Fagerstrom 1987, by kind permission of John Wiley & Sons, Inc.)

Environmental parameter	Azooxanthellate	Zooxanthellate
Salinity (ppt)	27–40 (optimal: 34–36)	27–40 (34–36)
Temperature (°C)	6–10	16 min., 18 for reef, optimal 25–29
Depth (m)	60–300, but up to 6200	Less than 100, thrive at low tide to 25
Latitude	?70°N–78°S	28°N–28°S
Nutrient levels	Medium–high	Low
Turbidity	—	Low
Circulation	—	High

xanthellate corals (Coates and Jackson 1985). This suggests that zooxanthellate corals are favoured in warm, shallow, low-productivity settings independent of their ability to form reefs (Table 8.3).

In contrast, the azooxanthellate coral *Tubastraea* dominates some shallow Indo-Pacific reef communities, outcompeting many zooxanthellate forms (Wellington and Trench 1985). *Tubastraea* shows fast rates of growth and is rarely attacked by predators, possibly due to the presence of chemical deterrents. Indeed, some species such as *T. micrantha* calcify at rates comparable to *Acropora cervicornis* (which itself calcifies up to 11 times faster than other zooxanthellate corals; Goreau and Goreau 1959), and can withstand current-exposed habitats (Schuhmacher 1984).

There has been a long history of confusion between the separate issues of the presence of zooxanthellae in corals and their construction of reefs. This confusion has been further promoted by the definition of the term 'hermatypic' to describe both reef-building and zooxanthellate corals (Wells 1933). The foregoing discussion has shown that criteria used to infer the presence of zooxanthellae need to be independent of those features that define reefs. Such a distinction is especially important in the knowledge that the environmental and ecological conditions under which reefs form both today and in the Ancient are far from uniform.

8.3 Attributes of living photosymbiotic reef organisms

In order to determine whether fossil reef-builders possessed photosymbionts, we need to assess the criteria available for recognizing photosymbiosis in the fossil record.

Endosymbionts inhabit soft tissue and so are never found preserved in fossil material (but see Lee and Hallock 1986). However, the former presence of symbionts in fossil organisms might be inferred—with varying degrees of precision—from morphological and geochemical characteristics of the skeleton, or from independent environmental indicators which correlate with photosymbiosis in the Modern.

Symbiosis can influence the form and anatomy of hosts, but the degree of modification may vary greatly. For example, the harvesting of light by the giant clam *Tridacna* is promoted by asymmetrical growth and a large mass of greatly expanded mantle tissue, whereas other bivalves—such as the freshwater forms *Unio* and *Andonta*—show no such substantial modifications associated with photosymbiosis. Moreover, many of the modifications associated with photosymbiosis do not involve the skeleton and so will not be recorded in the fossil record. Symbiotic hosts often show a wide range of behavioural, biochemical, and physiological adaptations in addition to morphological changes that promote, or are a consequence of, the association. All such modifications serve to enhance or maximize light-capture and subsequent penetration through their soft tissue, or to gain easy access to symbiont photosynthates.

8.3.1 Morphology

A major problem in the study of photosymbioses is to ascertain whether particular morphological attributes of symbiotic hosts are conferred by the symbiosis, or if they are preadaptations to that condition. In this context, *preadaptation* refers to a feature which, by virtue of its fortuitous suitability for a novel function, becomes co-opted as a new adaptation (Skelton 1985). For example, many photosymbioses occur in lower invertebrates which already have body plans with large surface areas of exposed soft tissue. But the possession of such an organization confers many advantages to sessile organisms that are quite independent of photosymbiosis (see Section 6.1.9). In addition, while the acquisition of photosymbiosis can select for substantial morphological changes in the host, even some supposedly unusual features of photosymbiotic hosts are known not to be a consequence of the symbiosis as they were already present in non-symbiotic ancestors. The only way to assess whether a morphological feature is a preadaptation to, or a consequence of, the photosymbiosis is by reference to the phylogeny of the host.

Once photosymbiosis is established, there will be a tendency for selection to optimize the benefits of the symbiosis. Here, knowledge of the ancestry of a photosymbiotic host becomes of particular interest and importance, in that both the morphological organization and mode of life of ancestral forms will place constraints upon the morphological pathways available to the host to achieve specialization to the symbiosis. In some cases, many solutions are possible, but as most involve the increased exposure of soft tissue to light, these will also increase the vulnerability of the host to predation. As we shall see, it is sometimes possible to disentangle those features which are antipredation traits from those that are a consequence of the photosymbiosis.

Thin-tissue syndrome

The maintenance of photosymbioses requires hosts to inhabit shallow, non-turbid waters and to have relatively stable habits which will withstand moderate levels of disturbance. Indeed, many photosymbioses are confined to tropical, shallow marine and pelagic surface environments. Host anatomy must expose a large area of symbiont-bearing tissue on light-receiving surfaces, and so photosymbioses are often found in hosts which optimize the photosynthetic surface area, frequently those with thin-tissue layers. Such an organization is known as *thin-tissue syndrome* (Cowen 1983).

Foraminiferans have extensive external tissue in which their symbionts are housed during daylight hours. We have already noted that the closely related bivalved molluscs *Tridacna* (superfamily Tridacnacea) and *Hippopus* (superfamily Cardiacea) show asymmetrical growth which produces a large mass (known as hypertrophied) of mantle tissue exposed to light immediately beneath a widely gaping shell aperture. Such modifications increase the hosts' vulnerability to predation, and many have specific defensive adaptations to counteract this, such as spines in foraminiferans and light-sensitive optical systems and siphonal water-jet expulsion in *Tridacna*.

Many photosymbiotic cnidarians, sponges, and ascidians show thin-tissue syndrome. Wilkinson (1983, 1987) has noted that living photosymbiotic sponges (which bear spicules but no calcified skeleton) tend to have multiserial encrusting, dish-shaped, or compact fan-like, morphologies. But whilst such an organization maximizes the surface area available to incident light, it may also be a mechanism to survive partial predation by minimizing the biomass available to predators and also maximizing the cost of collection, as well as being an additional benefit of the highly integrated, modular condition common in living sponges. Many photosymbiotic sponges contain toxins, and all cnidarians—with or without symbionts—bear a battery of nematocysts which also serve as highly effective defence structures.

The presence of thin tissue might be accompanied by skeletal modifications. Flattened shapes are common in benthic foraminifera and fungiid corals. Some cardiacean bivalves (*Corculum*) do not directly expose soft tissue to light, but have translucent windows (formed by bundles of needle-like crystals which radiate inwards from the external surface) on highly flattened posterior valve surfaces through which light can penetrate to the symbiont-rich tissues. In *Tridacna*, the shell-gape is maximized upwards with undulatory valve margins (Fig. 8.1(b)) to increase the area of exposed mantle tissue.

The possession of zooxanthellae by two remarkable burrowing species of the strawberry cockle *Fragum* (*F. fragum* and *F. loochooanum*) is made possible by the very low-light requirements (that is shade-adapted characteristics) of their zooxanthellae (Ohno *et al.* 1995). Unlike their semi-infaunal, zooxanthellate close relative *Fragum unedo*, these infaunal species do not bear hypertrophied mantle edges, but possess thin and semi-transparent shells which allow uniform illumination of the shell interior. Zooxanthellae are concentrated in the wholly internal soft tissues; the valves show a disproportionate increase in the length of the posterior shell-gape and a very rapid decrease of the angle between the posterior and ventral valve margins during growth, which ensures effective light harvesting by zooxanthellae. By adopting such shell modifications, these species enjoy the benefits of photosymbiosis without forfeiting the security of their infaunal, rapid-burrowing habit.

Coral colony growth form

Most living corals are either always zooxanthellate or always azooxanthellate; very few species appear to be facultatively photosymbiotic. Solitary corals can be either azooxanthellate, e.g. *Caryophyllia*, or zooxanthellate (and often clonal), e.g. *Fungia*; likewise species with phaceloid growth forms can be zooxanthellate or non-zooxanthellate—*Caulastraea*, *Lobophyllia*, *Eusmilia* are zooxanthellate, but *Dendrophyllia* and *Tubastraea* are not. Living coral species with complex, highly integrated multiserial organizations, however, are always zooxanthellate. For example, meandroid growth forms (see Fig. 6.23(g)) are known only from zooxanthellate corals. So while some growth forms correlate exclusively with photosymbiosis in living corals, many others offer no reliable indication of photosymbiotic status.

Coates and Jackson (1987) attempted to overcome this problem by considering the correlation between zooxanthellate status and the dominant morphological features within total living coral assemblages. They analysed four major well-known coral faunas from different settings: Jamaican reef corals; eastern Australian shallow-water corals; shallow-water eastern Pacific corals from Panama and the Galapagos; and deep-water corals (>200 m) from the Caribbean, Atlantic, and Mediterranean. They found that the occurrence of zooxanthellae strongly correlates with particular distributions of growth form, the degree of morphological integration between polyps (as expressed in the degree of connection between corallites; Section 6.1.1), and corallite size: that is with morphological characteristics associated with clonal growth.

Although zooxanthellate corals are represented by all growth forms, the vast majority possess multiserial morphologies (encrusting/massive or erect) irrespective of the ecology of the community (Fig. 8.6): Zooxanthellate corals tend to be multiserial in reef settings (Jamaica and eastern Australia; Figs 8.6(a) and (b)), and subtropical non-reef environments (eastern Pacific; Fig. 8.6(c)). Azooxanthellate corals also show the full range of growth forms, but solitary or phaceloid morphologies predominate.

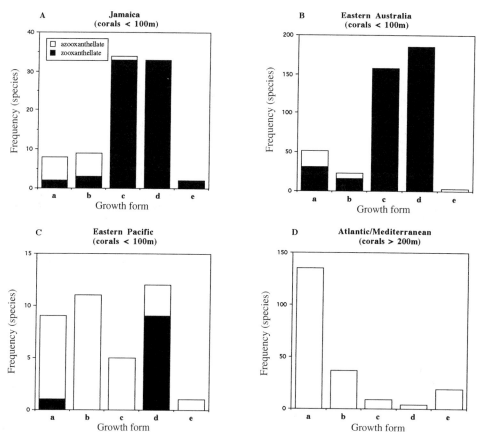

Fig. 8.6 Distribution of growth forms of Recent coral species (azooxanthellate and zooxanthellate) within communities from different environmental settings, showing total numbers of species for each growth form. (a) Solitary (clonal and aclonal); (b) uniserial encrusting and phaceloid; (c) multiserial encrusting; (d) multiserial erect; (e) uniserial erect. (Replotted from Coates and Jackson 1987.)

The degree of integration also correlates with zooxanthellate status, where the vast majority of zooxanthellate corals show high-integration states (Fig. 8.7). Azooxanthellate forms show both low- and high-integration states, but with low-integration predominating. When the ratio of high to low integration is compared between assemblages, the two coral reef communities show ratios >4 (Figs 8.7(a) and (b)), but the temperate, azooxanthellate community has a ratio of <1 (Fig. 8.7(d)). The Eastern Pacific non-reef community shows approximately the same numbers of high- and low- integration forms (Fig. 8.7(c)).

Coates and Jackson (1987) also considered the distribution of corallite diameter (a good proxy for original polyp size) within living coral assemblages. The majority of zooxanthellate corals possess very small (<5 mm diameter) corallites (Fig. 8.8). Small corallite size (especially 5–10 mm in diameter), however, is also favoured in azooxanthellate forms, suggesting that small corallite size is advantageous for reasons independent of the possession of zooxanthellae, especially at low latitudes (compare Figs 8.8(c) and (d)).

These analyses confirm that living, predominantly zooxanthellate, coral communities show particular distributions of morphological characters. But although the correlation is

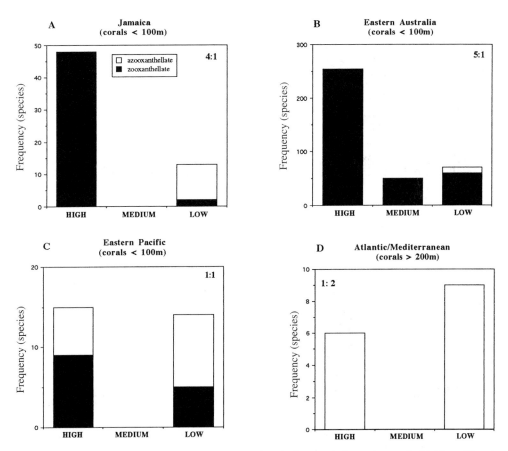

Fig. 8.7 Distribution of integration states among Recent coral species (azooxanthellate and zooxanthellate) from different environmental settings with the ratio of high:low integration. (Replotted from Coates and Jackson 1987.)

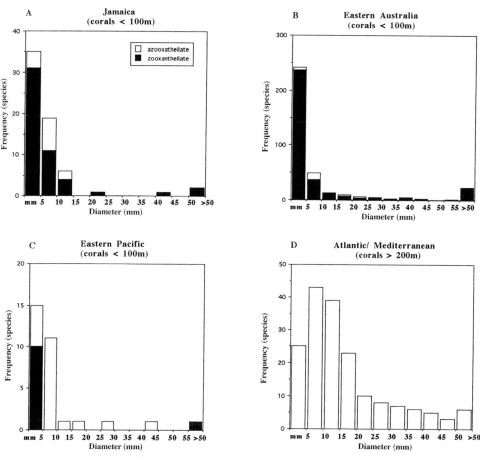

Fig. 8.8 Size distribution of corallite diameters of Recent coral species (azooxanthellate and zooxanthellate) from different environmental settings. Diameters for each species are the mean of all corallites measured. (Replotted from Coates and Jackson 1987.)

strong, it is not exclusive. Indeed, the distribution of growth forms within communities appears to correlate more with environmental demands than zooxanthellate status. Moreover, any shallow-water fauna is not composed exclusively of zooxanthellate species, and so the usefulness of morphological criteria in detecting the occurrence of photosymbiosis in fossil corals will depend on the proportions of species present in any assemblage that possessed photosymbionts (Coates and Jackson 1985). Also, while such an analysis might offer an indication of whether a particular community or assemblage was composed of predominantly photosymbiotic or non-photosymbiotic taxa, it does not allow us to determine whether a particular taxon or specimen bore photosymbionts. Most importantly, the application of such reasoning to fossil communities is based upon the assumption that the proportions of different growth forms, corallite size, and integration status of scleractinian corals have not changed over geological time. But, we have already established that there have been marked changes in the proportion of morphological types through the history of the Scleractinia (see Chapter 6).

Specialized skeletal structures

Some aspects of scleractinian coral morphology are clearly photoadaptive, such as the ledge-like structures found on septa, known as pennular or meniane structures, in the remarkable living coral *Leptoseris fragilis* which lives at depths of 95–145 m. Pennular structures support gastric ducts which radiate from the polyp centres, and are thought to aid photosymbiotic corals to feed heterotrophically in waters of highly reduced illumination (Schlichter 1991; Insalaco 1996). The zooxanthellae of *L. fragilis* also show many pigment modifications, in particular additional fluorescent pigments which alter the short-wavelength radiation into harvestable longer wavelengths.

Large benthic foraminifera show complex internal architecture, with the most intricate anatomical modifications for symbiosis. For example, the diatom symbionts in *Amphistegina lessonii* fit precisely into egg-shaped depressions formed by the pore rims of the test. Here, test sculpture is clearly designed to house one specific symbiont.

8.3.2 Large size and growth rate

We have noted that while zooxanthellate metazoans can reach large size, this is not always conferred by possession of algal symbionts. Although no compilation has been undertaken, it likely that while zooxanthellate corals usually show mean larger colony sizes than azooxanthellate species, these ranges overlap. Indeed, large size appears to correlate more with the environmental and ecological conditions that persist in the modern coral reef environment than possession of zooxanthellae *per se* (Section 8.2.4).

Skeletal extension in non-symbiotic corals occurs in the dark, and such forms grow in deep, cold waters that are undersaturated with respect to calcium carbonate. Skeletal extension rates in non-symbiotic corals are very variable, but they can grow rapidly, with branching forms showing rates of up to 50 mm year^{-1}.

Linear extension rates in symbiotic corals are also highly variable, and calcification rates can vary as much as an order of magnitude between species (Goreau and Goreau 1959). However, when colony calcification is normalized to a standard mean solid radius (Maragos 1978), the apparent variability is much reduced (Kinzie and Buddemeier 1996). Most corals growing under favourable conditions have an equivalent extension rate of between 5 and 10 mm yr^{-1}, similar to that of rapidly accumulating reef sediments (Buddemeier and Smith 1988).

However, heavy calcification does not always correlate with a zooxanthellate status, for example the azooxanthellate coral *Lophelia* is very heavily calcified. Photosymbiotic coral extension rate is strongly affected by environmental factors, especially temperature, water motion, and light. Warm temperatures and water motion enhance coral growth, which increase both light and dark calcification rates (Dennison and Barnes 1988). Water motion may act to increase the supply of carbonate, or remove CO_2. The marked influence of these environmental factors makes the symbiotic effect difficult to isolate.

Skeletal morphology changes in many species in response to light levels (see Fig. 6.3), and growth rates decrease with depth (Fig. 8.5(a)). Branching species grow fastest (up to 150 mm year^{-1}), followed by foliaceous, then massive colonies (Fig. 8.5(b)). Azooxanthellate forms, in particular branching species such as *Oculina varicosa* and *Tubastraea micrantha*, can achieve extension rates equal to zooxanthellate branching forms.

Large size, rapid growth and heavy calcification are terms which are only meaningful if used comparatively. Whilst the presence of large, calcified skeletons in reef-associated fossils might be supposed to be indicative of the former presence of symbionts, there is a real necessity for this assertion to be confirmed by good comparative growth-rate data for species with or without zooxanthellae within living faunas from the same environments. Although numerous experiments have shown that calcification rates plummet when zooxanthellae are artificially removed from living corals, such a 'deprived' coral should not be automatically compared to a true non-zooxanthellate example. Few studies have directly compared calcification rates between naturally occurring reef-dwelling zooxanthellate and azooxanthellate forms. Indeed there is some, albeit limited, experimental evidence to suggest that zooxanthellae in corals may not always boost calcification rates (A.T. Marshall 1996a; Section 8.2.3).

The coral skeleton shows marked alternating growth bands; dark-coloured bands of high-density skeletal growth correspond to the 'cool' season when light intensity and temperatures are lower; the high temperatures and light-intensity of the 'hot' season produce wider bands of slower growing, low-density skeletal material. Such growth bands enable the age of a coral colony to be determined as well as providing a record of relative extension rate, and may be detected in well-preserved fossil material.

8.3.3 Geochemical attributes

It is possible to detect the presence of an algal vital effect by analyses of the carbon and oxygen isotope chemistry of the host carbonate skeleton, but these effects produce distinct isotopic signatures in different symbiotic hosts due to dissimilar calcification pathways.

In general, photosynthesis preferentially sequesters the lighter ^{12}C isotope leaving a pool enriched in ^{13}C, whereas respired CO_2 is assumed to be depleted in both ^{13}C and ^{18}O relative to the CO_2 source (Land et al. 1975). This means that carbonate skeletons precipitated under the influence of photosynthesis are predicted to show enrichment in $\delta^{13}C$ composition compared to a skeletons of non-symbiotic organisms from the same environment.

Depletion of $\delta^{13}C$ composition has been found in the shell carbonate of *Tridacna maxima*, with a 2‰ depletion recorded compared to an non-symbiotic gastropod (D.S. Jones et al. 1986; Romanek et al. 1987). However, such a negative departure from equilibrium also appears to be common in living, non-symbiotic bivalves (Wefer and Berger 1991), such that the recognition of photosymbiosis from isotopic analyses of bivalve shell appears to be equivocal (D.S. Jones et al. 1988). Indeed, some have suggested that photosymbiont metabolism has a negligible influence on the isotopic composition of *Tridacna* skeletal material (Aharon 1991) because bivalves may draw their CO_2 directly from sea water.

In large benthic foraminifera, ^{13}C and ^{18}O depletion occurs with increased light intensity and through ontogeny only in larger rotaliines, not in larger miliolines which show a decrease. In contrast, the isotopic signatures of Neogene planktonic foraminifera show very negative $\delta^{18}O$, as well as enrichment in $\delta^{13}C$ due to the preferential removal of ^{12}C by photosymbionts (Norris 1996). There is therefore no clear photosymbiotic effect that can be applied to all benthic foraminifera.

Although living scleractinian coral skeletons show a wide range in carbon and oxygen isotope composition, there are also distinctive differences between zooxanthellate and azooxanthellate individuals (Fig. 8.9; Swart 1983; G.D. Stanley and Swart 1995). Azooxanthellate corals show a strong positive correlation between $\delta^{13}C$ and $\delta^{18}O$ isotopes

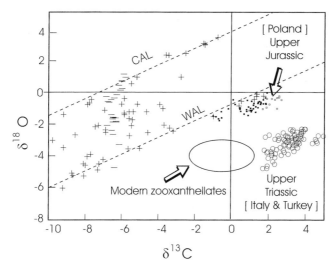

Fig. 8.9 Summary of isotopic data from Recent and Ancient scleractinian corals. There is a strong positive association between $\delta^{13}C$ and $\delta^{18}O$ in the living azooxanthellate corals, but none in living zooxanthellate forms. The spread of azooxanthellate data reflect the wide range of temperatures at which these forms grew, and define a Cold zooxanthellate Line (CAL) and Warm Azooxanthellate Line (WAL). Carbon isotopic compositions of the Upper Triassic samples, although enriched, are similar to Recent zooxanthellates, but the Upper Jurassic samples lie close to the Recent azooxanthellate field and show a strong positive correlation with oxygen. (Modified from G.D. Stanley and Swart 1995.)

within their skeletons and occupy a wide range of values. Zooxanthellate coral skeletons usually show no such correlation, and most occupy a limited field of values between –5 and –3‰ for oxygen and –2 and +2‰ for carbon showing an enrichment in ^{13}C (G.D. Stanley and Swart 1995). Temperature influences the $\delta^{18}O$ content of coral skeletons probably through normal equilibrium fractionation processes, but absolute $\delta^{18}O$ values of the skeletons are depleted by 2–4‰ for reasons as yet unclear (Weber and Woodhead 1972).

The rate of skeletal growth (extension) and its variation within a colony can also affect $\delta^{13}C$ content, and the seasonal periodicity in coral $\delta^{13}C$ records is likely to be a function of the relative growth rate where the faster depositing low-density bands are lighter in $\delta^{13}C$ relative to the slower depositing high-density bands. Again, there is a need for comparative data within single communities.

Valid isotopic analysis of skeletal carbonate requires diagenetically unaltered, pristine material. But aragonite is metastable and converts fairly rapidly to calcite, so that pristine scleractinian aragonite is rare in the geological record. However, notwithstanding the many inherent problems of diagenesis, analysis, and interpretation, isotopic measurements remain probably the most direct, if not wholly unequivocal, method of establishing the presence of ancient photosymbioses.

8.3.4 Environmental setting

The maintenance of algal symbiosis demands that most hosts occupy shallow, well-lit environments, and such limitations are assumed to have prevailed throughout the history of algal symbiosis. The unexpected absence of other photosymbiotic organisms or algae in a biota (suggesting low-light levels), or evidence of high-nutrient levels (e.g. in predicted areas of upwelling or coastal run-off), low water clarity (e.g. presence of fine-grained clastic sediments), or growth at depths below the photic zone, might all be counterindicative of such conditions. But these criteria are far from unequivocal, as some zooxanthellate corals—and indeed coral reefs—appear to be able to thrive in areas of relatively high-nutrient input, at depths greater than 30 m, and in areas of turbidity. Photosymbioses among benthic skeletal organisms, however, is predominantly a tropical phenomenon, and palaeolatitude reconstructions should confirm such a geographical distribution for abundant assemblages of putative photosymbiotic organisms.

Whilst the abundance and diversity of boring organisms is certainly an indicator of nutrification on modern coral reefs, many of these borers are geologically recent additions to the reef biota, appearing only from the Late Jurassic onwards. Their absence/presence in older communities will serve no useful purpose in assessing ancient nutrient levels. Moreover, any factor which decreases growth could lead to increased bioerosion within reef communities, such as changes in rates of sedimentation, temperature-related oxygen deficiency, or light limitation.

The relative abundance of algal growth is a good indicator of nutrification in recent seas, but many algae are non-calcified and so leave no trace in the fossil record. Moreover, the temporal distribution of calcified algae probably reflects the changing excavatory ability of herbivores, rather than ambient nutrient levels (see Section 7.3.1).

8.4   Photosymbiosis in fossil reef-builders

We have explored those characteristics of modern photosymbiotic hosts which might be indicative of photosymbiosis in the fossil record. In the following section, we will consider evidence for photosymbiosis in common, ancient reef organisms, by consideration of first the fossil record of potential symbionts, and second, morphological and geochemical aspects of the potential host skeletons.

8.4.1   Fossil record of potential photosymbionts

Cyanophytes, chlorophytes, acritarchs, and rhodophytes are ancient groups which probably appeared during the Proterozoic. Acritarchs were most abundant in the early Palaeozoic, but rapidly declined at the end of the Devonian. A second radiation, probably of unrelated forms, occurred in the Jurassic–Cretaceous. No living acritarchs (which may, however, be more correctly considered non-tabulate dinoflagellate cysts) are known to enter into symbioses. Diatoms appeared in the Cretaceous, and diversified during the Tertiary.

The oldest dinoflagellate fossils are known from the Middle Triassic (Anisian, ~240 Ma), recording those forms which encyst (only 7 of the 14 living orders are capable of producing fossilizable cysts). The fossil record of the non-cyst-forming dinoflagellate order Gymnodiniales is very poor. However, biomarkers for dinoflagellates (triaromatic dinosteroids, derived from dinosterols) are known from the late Neoproterozoic to the present day (Moldowan *et al.* 1996; Fig. 8.10). But dinosteroids found from the late Neoproterozoic to the Devonian are suggested to be the products of acritarchs, which are most probably closely related and possibly even ancestral to the dinoflagellates, and whose fossil record covers the same stratigraphic range. Dinosteroid abundance is, however, only ~9 to 60% relative to the levels in Mesozoic sediments (100%). No triaromatic dinosteroids are known

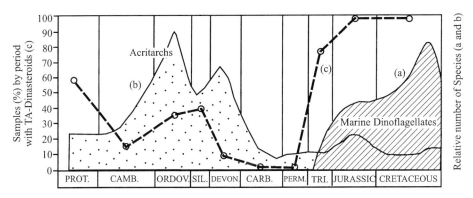

Fig. 8.10 The diversity and abundance of acritarchs and dinoflagellates over geological time. Relative numbers of species of (a) dinoflagellate cysts, and (b) acritarch cysts (Tappan and Loeblich 1973). Dashed line and circles (c) gives frequency of occurrence of detectable triaromatic dinosteroids in samples from each geological period. (Modified from Moldowan *et al*. 1996, by kind permission of the Geological Society of America.)

from the Carboniferous and the Permian, but the record resumes sporadically in the Early–Middle Triassic to reach abundant levels by the Upper Triassic which are maintained in sediments to the present day (Moldowan *et al*. 1996). These data show that detected tri-aromatic dinosteroid abundance correlates with the combined acritarch and dinoflagellate body fossil record through geological time. But whilst corroborating earlier suggestions that dinoflagellates are an ancient group, these data also suggest that dinoflagellate populations were not globally significant during the Early–Middle Triassic, and only underwent an increase in diversity/abundance to become abundant members of the plankton from the Upper Triassic onwards (Fig. 8.10).

So the fossil record suggests that dinoflagellates did not become widely available to enter into photosymbioses until the Upper Triassic. Moreover, due to the general dearth of dinoflagellate records in the Carboniferous and the Permian, we can speculate that they may not have been present—at least in any abundance—during this interval.

8.4.2 Evidence of photosymbiosis in fossil reef-builders

Calcified sponges

Like the vast majority of Cambrian sponges, archaeocyaths possessed a solitary or low-integration organization. Their morphology does not appear to have been adapted for photo-symbiosis: they were relatively small (up to 30 cm in length, but more usually 5–10 cm long), and had little externalized soft tissue (Wood 1990, 1993; Wood *et al*. 1992*a*). Moreover, many possessed a cryptic habitat (Zhuravlev and Wood 1995), and no algal fractionation signal has been detected in carbon isotope analyses of their skeletons, i.e. their skeletons were precipitated in isotopic equilibrium with sea water (Brasier *et al*. 1994; Surge *et al*. 1997).

Calcified sponges with sphinctozoan (chambered) organization show similar morphological attributes to archaeocyaths. Living representatives (which bear no photosymbionts) are often cryptic or deep-water dwellers, and many ancient sphinctozoans also occupied this habitat (Wood 1997).

Multiserial calcified sponges (stromatoporoids and chaetetids) possess thin veneers of soft tissue. Living representatives, e.g. *Astrosclera*, *Calcifibrospongia*, and *Acanthochaetetes*,

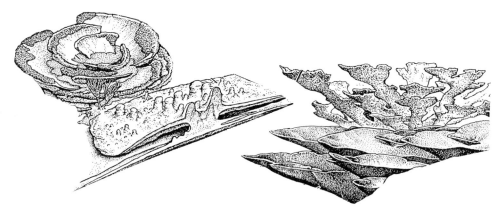

Fig. 8.11 Some of the platy morphologies of reef-builders which supported large areas of exposed soft tissue found in shallow-water; tropical Upper Devonian stromatoporoids (left) are reminiscent of those growth forms found in living zooxanthellate corals (right).

are generally cryptic and do not possess photosymbionts. They can reach up to 0.5 m in diameter, but show very slow growth rates, often only 0.2 mm year^{-1}. Shallow-water Palaeozoic stromatoporoids, by contrast, achieved tremendous diameters of up to 5 m, and in some environments, particularly well-agitated pure carbonate settings, they were able to outgrow other metazoans in competitive interactions (probably due to rapid lateral growth) even though only modest vertical extension rates of up to approximately 3 mm year^{-1} have been recorded to date (Gao and Copper 1997). Some Devonian stromatoporoids show complex, highly integrated, platy growth-forms that are reminiscent of those found in living zooxanthellate corals (Fig. 8.11).

Foraminifera

Large Permo-Carboniferous fusiline foraminifera with widespread abundance in shallow-marine carbonates probably contained algal symbionts (Ross 1974; Tappan 1982; Cowen 1983), although their wide geographical distribution beyond the tropics and subtropics makes it unlikely that all representatives were symbiotic (Ross 1979). These symbionts were presumably either chlorophytes, cyanophytes, or rhodophytes—groups that have high or broad nutrient tolerances.

A second phase of large benthic foraminiferan diversification involving 11 families occurred in the Late Cretaceous and early Tertiary (Ross 1977), presumably recording in part forms which became symbiotic with the newly evolved diatoms. Larger foraminifera evolved at least four times during this period at low latitudes, and all groups show complex test structures with multiple chambers, sculptured walls, and intricate interconnections (Cowen 1983). Modern diatom-bearing foraminiferans (rotaliines) are found predominantly in the oligotrophic to highly oligotrophic waters of the Indo-Pacific (Hallock 1988).

About 25% of the 40–50 extant species of planktic foraminifera are known to harbour algal symbionts (Hembleben *et al.* 1989). The exact number has proved difficult to determine as some species have facultative relationships. Data from extant taxa and isotopic studies of extinct species demonstrate that photosymbiosis has been present in different clades of planktic foraminifera throughout at least the last 75 million years (d'Hondt and Zachos 1995).

Bivalves and brachiopods

All known living photosymbiotic bivalves are closely related, and the acquisition of photosymbiosis is closely associated with an adaptive radiation of bivalve morphology and ecology. Here, the phylogeny of these clades are relatively well known such that we can trace the developmental pathways of symbiosis acquisition and subsequent modification. Both *Fragum* and *Tridacna*, and probably *Hippopus*, appeared in the Miocene. Fossil examples of *Tridacna* are known from the Caribbean and Europe, although living representatives are exclusively Indo-Pacific. *Corculum* is known only from the Recent.

Whilst most photosymbiotic species of the genus *Fragum* are epifaunal, it is probable that their ancestors were infaunal and, like the two extant infaunal photosymbiotic species of *F. fragum* and *F. loochooanum* did not possess hypertrophied mantles. The most parsimonious pathway for photosymbiosis acquisition in this bivalve might be that the association was initiated by the successful invasion of already shade-adapted *Symbiodinium* in an ancestral infaunal bivalve which was not preadapted to photosymbiosis (Ohno *et al.* 1995). Subsequently, *Fragum unedo* became specialized towards the semi-infaunal habit and developed hypertrophied mantles. An alternative explanation is that photosymbiosis appeared in preadapted forms, where features such as hypertrophied mantles had already evolved as a means for more efficient filter-feeding of zooplankters which subsequently became infested by photosymbionts. The ancestor of *Corculum* was also infaunal, but this genus has subsequently gained an epifaunal habit; like *F. fragum* and *F. loochooanum* it possesses a thin and transparent shell to illuminate algae, thus allowing the soft parts to remain protected.

The hypertrophied tissues and asymmetrical growth which promote light capture in *Tridacna* are not present in the ancestral genus *Avicularium* (Cowen 1983). It is likely that tridacnids—which probably evolved from a byssate, epifaunal ancestor—evolved hypertrophied mantle edges and shifted to epifaunal or semi-infaunal life habits (except the coral chasmolith *T. crocea*). The development of wide mantle edges for the placement of highly sensitive eyes may have been in response to increased predation pressure (Stasek 1962).

So we see that forms which had an infaunal ancestor were probably invaded by a shade-adapted *Symbiodinium*. Some species have retained the infaunal habit—so avoiding the vagaries of life on the open surface—but have promoted the association by the acquisition of thin, translucent shells, and an increase in the length of the posterior shell-gape to ensure efficient light-harvesting. In contrast, where ancestors were already epifaunal, the host evolved modifications for antipredatory defences such as light-sensitive optical systems and siphonal water-jet expulsion, and a stable habit gained by acquiring large size or a chasmolithic mode of life—which also reduces the ease of predator manipulation.

It has been noted that some non-photosymbiotic bivalves e.g. *Spondylus butleri*, contain dinoflagellates in their tissues. These are inferred to be a toxic defence against predation (Harada *et al.* 1982). This suggests a mechanism by which dinoflagellates can invade bivalves, whereby the host may gain increased fitness and only later farms the dinoflagellates as a source of carbon (Harper and Skelton 1993).

Several groups of extinct, large aggregating, epifaunal bivalves are known, especially from the Mesozoic. These include alatoconchids and rudists.

Alatoconchids were a Permian group of myalinid bivalves which formed gregarious communities in tropical reef lagoons. The largest examples could reach over 1 m in length and up to 10 kg in weight. On each valve, they bore high, narrow ridges of very thin shell which extended ventrally from umbo to commissure. Alatonconchids lay with their commissures vertical, such that these expanded ridges were horizontal to the sediment surface. Yancey (1982) proposed that these ridges were lined with symbiont-rich mantle tissues. The opening and closure of the commissure would certainly have been difficult in adult forms, also suggesting substantial dependence on photosymbionts for nutrition.

The Jurassic–Cretaceous rudists were remarkable producers of calcium carbonate: some of the largest examples (e.g. *Durania* aff. *nicholasi*) are estimated to have yielded up to 100 kg of carbonate per year, or approximately 0.1 m^3 within 10–20 years (Schumann 1995); the carbonate production of dense hippuritid associations has been estimated to be 21 kg m^{-2} year^{-1} (Steuber 1997). Rudists display a variety of morphologies, including tightly aggregating, upright-growing monospecific groups (hippuritids and radiolitids) which grew in abundance on both carbonate platforms and mixed carbonate–terrigenous settings in shallow, warm, tropical waters (estimated to be between 20–36 °C; Steuber 1996). These morphologies could reach several metres in length and growth rates of up to 54 mm year^{-1} have been recorded (the radiolitid *Gorjanovicia*: Steuber 1996).

Rudists are considered by some to have been photosymbiotic by dint of their high growth rates, prolific ability to produce calcium carbonate, and the presence of a porous outer layer on some upper (left) valves that covers a series of radial canals inferred to be used for farming photosymbionts (e.g. Cowen 1983; Kauffman and Johnson 1988). However, close morphological analysis has failed to secure any evidence for the extensive externalization of soft tissue, such that photosymbiosis is now thought to have been restricted to only a few non-aggregating, radiolitid taxa such as *Torreites*, that possess unusual morphological traits which do suggest the extensive exposure of the mantle tissue to light (Skelton and Wright 1987). This has been confirmed by analysis of well-preserved outer shell layers in these species which show similar isotopic compositions to those found in photosymbiotic scleractinian corals (Steuber 1994). In contrast, no such fractionation signature has been found in aggregating hippuritids and radiolitids (Steuber 1994, 1997). The fact that the hippuritids can show preferential orientation downstream of prevailing currents also supports a hypothesis of a predominantly heterotrophic, rather than phototrophic, mode of life. Moreover, hippuritid rudists are most abundant and diverse in mixed carbonate–terrigenous settings, suggesting that the pore system of the left valve may be better interpreted as a screen to reject unwanted sedimentary particles (Steuber 1997).

The acquisition of photosymbionts by some rudists has been proposed to be due to preadaptation. In one such scenario, the presence of an extremely narrow gape and small body/mantle cavity ratio in *Radiolites cf. angeoides* has been interpreted to indicate atrophy of the gills and the loss of effective filter-feeding (Skelton 1979). Skelton suggested that the species developed/expanded its tentacles and mantle margins to compensate for ineffective filter-feeding—which were later infected by zooxanthellae. However, an alternative scenario is to argue for a sciaphilous origin for the partnership, where infection of shade-adapted zooxanthellae occurred within the internal tissues of the ancestor. The expanded mantle margins could then be explained as being the result of optimized light-harvesting, which was only subsequently followed by the loss of efficient filter-feeding. This perhaps offers a more convincing explanation for the aberrant morphology of *Radiolites cf. angeoides*, as it removes the need to evoke the evolutionary stage of food collection with expanded and tentacled mantle margins.

Photosymbiosis in the minor Permian brachiopod groups, the lyttoniacids and richthofeniids, has been proposed by (Cowen 1983) on the basis of:

(1) evidence from the conical shell organization which provided exposure of the mantle to light;

(2) possession of a very large shell compared to body size;

(3) unusual feeding mechanisms;

(4) their preferred reef-associated habitats.

Others, however, have dismissed these lines of evidence as many aspects of the unusual morphology of these brachiopods can be explained as adaptations to a sessile, attached habit but with retention of a normal mode of brachiopod feeding associated with a wide gape between the valves (e.g. R.E. Grant 1975).

Cnidaria

Photosymbionts are present in all three living classes of Cnidaria (Scyphozoa, Hydrozoa, and Anthozoa). Molecular data clearly indicate a highly complex history of multiple acquisition (and probably loss) of zooxanthellae (Section 8.1.3).

Palaeozoic corals

Palaeozoic rugose corals were predominantly small, and 64% of genera were solitary (Scrutton 1998). Of the 104 colonial genera, 58 (53%) showed low-integration (phaceloid) colonies, with comparatively large corallites (diameters range from 1.5 to 8 mm; Coates and Jackson 1985). However, other more complex, colonial growth forms are known, including cerioid (34% of colonial forms) and asteroid–aphroid growth types (see Fig. 8.12). Only 13% of colonial rugosans were of high integration—5% of all genera (Scrutton 1998); one genus with meandroid organization is known from the Permian.

Rugosans occupied a range of shallow- and deep-water habitats, and show very variable growth rates. Rates are lowest in the most highly integrated colonies, and highest in solitary and phaceloid forms, although rates of up to 31 mm year^{-1} have been recorded in some phaceloid/cerioid species (Gao and Copper 1997; Fig. 8.13). On the basis of morphological comparison with the distribution of growth form and corallite size with living zooxanthellate scleractinian corals, Coates and Jackson (1987) concluded that they did not possess symbionts.

The vast majority of the Palaeozoic tabulate corals, however, show multiserial organizations, with small corallite sizes (mean diameter, 1–3 mm; Coates and Jackson (1985)). Some level of integration is seen in 85% of all genera, as well as growth rates equal to that of some modern massive zooxanthellate scleractinian corals (Fig. 8.13). Most tabulate corals show mean annual growth rates of less than 10 mm year^{-1} and, like the Rugosa, the most highly integrated corals show the slowest growth rates. Recorded rates give 2–6 mm year^{-1} for heliolitids, and more rapid rates of 5–18 mm year^{-1} for favositids (see Gao and Copper 1997; Scrutton 1998). The only data available for growth rate variation with depth is for *Heliolites megastroma*, which showed a mean growth rate decrease from 4.2 mm year^{-1} at a depth of ~30 m, to 3.4 mm year^{-1} at a depth of ~40–60 m in Silurian (Wenlock) environments in Shropshire (Powell 1980). Indeed, where intraspecific data are available, higher growth rates appear to correlate with higher rates of sedimentation in Palaeozoic corals.

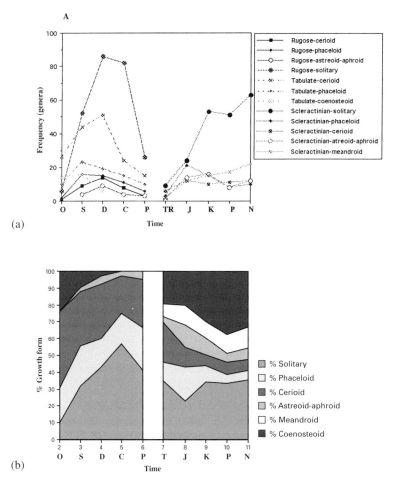

Fig. 8.12 Growth forms in tabulate, rugose, and scleractinian corals through the Phanerozoic. (a) Diversity; (b) relative proportion. Note how the proportion of meandroid, and particularly coenosteoid, growth forms has increased since the Jurassic, with other growth forms decreasing. O: Ordovician, S: Silurian, D: Devonian, C: Carboniferous, P: Permian, T: Triassic, J: Jurassic, K: Cretaceous, P: Palaeocene, N: Neogene. (Plotted from data in Coates and Oliver 1973.)

Based on morphological criteria, several authors have favoured the possession of symbionts in tabulate corals (Coates and Oliver 1973; Coates and Jackson 1987; Cowen 1988). However, colonial organizations and small corallite diameters are systematic characteristics of the group as a whole, and their utility as an indicator of photosymbiont possession is therefore questionable.

It is noteworthy that while growth rates for similar colony growth forms is slightly higher in zooxanthellate scleractinians than Palaeozoic corals, the differences are not marked—with the striking exception of the very rapid extension rates recorded for *Acropora*. However, no change of growth form in response to declining light levels with depth is found in Palaeozoic corals. The modest changes in mean growth rate with depth are more likely to be in response to declining temperature and sedimentation rates rather than illumination.

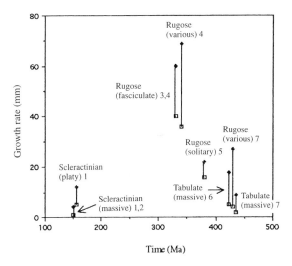

Fig. 8.13 Annual extension rates inferred from growth banding in some fossil corals. (Data from: 1, Ali 1984; 2, Insalaco 1996; 3, Ali 1984; 4, G.A.L. Johnson and Nudds 1975; 5, Scrutton 1965; 6, Scrutton and Powell 1981; 7, Gao and Copper 1997.)

No isotopic data are currently available to determine the presence of an algal fractionation signal in any Palaeozoic corals.

Scleractinian corals

About one-half of modern coral genera have been estimated to be photosymbiotic (Veron 1995). Scleractinian corals have a far greater range of colonial growth forms than Palaeozoic corals (see Fig. 6.2 and Fig. 8.12 for functional organizations), and have invaded a broader range of niches, including higher energy settings. They had occupied the full range of modern coral settings by the Toarcian (late Early Jurassic). Scleractinian corals also show very high species diversities: on the GBR alone, 252 species are described belonging to 70 genera. By comparison, the maximum number of species recorded from a shallow marine Palaeozoic coral fauna is 36 species (15 genera) (Webb 1990). These observations are consistent with a substantial adaptive radiation in the Scleractinia.

On the basis of the distribution of colony growth form, integration, and corallite size within communities (Fig. 8.14), it has been argued that scleractinian corals gained symbionts by the latest Triassic (Norian) and that photosymbiosis was widespread by the Lower Jurassic (Coates and Jackson 1987). While there is no palaeoecological evidence for framework formation in middle and early Late Triassic scleractinian corals, it has been suggested that the appearance of large branching individuals of the phaceloid–dendroid genus 'Thecosmillia', which reaches up to 10 m in height, together with faunal domination in shallow waters is indicative of photosymbiosis acquisition by this time (G.D. Stanley 1981, 1988). As discussed before, such inferences are based upon circular reasoning.

Zooxanthellate isotopic signatures have been sought in fossil scleractinian corals. To date, only material from two Upper Triassic (Norian: Italy and Turkey) sites, and one Upper Jurassic (Tithonian: Poland) and one Eocene (England) site have been deemed sufficiently unaltered to be examined geochemically (G.D. Stanley and Swart 1995; Mackenzie et al. 1997).

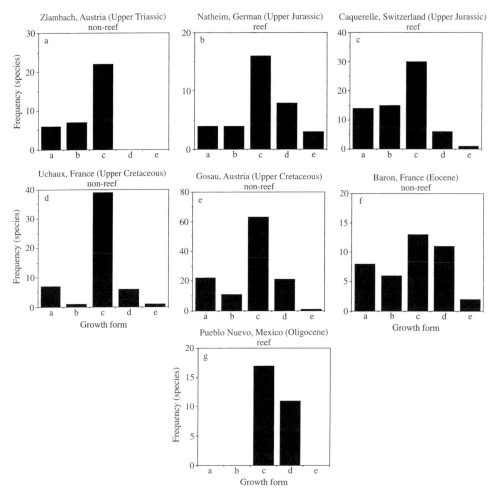

Fig. 8.14 Distribution of coral species growth forms within a variety of Mesozoic and Cenozoic communities showing total numbers of species for each growth form. (a) Solitary (clonal and aclonal); (b) uniserial encrusting and phaceloid; (c) multiserial encrusting; (d) multiserial erect; (e) uniserial erect. (Replotted from Coates and Jackson 1987).

The resultant data show a strong positive correlation between ^{13}C and ^{18}O isotopes in the Jurassic material (Fig. 8.9). Although these corals show multiserial morphologies, they occupied a temperate-latitude, and possibly cool-water, environment—and growth rates have been inferred to have been slow (Gruszczyñski *et al.* 1990; G.D. Stanley and Swart 1995); however, these have not been adequately quantified. In contrast, most (11 of 13) of the species analysed from the two Upper Triassic subtropical, shallow-water Tethyan localities showed isotopic signals similar to that of living zooxanthellates (Fig. 8.9), although carbon isotopes are 3–5‰ heavier than the Modern. No comparative growth or calcification rates are known for this analysed material. Specimens of the Eocene *Goniopora websteri* from a high latitude, but warm, non-reef setting also show a signal similar to living zooxanthellate corals. These corals show growth rates of 2 mm year^{-1} (Mackenzie *et al.* 1997).

Of most note, however, is the distribution of growth forms in the coral species which show a zooxanthellate signature. Of the 11 species, 4 are solitary, 3 phaceloid, and 4 bear multiserial morphologies: a distribution and proportion quite different to that of living zooxanthellate corals.

Some Mesozoic platy scleractinian corals, such as microsolenids, bear pennular structures similar to those found in *Leptoseris fragilis* (see Section 8.3.1—Specialized skeletal structures). These corals form tabular beds of low diversity, and are known from a variety of Jurassic (Oxfordian), Early Cretaceous (Barremian–Early Aptian, and Albian), and Late Cretaceous localities (see Insalaco 1996; Rosen 1998; Fig. 8.15). These assemblages have been inferred to have been zooxanthellate, but are suggested to have occupied deeper water settings, but still within the photic zone. Interesting, many Jurassic and Cretaceous coral seem to show slow growth rates. No growth rates of more than 20 mm year^{-1} have been recorded (E. Insalaco, personal communication).

Members of living genera such as *Montastraea* are known from the Cretaceous (Budd and Coates 1992), and by the Eocene a substantial proportion of other living genera had appeared that might be inferred to have been zooxanthellate on the basis of taxonomic uniformitarianism (Rosen 1998). On such grounds we might conclude that algal symbiosis was

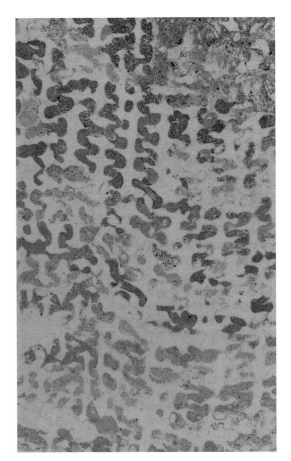

5 mm

Fig. 8.15 Pennular structures in a Jurassic microsolenid coral. These structures are thought to aid photosymbiotic corals to feed heterotrophically in waters of highly reduced illumination. (Photograph: E. Insalaco.)

present throughout the Cenozoic. (Although it should be noted that some modern coral genera bear both zooxanthellate and non-zooxanthellate species.)

Bryozoans

While no living bryozaons are known to possess photosymbionts, there is some evidence to suggest that the mid-Permian trepostome bryozoan *Tabulipora* may have possessed symbionts (Håkansson and Madsen 1991):

1. The branch width is three times that of other stenoporoid trepostomes.
2. The growing tips of branches were covered in substantial areas of soft tissue which lacked zooids.
3. Isotopic analysis of material with excellent preservation of skeletal microstructure shows depleted $\delta^{13}C$ values compatible with the influence of a 'vital effect'.

Tabuliopora was not a major reef-builder.

Further isotopic sampling of both fossil material, excluding diagenetic cement, and living bryozoan skeletal material is required to confirm these observations.

8.4.3   Dating the origin of symbioses

Dating the origin of photosymbioses can be approached from two angles: using morphological and isotopic data from potential hosts; and analysing the fossil record for potential symbionts.

We have seen that inferring photosymbiosis in hosts can be highly problematic on morphological grounds alone, except where taxonomic uniformitarianism can be safely evoked. Pennular structures do, however, appear to present an unequivocal indicator of photosymbiosis in scleractinian species adapted to low-light levels. Reasoning based on morphology and growth form relies heavily upon modern analogies, but not only are none available for several important extinct groups of reef-associated metazoans, such uniformitarian reasoning may also be invalid. As low-integration growth forms in Triassic scleractinian corals show apparent zooxanthellate isotopic signatures (and indeed living zooxanthellate corals can, with solitary growth forms also known), inferences derived from the analysis of typical growth morphology distribution in living zooxanthellate corals would appear to be a poor predictor of zooxanthellate status in Ancient representatives. Indeed, we have seen that the distribution of growth forms has changed through time in all groups of benthic metazoans associated with reefs (Chapter 6). If the solitary habit then is a poor indicator of the absence of photosymbiosis, then this criteria cannot be used to infer zooxanthellate status in rugosan corals. We are then left with the conclusion of Hill (1981), who regarded photosymbiosis in Rugosa as incapable of proof. Coral growth morphology is clearly not determined entirely by the presence or absence of zooxanthellae, but is a result of many other ecological demands.

The need for more comparative growth rate and isotopic data within and between fossil assemblages is critical to gaining any further insight into photosymbiotic versus non-photosymbiotic calcification, and the rarity of pristine skeletal material makes valid isotopic data difficult to acquire. (Although analysis of the exceptionally well-preserved Anisian coral faunas from south China (Qi 1984) would be most fruitful.) The only way forward in many fossil groups may be to assemble data from as many independent methodologies as possible, but for many—as for rugosan corals—the unequivocal presence or absence of photosymbiosis might be incapable of proof.

Table 8.4 outlines the inferred timing of the appearance of reef-associated photosymbiotic associations through the Phanerozoic, but these details are very poorly constrained. We can be relatively secure in our assertion that some fossil scleractinian corals, many foraminifera, and tridacnid bivalves possessed photosymbionts. If the isotopic data are correct, scleractinian corals gained photosymbionts at least by the latest Triassic; to date older material has not been analysed isotopically. Pennular structures, known from the Jurassic, indicate that not only were some corals photosymbiotic at this time, they were also adapted to low-light levels. A few, non-aggregating rudists were probably photosymbiotic but the evidence for the Palaeozoic tabulate corals and stromatoporoids is equivocal, although the platy growth forms of some Devonian stromatoporoids is suggestive of extensive exposure of soft tissues to light. That photosymbiosis was present in other extinct reef-building groups is at best unconfirmed; it is not necessary to conclude that by virtue of their reef-building abilities all major Phanerozoic reef-building groups possessed symbionts.

The history of dinoflagellate body fossils and biomarkers may provide less equivocal data for inferring the appearance of photosymbiosis in reef communities. This record suggests

Table 8.4 Inferred symbiont appearance, host appearance, and proposed timing of symbiont acquisition in extinct and living reef-associated organisms

Host	Symbiont	First appearance of symbiont	First appearance of host	Inferred timing of acquisition by host
Living groups				
Scleratinian corals	Dinoflagellates	?Proterozoic abundant by Upper Triassic (Norian)	Middle Triassic (Anisian)	At least by upper Triassic (Norian)
Rotaliine foraminiferans	Diatoms	Late Cretaceous	Late Cretaceous	Late Cretaceous
Milioline foraminiferans	Dinoflagellates	?Proterozoic, abundant by Upper Triassic (Norian)	Upper Triassic	Upper Triassic
Tridacnid bivalves	Dinoflagellates	?Proterozoic, abundant by Upper Triassic (Norian)	Eocene	Eocene
Extinct groups				
Fusiline foraminiferans	?Chlorophytes ?Rhodophytes	Proterozoic	Late Carboniferous	Late Carboniferous
Rudist bivalves	?Dinoflagellates	?Proterozoic, abundant by Upper Triassic (Norian)	Late Cretaceous	Late Cretaceous
?Tabulate corals	?Chlorophytes ?Rhodophytes ?Acritarchs	Proterozoic	Cambrian	?
?Stromatoporoid sponges	?Chlorophytes ?Rhodophytes ?Acritarchs	Proterozoic	Ordovician	?

that dinoflagellates did not become widely available to enter into photosymbiosis until the Upper Triassic. The widespread occurrence of the *Symbiodinium*-complex in photosymbioses indicates that this alga is particularly successful at host invasion and manipulation, that is capable of overcoming the self-defence system of potential hosts. Indeed there is evidence to suggest that some types of *Symbiodinium* actively seek new hosts (Fitt *et al.* 1981), or respond to positive chemotaxis (attraction to water-borne chemicals) towards prospective coral hosts. The singular success of the *Symbiodinium*-complex indicates that once a suitable symbiont has evolved, it might infect many unrelated organisms. This ability might also be related to its long symbiotic history. Although it has been suggested that prospective hosts may be required to be in some way preadapted to acquire photosymbiosis (Cowen 1983), an intriguing alternative scenario might be that some photosymbioses may have originated in low-light conditions made possible by shade-adapted *Symbiodinium*, and only later invaded well-lit environments as a result of morphological modifications that optimized the partnership.

8.5 Photosymbiosis and the history of reef-building

Contrary to received opinion (Cowen 1988; Talent 1988), current evidence suggests that photosymbiotic metazoans have not always been present in reef communities. Indeed, there are no clear data to support the presence of photosymbiotic reef-associated faunas before the Upper Triassic, with the possible exception of fusilinid foraminifera and alatoconchid bivalves. Moreover, these groups were presumably symbiotic with either ancestral dinoflagellates, chlorophytes, rhodophytes, or cyanophytes, and such symbioses would not necessarily have conferred the same metabolic capabilities, or indeed adaptation to the same environments, as do modern photosymbioses with dinoflagellates and diatoms (Table 8.4).

8.5.1 Trophic organization of Palaeozoic reefs

The lack of evidence for photosymbiosis in many Palaeozoic communities has far-reaching consequences. Most Palaeozoic reefs were constructed by microbes, calcified algae, and invertebrate heterotrophs (suspension-feeders). So unlike modern coral reefs, where the main primary producers are also the reef constructors—and indeed the main source of carbonate—the primary producers in Palaeozoic reefs can be inferred to have been either microbial communities, or soft-bodied alga (for which we have no record), or phytoplankton (compare the trophic webs shown in Fig. 7.8). Indeed, to support the sometimes substantial heterotrophic communities found in Palaeozoic reefs, we need to infer the presence of considerable suspended food matter in the water column. The trophic and environmental requirements of many ancient reef communities were therefore very different to that of modern coral reefs, and as a result they might have occupied quite different geographical settings.

That Palaeozoic reef communities show no evidence of adaptation to low-nutrient conditions is in some cases supported by independent sedimentological and palaeogeographical observations. For example, the Permian Capitan reef (Texas and New Mexico) grew around the small, probably nutrified, intracratonic Delaware basin (see Case study 3.9) and was built by relatively fragile frondose bryozoans, laminar sponges, and problematic ?algal encrusters. Here, the lack of photosymbionts did not prevent the construction of a massive reef framework to sea level with sustained growth over 2.5 Myr. Indeed, periods of global reef formation during the Phanerozoic—such as the Late Devonian—clearly do not correlate with the widespread availability of confirmed photosymbiotic reef-builders, but rather with the widespread distribution of environmental conditions that were conducive to tropical carbonate production and reef-formation.

8.5.2 Photosymbiosis and carbonate-platform accumulation rates

Many Palaeozoic reefs and carbonate platforms, e.g. those from the Late Devonian and the Permian, achieved rates comparable to those recorded from modern carbonate platforms (Fig. 4.7). Notwithstanding the fact that the sources of this carbonate were profoundly different, with many Palaeozoic reefs possessing very large volumes of inorganic cement, these data demonstrate that while short-term rates of deposition may have varied, carbonate accumulation rates measured over geological timescales are largely independent of the ability of reef metazoans to photosynthesize.

8.5.3 Photosymbiosis and the history of reef-building by scleractinian corals

Scleractinian corals appeared in the Anisian with a high-standing diversity as well as complex skeletal features and colonial morphologies. The synchronous onset of normal marine conditions at this time may have provided an environmental or ecological cue that initiated skeletonization in the two clades of zooantharians which independently acquired calcareous skeletons (Section 5.3.4).

Most of the earliest scleractinians, at least until the Carnian, were small solitary or phaceloid forms which inhabited soft substrates in deep or otherwise protected settings (Fois and Gaetani 1984). Higher energy settings continued to be occupied by calcareous sponges, bryozoans, and solenoporacean algae. It is possible then, that photosymbiosis appeared first in such forms and that the invasion of higher energy settings was a result of coevolved morphological modifications which optimized the partnership.

A marked faunal changeover occurred from the Carnian to the Norian, with a rapid extinction and adaptive radiation which resulted in the loss of the Permian reef 'holdovers' and the rise in dominance of scleractinian corals. This reorganization represents the start of the assembly of the modern coral reef community. During the Norian, corals, as other reef benthos, increase in both diversity and in ecological roles. Isolated, shallow reefs are dominated by thickets of the substantial, but delicately branching, genus '*Thecosmillia*'.

The most striking aspect of the foregoing analysis of ancient photosymbioses, are the predominantly solitary and low-integration growth forms of Upper Triassic scleractinian corals which exhibit a supposedly zooxanthellate isotopic signal. These corals are all small, with colony sizes rarely exceeding 6 cm in any dimension. Although no good comparative data are available to assess the relative growth rates of supposed zooxanthellate and azooxanthellate forms, what is clear is that possession of zooxanthellae did not confer either noticeably large size or reef-building ability to these Norian corals.

We have already established that from their first appearance, there was a steady trend of increase in the proportion of high-integration multiserial scleractinian corals, and that a spectacular rise of multiserial branching forms occurred from the Late Cretaceous onwards. From these data we might conclude that the predominantly high proportion of multiserial organizations found in zooxanthellate corals on modern reefs is a phenomenon which postdates the

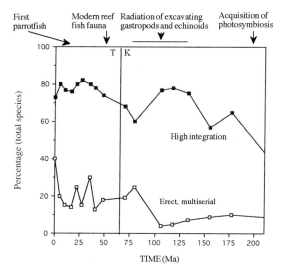

First parrotfish   Modern reef fish fauna   Radiation of excavating gastropods and echinoids   Acquisition of photosymbiosis

Fig 8.16 The high proportion of multiserial colony organizations, particularly erect growth forms, found in zooxanthellate corals on modern reefs is a phenomenon which postdates the proposed time of photosymbiosis acquisition.

proposed time of photosymbiosis acquisition (Fig. 8.16). Moreover, an analysis of the distribution of shallow-water scleractinian corals through time—excluding solitary forms that are clearly not found associated with reefs—while showing an increase in diversity of all growth forms, demonstrates that only meandroid, and particularly coenosteoid, growth types show a proportional increase (Fig. 8.12(b)). We can speculate that this marks the burgeoning and adaptive radiation of the zooxanthellae–scleractinian coral partnership.

Notwithstanding the probable presence of the algal–coral symbiosis, however, some Mesozoic shallow marine environments were not especially favourable to large-scale reef-building by scleractinian corals. Only in the Jurassic are reefs found which are dominated by scleractinian corals, although the rapid growth rates shown by some modern corals have not been recorded. During the Cretaceous—although the taxonomic diversity of corals was globally high—growth forms were predominantly multiserial, and while many species probably possessed zooxanthellae, colonies rarely reached more than 0.5 m in diameter and did not aggregate or build extensive shallow-water reefs even though important modern reef-building genera were present, e.g. *Montastraea* (Budd and Coates 1992). For example, 115 coral species are known from the celebrated Gosau faunas from the Upper Cretaceous of Austria, and although 73% of species are multiserial and had morphological features similar to modern corals, these communities did not form reefs (Coates and Jackson 1987). Some Cretaceous coral communities occupied deeper water, outer-shelf settings, rather than the inner platforms which may have been dominated by shifting, often muddy, substrates. With their pennular structures, these faunas appear to have been photosymbiotic but adapted to low-light conditions.

All these observations suggest that independent of the possession of photosymbiosis—reef-building by scleractinian corals, or indeed any group of metazoans—can only occur when environmental conditions are favourable. Possible explanations as to why zooxanthellate scleractinians have not dominated reef communities since their first appearance might include:

- Suitable environments (with temperature fluctuations within the thresholds for scleractinian corals, low clastic input, stable substrates) were not widespread in the shallow marine tropics. Sediment accumulation, for example, is known to inhibit coral larvae settlement.

- Periods of high pCO_2 may not have favoured groups with aragonitic skeletons. The Cretaceous was such a period, and indeed it was at this time that molecular phylogenies show that some living non-skeletal hexacorals (actinians and corallimorpharians) evolved from the parental scleractinian stock (Buddemeier and Fautin 1996a).

- Fast-growing zooxanthellate types—or host–symbiont associations—that enabled a high degree of autotrophy and very rapid host growth had not evolved. No extension rates exceeding 20 mm year^{-1} have been recorded for pre-Tertiary scleractinian corals; most grow in the order of a few millimetres (Fig. 8.13).

- Specialist predators and herbivores that promote the growth of reef-building corals had not evolved (Figure 8.16).

To test whether the acquisition of photosymbiosis provided a trigger for the appearance of new clades requires knowledge of phylogenetic relationships and photosymbiosis distribution in all members of both host and symbiont clades, and, in particular, comparison of the relative fates of symbiotic and asymbiotic sister groups. None of these data are currently available for the scleractinian corals, and so the actual mechanisms for the formation of symbioses, as well as the rate of formation, are unknown. We cannot at present determine the order of appearance of novelties associated with symbiosis and so test the macroevolutionary significance of the association.

But in groups such as planktic foraminifera where both sets of data are well constrained, photosymbiont acquisition has been shown to have triggered the initial adaptive radiations (Norris 1996). However, after their initial appearance, possession of photosymbionts does not appear to control long-term morphological or ecological patterns, or rates of species-turnover to their hosts. Different symbiotic groups of planktic foraminifera display independent patterns of diversification, which suggest that traits due to membership of a particular clade (such as characteristic growth rates, methods of feeding, temperature tolerances, and predator resistance) as well as changes in the geographical extent of suitable environments, are more important determinants of diversification than shared symbiotic ecology (Norris 1996).

So while photosymbiosis may provide the catalyst for the differentiation of major groups by allowing entry into a new adaptive zone, this does not necessarily confer a similar evolutionary fate. Subsequent evolution after the acquisition of symbiosis appears to be determined both by the inherent characteristics of the clade and by changes in the environment.

8.5.4   Effect of environmental perturbation and mass extinction events on symbioses

Although it has been suggested that a lengthy period of millions of years is needed for the acquisition of the complex symbiosis relationship, and has thus been the major constraint on the re-establishment of a reef-framework building biota after mass extinction (e.g. Copper 1988, 1989; Cowen 1988), there is little evidence to support this assertion. The long and highly complex history of multiple acquisition, and the presence of at least some multi-species complexes present in the coral–algal partnership, implies that host and symbiont can recombine with relative ease. Indeed, it seems likely that the seemingly labile, highly dynamic nature of host–symbiont partnerings, might provide for great stability of the symbiosis on an evolutionary timescale.

If bleaching episodes allow recombination, then we might conclude that given the right environmental conditions, the scleractinian–algal symbiosis does not take a long time to reform. In such a scenario, the availability of a suitable symbiont from the environment becomes of paramount importance. Sadly, reliable data are so patchy that little can be currently gleaned from the fossil record as to the temporal and geographical distribution of photosymbiotic coral reef communities. However, the continuous record of reef-building by scleractinian corals through the strong climatic upheavals of the Neogene is testament to the persistence of scleractinian–algal symbiosis. Symbiosis can be maintained through cladogenesis of the host species.

Several extinction events are apparent during the history of the scleractinian corals: there has been a complete turnover of coral species since the Carnian (Late Triassic); and coral species losses were high during the Lower Jurassic, Early Cretaceous, early Late Cretaceous, and the Cretaceous/Tertiary (KT) mass extinction event. At present, there are no data to support the wholesale loss of photosymbioses at the KT boundary, although inferred zooxanthellate species appear to have been more prone to extinction (70% compared to 40%; Rosen and Turnšek 1989) as might be predicted by the onset of cooler climatic conditions and loss of carbonate-platform habitat. Modern coral faunas may have their origins in the Late Cretaceous, as nine coral genera survived the extinction event into the Palaeocene— *Diploria*, *Favia*, *Goniopora*, *Hydnophora*, *Leptoria*, *Montastraea*, *Pachypyra*, *Siderastrea* and *Stephanocoenia*—which today are zooxanthellate. All either became important reef-builders during the Cenozoic, or were closely related to reef builders (Rosen 1998). However, while reefs appeared after only between 2–3 and 5 Myr after the mass extinction, inferred zooxanthellate corals may have taken far longer to recover their former species' richness. Rosen (1998) suggests that this was not achieved until the Oligocene.

Diversification of photosymbiotic planktic foraminifera proceeded soon after the KT crisis: photosymbiotic foraminiferans are known some 3.5 Myr after the extinction event boundary (Norris 1996). The Palaeogene is characterized by the transition from a globally equable climate to the strong thermal gradients of the glacial world, and such changes certainly involved shifts in oceanic circulation rates and patterns of nutrient flux to open surface waters.

8.6 Summary

It is possible to conclude that reef formation is the consequence of the long-term growth of organisms under particular environmental and biological conditions which are quite independent of the ability to photosynthesize (Coates and Jackson 1985). Photosymbiosis has not been a necessary prerequisite for reef-building in the past, nor has its evolutionary appearance notably increased carbonate-platform accumulation rates on geological timescales. In the Modern, the ability to achieve fast rates of growth and large size are more important determinants of reef-building capability in shallow tropical waters, together with resistance to partial predation, fouling, or smothering by sediments. Many of these features are associated with clonal growth, which is clearly preadapted for photosymbiosis acquisition.

Acquisition of symbionts allows a host to gain novel metabolic capabilities. Photosymbiosis can confer a variety of additional advantages to reef-building organisms, including rapid rates of growth and calcification, the potential to reach large size, possibly the ability to occupy low-nutrient settings where competition from macroalgae is reduced, and the potential for host selection of optimal symbionts in a highly fluctuating environment. The apparent fragility of the photosymbiotic relationship, however, also holds the key to its success: the dynamic nature of symbiotic combinations in corals has allowed them to persist through 240 Myr, and through periods of rapid, and sometimes extreme, environmental change.

Many ancient reefs had profoundly different trophic organizations to modern coral reefs, where neither the primary producers nor framebuilders were photosymbiotic metazoans. In Proterozoic and Palaeozoic reefs, the main sources of carbonate were either inorganic or microbial. Moreover, few show the same style of pronounced zonation as found on modern reefs. In addition, the loss of photosymbioses cannot alone account for periods in Earth's history when there was no widespread reef-building.

The appearance and diversification of symbioses may be determined by the availability of an appropriate symbiont, together with the selection pressure to acquire the metabolic capability of that symbiont. It is highly likely that dinoflagellate symbionts—particularly the *Symbiodinium* group which is so successful at overcoming the defence systems of a host—did not become widely available until the Upper Triassic. But the need to expose large areas of exposed soft tissue makes sessile photosymbiotic metazoans even more vulnerable to predation, and symbiosis may have favoured—at least in part—the proliferation of antipredation traits.

Some scleractinian corals exhibit traits that are photoadaptive and genetic, suggesting some coevolution with algal symbionts. Although the fossil record is not adequate to record the detail, it seems likely that the acquisition of photosymbiosis provided a catalyst for an adaptive radiation: the first unequivocal skeletal photoadaptive traits in scleractinian hosts were present by the Late Jurassic (~155 Ma). The highly dynamic coral–algal partnership—as well as extensive intraspecific variation within both hosts and symbionts—has enabled corals to invade a variety of dynamic settings, including adaptation to extreme environments.

Appendix

Geological timescale

Era	Period	Epoch	Stage	Age Ma
Palaeozoic	Permian	Tatarian	Changxingian	251
			Diulfian	
		Guadalupian	Capitanian	
			Wordian	
			Roadian	
		Early	Leonardian	
			Artinskian	
			Sakmarian	
			Asselian	290
	Carboniferous — Pennsylvanian	Gzelian		
		Kazimovian		
		Moscovian		
		Bashkirian		
	Carboniferous — Mississipian	Serpukhovian		
		Viséan		
		Tournaisian		353.7
	Devonian	Late	Famennian	
			Frasnian	
		Middle	Givetian	
			Eifelian	
		Early	Emsian	
			Pragian	
			Lochkovian	408.5
	Silurian	Pridoli		
		Ludlow		
		Wenlock		
		Llandovery		439
	Ordovician	Ashgill		
		Caradoc		
		Llandeilo		
		Llanvirn		
		Arenig		
		Tremadoc		500
	Cambrian	Late	Franconian	
		Middle		520
		Early	Toyonian	
			Botoman	
			Atdabanian	
			Tommotian	531
			Nemakit-Daldyn	543
	Vendian			565
Proterozoic				2500
Archaean				3600

Era	Period	Epoch	Stage	Age Ma
Cenozoic	Quaternary	Holocene		
		Pleistocene		1.8
	Tertiary (Neogene)	Pliocene	Placenzian	
			Zanclian	5.2
		Miocene	Messinian	
			Tortonian	
			Serravalian	
			Langhian	
			Burdigalian	
			Aquitanian	23.8
	Tertiary (Palaeogene)	Oligocene	Chattian	
			Rupelian	33.5
		Eocene	Priabonian	
			Bartonian	
			Lutetian	
			Ypresian	55.6
		Palaeocene	Thanetian	
			Danian	65.0
Mesozoic	Cretaceous	Late	Maastrichtian	
			Campanian	
			Santonian	
			Coniacian	
			Turonian	89
			Cenomanian	98.9
		Early	Albian	
			Aptian	
			Barremian	
			Hauterivian	127
			Valanginian	
			Berriasian	144
	Jurassic	Late	Tithonian	
			Kimmeridgian	
			Oxfordian	160
		Middle	Callovian	
			Bathonian	
			Bajocian	
			Allenian	180
		Early	Toarcian	
			Pliensbachian	
			Sinemurian	
			Hettangian	206
	Triassic	Late	Rhaetian	
			Norian	
			Carnian	228
		Middle	Ladinian	
			Anisian	242
		Scythian	Spathian	
			Nammalian	
			Griesbachian	

References

Adey, W.H. (1975). The algal ridges and coral reefs of St. Croix, their structure and Holocene development. *Atoll Research Bulletin*, **187**, 1–67.

Adey, W.H. (1978). Coral reef morphogenesis: a multidimensional model. *Science*, **202**, 831–7.

Adey, W.H. and Burke, R.B. (1977). Holocene bioherms of the Lesser Antilles—geologic control of development. In *Reefs and related carbonates—ecology and sedimentology*, Vol. 4 (ed. S.H. Frost, M.P. Weiss, and J.B. Saunders), pp. 67–81. American Association of Petroleum Geologists, Studies in Geology.

Adey, W.H. and Steneck, R.S. (1985). Highly productive eastern Caribbean reefs: synergistic effects of biological, chemical, physical and geological factors. In *The ecology of deep and shallow coral reefs*, Vol. 2 (ed. M.L. Reaka), pp. 163–87. Office of Undersea Research, NOAA, Rockville, Maryland.

Aharon, P. (1991). Recorders of reef environmental histories: stable isotopes in corals, giant clams, and calcareous algae. *Coral Reefs*, **10**, 71–90.

Ahr, W.M., Webb, G., and Yang, Z.D. (1991). Ramp-to-shelf evolution of Lower Carboniferous platforms. *Dolomieu Conference on Carbonate Platforms and Dolomitization, Abstracts*, p. 1.

Aitken, J.D. (1989). Giant 'algal' reefs, Middle/Upper Proterozoic Little Dal Group (>770, <1200 Ma), Mackenzie Mountains, N.W.T. Canada. In *Reefs. Canada and adjacent area*, Vol. 13 (ed. H.H.J. Geldsetzer, N.P. James, and G.E. Tebbutt), pp. 13–23. Canadian Society of Petroleum Geologists.

Ali, O.E. (1984). Sclerochronology and carbonate production in some Upper Jurassic reef corals. *Palaeontology*, **27**, 537–48.

Alldredge, A.L. and King, J.M. (1977). Distribution, abundance and substrate preference of demersal zooplankton. *Journal of Experimental Marine Biology and Ecology*, **44**, 133–56.

Aller, R.C. (1982). Carbonate dissolution in nearshore terrigenous muds: the role of physical and biological reworking. *Journal of Geology*, **90**, 75–9.

Aller, R.C. and Dodge, R.E. (1974). Animal–sediment relations in a tropical lagoon, Discovery Bay, Jamaica. *Journal of Marine Research*, **32**, 295–314.

Alvarez, L., Alvarez, W., Asaro, F., and Michel, H.V. (1980). Extraterrestrial cause for the Cretaceous–Tertiary extinction: experimental results and theoretical interpretation. *Science*, **208**, 1095–108.

Armstrong, H.A. (1996). Biotic recovery after mass extinction: the role of climate and ocean-state on the post-glacial (Late Ordovician–Early Silurian) recovery of the conodonts. In *Biotic recovery from mass extinction events* (ed. M.B. Hart), pp. 105–17. Geological Society of London Special Publication, **102**.

Atkinson, M.M. (1992). Productivity of Enewetak Atoll reef flats predicted from mass transfer relationships. *Continental Shelf Research*, **12**, 799–807.

Ausich, W.I. and Bottjer, D.J. (1982). Tiering in suspension-feeding communities on soft-substrata throughout the Phanerozoic. *Science*, **216**, 173–4.

Ausich, W.I. and Meyer, D.L. (1990). Origin and composition of carbonate buildups and associated facies in the Fort Payne Formation (Lower Mississippean, south-central Kentucky): an integrated sedimentological and paleoecologic analysis. *Bulletin of the Geological Society of America*, **102**, 129–46.

Awramik, S.M. (1971). Precambrian columnar stromatolite diversity: reflection of metazoan appearance. *Science*, **174**, 825–6.

Ayre, D.J. (1984). The effects of sexual and asexual reproduction on geographic variation in the sea anemone *Actinia tenebrosa*. *Oecologia*, **62**, 222–9.

Babcock, R.C. (1984). Reproduction and distribution of two species of *Goniastrea* (Scleractinia) from the Great Barrier Reef province. *Coral Reefs*, **2**, 187–204.

Babcock, R.C., Bull, G., Harrison, P.L., Heywood, A.J., Oliver, J.K., Wallace, C.C., *et al.* (1986). Synchronous multispecific spawnings of 107 scleractinian coral species on the Great Barrier Reef. *Marine Biology*, **90**, 379–94.

Bak, R.P.M. (1976). The growth of coral colonies and the importance of crustose coralline algae and burrowing sponges in relation with carbonate accumulation. *Journal of Sea Research*, **10**, 285–337.

Bak, R.P.M. (1983). Neoplasia, regeneration and growth in the reef-building coral *Acropora palmata*. *Marine Biology*, **77**, 221–7.

Bak, R.P.M. and Engel, M.S. (1979). Distribution, abundance and survival of juvenile hermatypic corals (Scleractinia) and the importance of life history strategies in the parent coral community. *Marine Biology*, **54**, 341–52.

Bak, R.P.M. and Steward-Van Es, Y. (1980). Regeneration of superficial damage in the scleractinian corals *Agaricia agaricites* F. *Purpurea* and *Porites asteroides*. *Bulletin Marine Science*, **30**, 883–7.

Bak, R.P.M., Brouns, J.J.W.M., and Heys, F.M.L. (1977). Regeneration and aspects of spatial competition in the scleractinian corals (ed. D.L. Taylor). *Proceedings of the 3rd International Coral Reef Symposium*, Miami, USA, May 1977, pp. 143–8.

Baker, A., Rowan, R., and Knowlton, N. (1997). Symbiosis ecology of two Caribbean acroporid corals. *Proceedings of the 8th International Coral Reef Symposium*, (ed. H.G. Lessios and I.G. Macintyre). Panama City, Panama, **2**, 1295–300.

Bakus, G.J. (1974). Toxicity in holothurians: a geographic pattern. *Biotropica*, **6**, 229–36.

Bakus, G.J. (1981). Chemical defence mechanisms on the Great Barrier Reef, Australia. *Science*, **211**, 497–9.

Ball, M.M., Shinn, E.A., and Stockman, K.W. (1967). The geological effects of Hurricane Donna in south Florida. *Journal of Geology*, **15**, 583–97.

Barnaby, R.J. and Reed, J.F. (1990). Carbonate ramp to rimmed shelf evolution. Lower to Middle Cambrian continental margin: Virginian Appalachians. *Geological Society of America Bulletin*, **102**, 391–404.

Barnes, C.R., Fortey, R.A., and Williams, S.H. (1995). The pattern of global bio-events during the Ordovician period. In *Global events and event stratigraphy* (ed. O.H. Walker), pp. 139–72. Springer-Verlag, Berlin.

Barnes, D.J. and Chalker, B.E. (1990). Calcification and photosynthesis in reef-building corals and algae. In *Coral Reefs* (ed. Z. Dubinsky), pp. 109–31. *Ecosystems of the world*, Vol. 25. Elsevier, New York.

Barnes, D.J. and Devereux, M.J. (1988). Variations in skeletal architecture associated with density banding in the hard coral *Porites*. *Journal of Experimental Marine Biology and Ecology*, **121**, 37–54.

Barnes, D.J. and Lough, J.M. (1996). Coral skeletons: storage and recovery of environmental information. *Global Change Biology*, **2**, 569–82.

Barnes, R.S.K. and Hughes, R.N. (1988). *An introduction to marine ecology* (2nd edn). Blackwell, Oxford.

Barnola, J.M., Raynaud, D., Korotkevich, Y.S., and Lorius, C. (1987). Vostok ice core provides 160,000 year record of atmospheric CO_2. *Nature*, **239**, 408–14.

Barrera, E. (1994). Global environmental changes preceding the Cretaceous–Tertiary boundary: Early–Late Maastrichtian transition. *Geology*, **22**, 877–80.

Barron, E.J., Thompson, S.L., and Schneider, S.H. (1981). Cretaceous 'oceanic events' as causal factors in development of reef-reservoired giant oil fields. *American Association of Petroleum Geologists Bulletin*, **63**, 870–85.

Bathurst, R.G.C. (1982). Genesis of stromatactis cavities between submarine crusts in Palaeozoic carbonate mud buildups. *Journal of the Geological Society of London*, **139**, 165–81.

Bebout, D. G. and Kerans, C. (ed.) (1993). Guide to the Permian Reef Geology Trail, McKittrick Canyon, Guadalupe Mountains National Park, West Texas. *Bureau of Economic Geology Guidebook*, *Bureau of Economic Geology*, *Austin*, **26**.

Beauvais, L. (1984). Evolution and diversification of the Jurassic Scleractinia. *Paleontographica Americana*, **54**, 219–24.

Beauvais, L. (1986). Evolution paléobiogéographique des formations a Scléractiniaires du Bassin téthysien au cours due Mésozoique. *Bulletin de la Société Géologique de France*, **8**, 499–509.

Beauvais, L. and Beauvais, M. (1974). Studies on the world distribution of the Upper Cretaceous corals. *Proceedings of the 2nd International Symposium on Coral Reefs*, 22 June–2 July, 1973 (ed. A.M. Cameron, J.S. Jell *et al.*), **1**, 475–94.

Becker, R.T., House, M.R., Kirchgasser, W.T., and Playford, P.E. (1991). Sedimentary and faunal changes across the Frasnian/Famennian boundary at the Canning Basin of Western Australia. *Historical Biology*, **5**, 183–96.

Beklemishev, W.N. (1964). *Principals of comparative anatomy of invertebrates*, Vol. 1 Promorphology (trans. J.M. MacLennan). University of Chicago Press, Chicago.

Belda, C.A. and Yellowlees, D. (1995). Phosphate acquisition in the giant clam–zooxanthellae symbiosis. *Marine Biology*, **121**, 261–6.

Belda, C.A., Cuff, C. and Yellowlees, D. (1993*a*). Modification of shell fomation in the giant clam *Tridacna gigas* at elevated nutrient levels in sea water. *Marine Biology*, **117**, 251–7.

Belda, C.A., Lucas, L.S., and Yellowlees, D. (1993*b*). Nutrient limitation in the giant clam–zooxanthellae symbiosis: effects of nutrient supplements on growth of the symbiotic partners. *Marine Biology*, **117**, 644–55.

Belle, R.A. van. (1977). Sur la classification des Polyplacophora: III. Classification systématique des Subterenochitonidae et des Ischnochitonidae (Neoloricata: Chitonina). *Inf. Societé Belgique Malacologique*, **5**, 15–40.

Bellwood, D.R. (1994). A phylogenetic study of the parrotfishes family Scaridae (Pisces: Labroidei) with a revision of genera. *Records of the Australian Museum Supplement*, **20**, 1–86.

Bellwood, D. R. (1995). Carbonate transport and within-reef patterns of bioerosion and sediment release by parrotfishes (family Scaridae) on the Great Barrier Reef. *Marine Ecology Progress Series*, **117**, 127–36.

Bellwood, D. R. (1996*a*). Coral reef crunchers. *Nature Australia*, Autumn, 49–55.

Bellwood, D. R. (1996*b*). The Eocene fishes of Monte Bolca: the earliest coral reef fish assemblage. *Coral Reefs*, **15**, 11–19

Bellwood, D. R. (1997). Reef fish biogeographgy; habitat associations, fossils and phylogenies. *Proceedings of the 8th International Coral Reef Symposium*, Panama City, Panama, June 24–29 1996 (ed. H.G. Lessios and I.G. Macintyre), **2**, 1295–1300.

Bellwood, D.R. and Schulz, O. (1991). A review of the fossil record of the parrotfishes (family Scaridae) with a description of a new *Calatomus* species from the Middle Miocene (Badenian) of Austria. *Naturhistorische Museum Wein*, **92**, 55–71.

Benayahu,Y. and Loya, Y. (1977). Space partitioning by stony corals, soft corals and benthic algae on the coral reefs of the northern Gulf of Eilat (Red Sea). *Helgoländer Wissenschaftliche Meersuntersuchungen*, **30**, 362–82.

Bengston, S. and Zhao, Y. (1992). Predational borings in Late Precambrian mineralized skeletons. *Science*, **257**, 367–9.

Benjamini, C. (1981). Limestone and chalk transitions in the Eocene of the western Negev, Israel. In *Microfossils from Recent and fossil shelf seas* (ed. J.W. Neale and M.D. Brasier), pp. 205–13. British Micropalaeontological Society, Ellis Horwood, Ltd., Chichester.

Benton, M.J. (1988). Mass extinctions and the fossil record of reptiles: paraphyly, patchiness, and periodicity. In *Extinction and survival in the fossil record* (ed. G.P. Larwood), pp. 269–94. Systematics Association Special Volume, **34**.

Benton, M.J. (1995). Diversification and extinction in the history of life. *Science*, **268**, 52–8.

Bergstom, J. (1979). Morphology of fossil arthropods as a guide to phylogenetic relationships. In *Arthropod phylogeny* (ed. A.P. Gupta), pp. 3–58. Van Nostrand Reinhold, New York.

Berner, R.A. (1994). 3Geocarb II: a revised model of atmospheric CO_2 over Phanerozoic time. *American Journal of Science*, **294**, 56–91.

Berry, W.B.N. and Boucot, A.J. (1973). Glacioeustatic control of Late Ordovician–Early Silurian platform sedimentation and faunal change. *Bulletin of the Geological Society of America*, **84**, 275–84.

Best, B.A. and Winston, J.E. (1984). Skeletal strength of encrusting cheilostome bryozoans. *Biological Bulletin*, **167**, 390–409.

Bhandari, N., Shukla, P.N., Ghevariya, Z.G., and Sundaram, S.M. (1995). Impact did not trigger Deccan volcanism: evidence from Anjar K/T boundary intertrappen sediments. *Geophysical Research Letters*, **22**, 43–6.

Billinghurst, Z., Douglas, A.E., and Trapido-Rosenthal, H.G. (1997). On the genetic diversity of the symbiosis between the coral *Monastraea cavernosa* and zooxanthellae in Bermuda. *Proceedings of the 8th International Coral Reef Symposium*, Panama City, Panama, (ed. H.G. Lessios and I.G. Macintyre) **2**, 1291–4.

Birkeland, C. (1977). The importance of rate of biomass accumulation in early successional stages of benthic communities to the survival of coral recruits. *Proceedings of the 3rd International Coral Reef Symposium, Miami, USA* (ed D.L. Taylor), **1**, 16–21.

Birkeland, C. (1987). Nutrient availability as a major determinant of differences among coastal hard-substratum communities in different regions of the tropics. In *Comparisons between Atlantic and Pacific tropical coastal marine ecosystems:community structure, ecological processes, and productivity* (ed. C. Birkeland), pp. 45–97. UNESCO Reports in Marine Sciences, **46**.

Birkeland, C. (1988). Geographic comparisons of coral-reef community processes. *Proceedings of the 6th International Coral Reef Symposium*, Townsville, Australia, **1**, 211–20.

Birkeland, C. (1997). Introduction. In *Life and Death of coral reefs* (ed. C Birkeland), pp. 1–12. Chapman and Hall, New York.

Birkeland, C. and Randall, R.H. (1982). Facilitation of coral recruitment by echinoid excavation. *Proceedings of the 4th International Coral Reef Symposium*, Manila, Phillipines, **1**, 695–98.

Birkeland, C., Nelson, S.G., Wilkins, S., and Gates, P. (1985). Effects of grazing of herbivorous fishes on coral reef community metabolism. *Proceedings of the 5th International Coral Reef Congress*, Tahiti, French Polynesia, **4**, 47–51.

Birkelund, T. and Hakansson, E. (1982). The terminal Cretaceous extinction in boreal shelf seas—a multicausal event. *Geological Society of America Special Paper*, **190**, 373–84.

Blanchon, P. and Shaw, J. (1995). Reef drowning during the last deglaciation: evidence for catastrophic sea-level rise and ice-sheet collapse. *Geology*, **23**, 4–8.

Bless, M.J.M., Becker, R.T., Higgs, K.T., Paproth, E., and Streel, M. (1992). Eustatic cycles around the Devonian–Carboniferous boundary and the sedimentary and fossil record in Sauerland (Federal Republic of Germany). *Annales de la Société Géologique de Belgique*, **115**, 689–702.

Blot, J. (1980). La fauna ichthyologique des gisements du Monte Bolca (Province de Vérone, Italy). Catalogue systematique présentant l'état actual des researches concernant cette fauna. *Bulletin Muséum Natural Histoire* **2(C4)**, 339–96.

Boardman, R.S., Cheetham, A.H., and Oliver, W.A., Jr. (1973). Introducing coloniality. In *Animal colonies, developments and function through time* (ed. R.S. Boardman, A.H. Cheetham, and W.A. Oliver), pp. v–ix. Dowden, Hutchinson and Ross, Stroudsburg, Pa.

Bolser, R.C. and Hay, M.E. (1997). Are tropical plants better defended? Palatability and defences of temperate vs. tropical seaweeds. *Ecology*, **78**, 2269–86.

Boreen, J.N. and James, N.P. (1993). Holocene sediment dynamics in a cool-water carbonate shelf: Otway, southeastern Australia. *Journal of Sedimentary Petrology*, **63**, 574–88.

Bosence, D.W.J. (1983). Coralline algal reef frameworks. *Journal of the Geological Society*, **140**, 365–76.

Bosence, D.W.J. and Bridges, P.H. (1995). A review of the origin and evolution of carbonate mud-mounds. In *Carbonate mud-mounds. Their origin and evolution* (ed. C.L.V. Monty, D.W.J. Bosence, P.H. Bridges, and B.R. Pratt), pp. 3–9. Special Publications of the International Association of Sedimentologists, **23**. Blackwell Science, Oxford.

Bosscher, H. and Schlager, W. (1992). Computer simulation of reef growth. *Sedimentology*, **39**, 503–12.

Bosscher, H. and Schlager, W. (1993). Accumulation rates of carbonate platforms. *Journal of Geology*, **10**, 345–55.

Bottjer, D. J. and Ausich, W.I. (1986). Phanerozoic development of tiering in soft substrata suspension-feeding communities. *Paleobiology*, **12**, 400–20.

Boulter, M.C., Spicer, R.A., and Thomas, B.A. (1988). Patterns of plant extinction from some palaeobotanical evidence. In *Extinction and survival in the fossil record*, Vol. 34 (ed. G.P. Larwood), pp. 1–36. *Systematics Association Special Volume.*

Bourque, P.-A. (1989). Silurian reefs. In *Reefs. Canada and adjacent area* (ed. H.H.J. Geldsetzer, H.H.J., N.P. James, and G.E. Tebbutt), pp. 245–50. *Canadian Society of Petroleum Geologists*, **13**.

Bourque, P.-A. and Gignac, H. (1983). Sponge-constructed Stromatactis mud mounds, Silurian of Gaspé, Quebec. *Journal of Sedimentary Petrolology*, **53**, 521–32.

Bowring, S.A., Grotzinger, J.P., Isachsen, C.E., Knoll, A.H., Pelechaty, S.M., and Kolosov, P. (1993). Calibrating rates of Early Cambrian evolution. *Science*, **261**, 1293–8.

Brachert, T.C., Buggisch, W., and Flügel, E. *et al.* (1992). Controls of mud mound formation: the Early Devonian Kess–Kess carbonates of the Hamar Laghdad, Antiatlas, Morocco. *Geologisches Rundschau*, **81**, 15–44.

Brandner, R. (1984). Meeresspiegelschwankungen und Tektonik in der Trias der NW-Tethys. *Jahrbuch der Geologischen Bundesanstalt*, **126**, 435–75.

Brandner, R. and Resch, W. (1981). Reef development in the Middle Triassic (Ladinian and Cordevolian) of the Northern Limestone Alps near Innsbruck. In *European reef models* (ed. D.F. Toomey), pp. 203–31. *Special Publication of the Society of Economic Paleontologists and Mineralogists*, **30**.

Brasier, M. (1988). Foraminiferal extinction and ecological collapse during global biological events. In *Extinction and survival in the fossil record* (ed. G.P. Larwood), pp. 37–64. Systematics Association Special Volume, **34**.

Brasier, M. (1995). The basal Cambrian transition and Cambrian bio-events. In *Global events and event stratigraphy* (ed. O.H. Walliser), p. 113–38. Springer-Verlag, Berlin.

Brasier, M., Corfield, R., Derry, L., Rozanov, A., and Zhuravlev, A.Yu. (1994). Multiple δ^{13}C excursions spanning the Cambrian explosion to the Botomian crisis in Siberia. *Geology*, **22**, 455–8.

Brasier, M., Green, O., and Sheilds, G. (1997). Ediacaran sponge spicule clusters from southwestern Mongolia and the origins of the Cambrian fauna. *Geology*, **25**, 303–6.

Bratton, J.F. (1996). Brachiopods and oxygen levels during the survival interval of the Late Devonian mass extinction recovery in the Great Basin, western USA. 6th North American Paleontological Convention, Abstract Volume. (ed. J.E. Repetski). *Paleontological Society Special Publication*, Washington D.C. **8**, 44.

Brenchley, P. J. (1989). The Late Ordovician extinction. In *Mass extinctions: processes and evidence* (ed. S.K. Donovan), pp. 104–32. Belhaven, London.

Bridges, P.H. and Chapman, A.J. (1988). The anatomy of a deep-water mud-mound complex to the southwest of the Dinantian platform in Derbyshire, UK. *Sedimentology*, **35**, 139–62.

Bridges, P.H., Gutteridge, P. and Pickard, N.A.H. (1995). The environmental setting of Early Carboniferous mud-mounds. *Special Publications International Association Sedimentologists*, **23**, 171–90.

Briggs, D.E.H., Fortey, R.A., and Clarkson, E.N.K. (1988). Extinction and the fossil record of the arthropods. In *Extinction and survival in the fossil record* (ed. G.P. Larwood), pp. 171–209. Systematics Association Special Volume, **34**.

Broeker, W.S. and Pang, T.-H. (1982). *Tracers in the sea*. Lamont-Doherty Geological Observatory, Columbia University, Palisades, New York.

Bromley, R. G. (1975). Comparative analysis of fossil and recent echinoid bioerosion. *Palaeontology*, **18**, 725–39.

Bromley, R.G. (1992). The palaeoecology of bioerosion. In *The palaeobiology of trace fossils* (ed. S.K. Donovan), pp. 134–54. Wiley, Chichester.

Brown, B.E. (1997*a*). Adaptations of reef corals to physical environmental stress. *Advances in Marine Biology*, **31**, 221–99.

Brown, B.E. (1997*b*). Coral bleaching: causes and consequences. *Coral Reefs*, **16** (Suppl.), 129–38.

Brown, B.E. and Ogden, J.C. (1992). Coral bleaching. *Scientific American*, **268**, 64–70.

Brown, B.E., Le Tissier, M.D.A., and Dunne, R.P. (1994). Tissue retraction in the scleractinian coral *Coeloseris mayeri*, its effect upon pigmentation, and preliminary implications for heat balance. *Marine Ecology Progress Series*, **105**, 209–18.

Bruggeman, J.H., van Kessel, A.M., van Rooij, J.M., and Breeman, A.M. (1996). Bioerosion and sediment ingestion by the Caribbean parrotfishes *Scarus vetula* and *Sparisoma viride*: implications of fish size, feeding mode and habitat use. *Marine Ecology Progress Series*, **134**, 59–71.

Brunton, F.R. and Copper, P. (1994). Paleoecological, temporal, and spatial analysis of Early Silurian reefs of the Chicotte Formation, Anticosti Island, Quebec, Canada. *Facies*, **31**, 57–80.

Brunton, F.R., Copper, P., and Dixon, O.A. (1997). Silurian reef-building episodes. *Procedings of the 8th International Coral Reef Symposium*, (ed. H.G. Lessios and I.G. Macintyre), Panama City, Panama, **2**, 1643–50.

Bryan, J.R. (1991). A Paleocene coral–algal–sponge reef from southwestern Alabama and the ecology of Early Tertiary reefs. *Lethaia*, **24**, 423–38.

Budd, A.F. and Coates, A.G. (1992). Nonprogressive evolution in a clade of Cretaceous *Montastrea*-like corals. *Paleobiology*, **18**, 425–46.

Budd, A.F. and Kievmann, C.M. (1994). Coral assemblages and reef environments in the Bahamas Drilling Project cores. In *Final draft report of the Bahamas Drilling Project*, **3**. Coral Gables, Florida. Rosenstiel School of Marine and Atmospheric Sciences, University of Miami.

Budd, A.F., Johnson, K.G., and Stemann, T.A. (1993). Plio-Pleistocene extinctions and the origin of the modern Caribbean reef-coral fauna. In *Proceedings of the Colloquium on Global Aspecs of Coral Reefs: Health, Hazards and History* (ed. R.N. Ginsburg) *Rosenstiel School of Marine and Atmospheric Science, University of Miami*, 420 pp.

Budd, A.H., Stemann, T.A., and Johnson, K.G. (1994). Stratigraphic distribution of genera and species of Neogene to Recent Caribbean reef corals. *Journal of Paleontology*, **68**, 951–77.

Budd, A.H., Johnson, K.G., and Stemann, T.A. (1996). Plio-Pleistocene turnover and extinctions in the Caribbean reef-coral fauna. In *Evolution and environment in tropical America* (ed. J.B.C. Jackson, A.F. Budd, and A.G. Coates), pp. 168–204. University of Chicago Press, Chicago.

Budd, A.F., Petersen, R.A., and McNeill, D.F. (1998). Stepwise faunal change during evolutionary turnover: a case study from the Neogene of Curaçao, Netherlands Antilles. *Palaios*, **13**, 170–88.

Buddemeier, R.W. (1997). Making light work of adaptation. *Nature*, **388**, 229–30.

Buddemeier, R.W. and Fautin, D.G. (1993). Coral bleaching as an adaptive mechanism. *BioScience*, **43**, 320–6.

Buddemeier, R.W. and Fautin, D.G. (1996*a*). Saturation state and the evolution and biogeography of symbiotic calcification. *Bulletin de L'Institute Océanographique*, **14**, 23–32.

Buddemeier, R.W. and Fautin, D.G. (1996*b*). Global CO_2 and evolution among the Scleractinia. *Bulletin de L'Institute Océanographique*, **14**, 33–8.

Buddemeier, R.W. and Hopley, D. (1988). Turn-ons and turn-offs: causes and mechanisms of the initiation and termination of coral reef growth. *Proceedings of the 6th International Coral Reef Symposium*, Townsville, Australia, **1**, 253–61.

Buddemeier, R.W. and Kinsey, R.A. (1976). Coral growth. *Oceanographical Marine Biology and Ecology*, **14**, 183–225.

Buddemeier, R.W. and Oberdorfer, J.A. (1986). Internal hydrology and geochemistry of coral reefs and atoll islands: key to diagenetic variations. In *Reef diagenesis* (ed. J.H. Schroeder and B.H. Purser), pp. 91–111. Springer-Verlag, Berlin.

Buddemeier, R.W. and Smith, S.V. (1988). Coral reef growth in an era of rapid rising sea level: predictions and suggestions for long-term research. *Coral Reefs*, **7**, 51–6.

Budyko, M.I. (1974). *Climate and life*. Academic Press, New York.

Buggisch, W. (1991). The global 'Kellwasser event'. *Geologische Rundschau*, **80**, 49–72.

Burchette, T. P. and Riding, R. (1977). Attached vermiform gastropods in Carboniferous marginal marine stromatolites and biostromes. *Lethaia*, **10**, 17–28.

Buss, L.W. (1986). Competition and community organisation on hard surfaces in the sea. In *Community ecology* (ed. T.J. Case and J. Diamond), pp. 517–36. Harper and Row, New York.

Buss, L.W. (1987) *The evolution of individuality*. Princeton, New Jersey.

Buss, L.W. and Iverson, E.W. (1981). A new genus and species of Sphaeromatida (Crustacea: Isopoda) with experiments and observations on its reproductive biology, interspecific interactions and color polymorphisms. *Postilla*, **184**, 1–23.

Buss, L.W. and Jackson, J.B.C. (1979). Comptetive networks: nontransitive competitive relationships in cryptic coral reef environments. *American Naturalist*, **113**, 223–34.

Butterfield, N.J. (1994). Burgess-shale type fossils from a Lower Cambrian shallow-shelf sequence in northwestern Canada. *Nature*, **369**, 477–9.

Buzas, M.A. and Culver, S.J. (1994). Species pool and dynamics of marine paleocommunities. *Science*, **264**, 1439–41.

Bythell, J.C., Douglas, A.E., Sharp, V.A., Searle, J.B., and Brown, B.E. (1998). Algal genotype and photoacclimatory response of the symbiotic alga *Symbodinium* in natural populations of the sea anemone *Anemonia viridis*. *Proceedings of the Royal Society* (in press).

Callander, B.A. (1995). *IPCC Working Group I 1995 Summary for Policy Makers*. Electronic message on the Internet, December 1995.

Camoin, G.F. (1995). Nature and origin of Late Cretaceous mud-mounds, north Africa. In *Carbonate mud mounds: their origin and evolution* (ed. C.L.V. Monty, D.W.J. Bosence, P.H. Bridges, and B.R. Pratt), pp. 385–400. International Association of Sedimentologists, Special Publication, **23**. Blackwell Science, Oxford.

Cañas, F. (1995). Early Ordovician carbonate platform facies of the Argentine Precordillera: restricted shelf to open platform evolution. In *Ordovician odyssey* (ed. J.D. Cooper, M.L. Droser, and S.C. Finney), pp. 221–4. Society of Economic Paleoentologists and Mineralogists.

Caputo, M.V. and Crowell, J.C. (1985). Migration of glacial centres across Gondwana during the Paleozoic era. *Bulletin of the Geological Society of America*, **96**, 1020–36.

Carpenter, R.C. (1986). Partioning herbivory and its effects on coral reef algal communities. *Ecological Monographs*, **56**, 345–63.

Carroll, R.L. (1987). *Vertebrate paleontology and evolution*. Freeman, New York.

Chafetz, H.S. (1986). Marine peloids: a product of bacterially induced precipitation of calcite. *Journal of Sedimentary Petrology*, **56**, 812–17.

Chappell, J. (1980). Coral morphology, diversity and reef growth. *Nature*, **286**, 249–52.

Chappell, J. (1981). Relative and average sea level changes, and endo-, epi- and exogenic processes on the earth. In *Sea level, ice and climatic change*. International Association of Hydrology, Scientific Publications, **131**.

Chappell, J. (1983). Sea-level changes and coral reef growth. In *Perspectives on coral reefs* (ed. D.J. Barnes), pp. 46–55. Australian Institute of Marine Science, Brian Clouston, Manuka.

Chappell, J. and Polach, H. (1991). Post-glacial sea-level rise from a coral record at Huon Pennisula, Papua New Guinea. *Nature*, **349**, 147–9.

Chatterton, B.D.E. and Speyer, S.E. (1989). Larval ecology, life history strategies, and patterns of extinction and survivorship among Ordovician trilobites. *Paleobiology*, **15**, 118–32.

Chevalier, J.P. (1971). Les scléractiniares de la Mélanésie française (Nouvelle Calédonie, Isles Chesterfield, Isles Loyauté, Nouvelles Hebrides). In *Expedition Française récifs Coralliens Nouvelle Calédonie*, pp. 5–307. Fondation Singer-Polignac, Paris, France.

Choat, J.H. (1991). The biology of herbivorous fishes on coral reefs. In *The ecology of fishes on coral reefs* (ed. P.F. Sale), pp. 120–55. Academic Press, London.

Choat, J.H. and Bellwood, D.R. (1991). Reef fishes: their history and evolution. In *The ecology of fishes on coral reefs* (ed. P.F. Sale), pp. 39–66. Academic Press, London.

Chornesky, E.A. (1983). Induced development of sweeper tentacles on the reef coral *Agaricia agaricites*: a response to direct competition. *Biological Bulletin*, **165**, 569–81.

Chuvashov, B. and Riding, R. (1984). Principal floras of Palaeozoic marine algae. *Palaeontology*, **27**, 487–500.

Clough, J.G. and Blodgett, R.B. (1989). Silurian–Devonian algal reef mound complex of Southwest Alaska. In *Reefs. Canada and adjacent area* (ed. H.H.J. Geldsetzer, N.P. James, and G.E. Tebbutt), pp. 404–7. *Canadian Society of Petroleum Geologists*, **13**.

Coates, A.G. and Jackson, J.B.C. (1985). Morphological themes in the evolution of clonal and aclonal marine invertebrates. In *Population biology and evolution of clonal organisms* (ed. J.B.C. Jackson, L.W. Buss, and R.E. Cook), pp. 67–106. Yale University Press, New Haven, CT.

Coates, A.G. and Jackson, J.B.C. (1986). Life cycles and evolution of clonal (modular) organisms. *Philosophical Transactions of the Royal Society London*, **B313**, 7–22.

Coates, A.G. and Jackson, J.B.C. (1987). Clonal growth, algal symbiosis and reef formation by corals. *Paleobiology*, **13**, 363–78.

Coates, A.G. and Oliver, W.A., Jr. (1973). Coloniality in zooantharian corals. In *Animal colonies, developments and function through time* (ed. R.S. Boardman, A.H. Cheetham, and W.A. Oliver), pp. 3–27. Dowden, Hutchinson and Ross, Stroudsburg, PA.

Cocks, L.R.M. (1988). Brachiopods across the Ordovician–Silurian boundary. *Bulletin of the British Museum* (*Natural History*) (*Geology*), **43**, 311–16.

Connell, J. H. (1973). Population ecology of reef-building corals. In *Biology and geology of coral reefs*. 2. *Biology*, 1. (ed. O.A. Jones and R. Endean), pp. 205–45. Academic Press, New York.

Connell, J.H. (1978). Diversity in tropical rain forests and coral reefs. *Science*, **199**, 1302–10.

Connell, J.H. (1980). Diversity and coevolution of competitors, or the ghost of competition past. *Oikos*, **35**, 131–8.

Conway Morris, S. (1977). Fossil priapulid worms. *Special Papers in Palaeontology*, **20**, 1–95.

Conway Morris, S. (1979). Middle Cambrian polychaetes from the Burgess Shale of British Columbia. *Philosophical Transactions of the Royal Society*, **B285**, 227–391.

Conway Morris, S. (1993). The fossil record and the early evolution of the Metazoa. *Nature*, **361**, 219–25.

Copper, P. (1988). Ecological succession in Phanerozoic reef ecosystems: is it real? *Palaios*, **3**, 136–52.

Copper, P. (1989). Enigmas in Phanerozoic reef development. In *Fossil Cnidaria* 5 (ed. P.A. Jell and J.W. Pickett), pp. 371–85. *Memoir of the Association of Australasian Palaeontologists*, **8**.

Copper, P. (1994). Ancient reef ecosystem expansion and collapse. *Coral Reefs*, **13**, 3–11.

Copper, P. (1996). *Davidsonia* and *Rugodavidsonia* (new genus), cryptic Devonian atrypid brachiopods from Europe and south China. *Journal of Paleontology*, **70**, 588–602.

Copper, P. (1997). Reefs and carbonate productivity: Cambrian through Devonian. *Proceedings of the 8th International Coral Reef Symposium*, (ed. H.G. Lessios and I.G. Macintyre), Panama City, Panama, **2**, 1623–30.

Copper, P. and Brunton, F.H. (1991). A global review of Silurian reefs. *Special Papers in Palaeontology*, **44**, 225–59.

Cornell, H.V. and Karlson, R.H. (1996). Species richness of reef-building corals determined by local and regional processes. *Journal of Animal Ecology*, **65**, 233–41.

Coudray, J. and Mantaggioni, L. (1982). Coraux et récifs coralliens de la province indo-pacifique; répartition geographique et altitudinale en relation avec la tectonique globale. *Bulletin de la Societé Géologique de France* (Series 7), **24**, 981–93.

Cowen, R. (1983). Algal symbiosis and its recognition in the fossil record. In *Biotic interactions in recent and fossil benthic communities* (ed. M.J. S. Tevesz and P.L. McCall), pp. 432–78. Plenum Press, New York.

Cowen, R. (1988). The role of algal symbiosis in reefs through time. *Palaios*, **3**, 221–7.

Cronin, G. and Hay, M.E. (1997). Induction of seaweed chemical defences by amphipod grazing. *Ecology*, **78**, 2287–301.

Cuffey, R.J. (1974.) Delineation of bryozoan constructional roles in reefs from comparison of fossil bioherms and living reefs. *Proceedings of the 2nd International Coral Reef Symposium*, (ed. A.M. Cameron, J.S. Jell *et al.*), Aboard M.V. Marco Polo, GBR Australia, **1**, 357–64.

Cutchis, P. (1982). A formula for comparing annual damaging ultraviolet (DUV) radiation doses at tropical and mid-latitude sites. In *The role of solar ultraviolet radiation in marine ecosystems* (ed. J. Chalkins), pp. 213–28. NATO Conference Series IV. Marine Sciences, **7**. Plenum Press, New York.

Darwin, C. (1872). *The structure and distribution of coral reefs* (2nd edn). Smith, Elder and Company, London.

Davies, G.R. and Nassichuk, W.W. (1990). Submarine cements and fabrics in Carboniferous to Lower Permian reefs, shelf margin and slope carbonates, north-western Ellesmee Island, Canadian Arctic Archipelago. *Geological Survey of Canada Bulletin*, **399**, 1–77.

Davies, G.R., Nassichuk, W.W., and Beauchamp, B. (1989). Upper Carboniferous 'Waulsortian' reefs, Canadian Arctic Archipelago. In *Reefs, Canada and adjacent areas* (ed. H.H.J. Geldsetzer and N.P.James), pp. 565–74. Memoir of the Canadian Society of Petroleum Geologists, **13**.

Davies, P.J. and Hopley, D. (1983). Growth fabrics and growth rates of Holocene reefs in the Great Barrier Reef. *BMR J of Australian Geologists and Geophysicists* **8**, 237–51.

Davies, T.A. and Worsley, T.R. (1981). Paleoenvironmental implications of oceanic carbonate sedimention rates. In *The deep sea drilling project: a decade of progress* (ed. J.E. Warme, R.G. Douglas, and E.L. Winterer), pp. 169–79. Special Publication of the Society of Economic Paleontologists and Mineralogists, **32**.

Dawson, E.Y. (1966). *Marine botany*. Holt, Rinehard and Winston, New York.

Day, R.W. (1983). Effects of benthic algae on sessile animals: observational evidence from coral reef habitats. *Bulletin Marine Science*, **33**, 597–605.

Debrenne, F. and James, N.P. (1981). Reef-associated archaeocyathans from the Lower Cambrian of Labrador and Newfoundland. *Palaeontology*, **24**, 343–78.

Debrenne, F. and Wood, R. (1990). A new Cambrian sphinctozoan from North America, its relationship to archaeocyaths and the nature of early sphinctozoans. *Geological Magazine*, **127**, 435–43.

Debrenne, F. and Zhuravlev, A. Yu. (1992). Irregular archaeocyaths. Morphology. Ontogeny. Systematics. Biostratigraphy. Palaeoecology. *Cahiers de Paléontologie*. Editions du CNRS, Paris.

Debrenne, F. and Zhuravlev, A. Yu. (1994). Archaeocyathan affinity: How deep can we go into the systematic affiliation of an extinct group? In *Sponges in space and time* (ed. E. Balkema), pp. 3–10. Proceedings of the 4th International Poriferan Congress, Amsterdam, 1993.

Debrenne, F., Rozanov, A. Yu., and Webers, G.F. (1984). Upper Cambrian Archaeocyatha from Antarctica. *Geological Magazine*, **121**, 291–9.

Debrenne, F., Gandin, A., and Rowland, S. (1989). Lower Cambrian bioconstructions in Northwestern Mexico (Sonora). Depositional setting, palaeoecology and systematics of archaeocyaths. *Geobios*, **2**, 137–95.

Debrenne, F., Zhuravlev, A. Yu., and Rozanov, A. Yu. (1990). Regular archaeocyaths: morphology, systematics, biostratigraphy, palaeoecology, biological affinities. *Cahiers de Paléontologie*. Editions du CNRS, Paris.

Dennison, W.C. and Barnes, D.J. (1988). Effect of water motion on coral photosynthesis and calcification. *Journal of Experimental Marine Biology and Ecology*, **115**, 67–77.

d'Hondt, S. and Arthur, M.A. (1996). Late Cretaceous oceans and the cool tropics paradox. *Science*, **271**, 1838–41.

D'Hondt, S. and Zachos, J.C. (1995). 75 million years of photosymbiosis in planktic foraminifera. *Geological Society of America Abstracts with Programs*, **27**, A244.

Digerfeldt, G., and Hendry, M.D. (1987). An 8000 year Holocene sea-level record from Jamaica: implications for interpretation of Caribbean reef and coastal history. *Coral Reefs*, **5**, 165–70.

Done, T.J. (1982). Patterns in the distribution of coral communities across the central Great Barrier Reef. *Coral Reefs*, **1**, 95–107.

Done, T.J. (1991). The debate continues—robust vs. fragile reefs. *Reef Encounters*, **9**, 5–7.

Donovan, D.T. and Jones, E.J.W. (1979). Causes of worldwide changes in sea-level. *Journal of the Geological Society*, **136**, 187–92.

Douglas, A.E. (1992). Symbiosis in evolution. *Oxford Surveys in Evolutionary Biology*, **8**, 347–82.

Douglas, A.E. (1994). *Symbiotic interactions*, 148pp. Oxford Science Publications, Oxford University Press, Oxford.

Douglas, A.E. (1995). The ecology of symbiotic micro-organisms. *Advances in Ecological Research*, **26**, 69–102.

Douglas, A.E. and Smith, D.C. (1989). Are endosymbioses mutalistic? *Trends in Ecology and Evolution*, **4**, 350–2.

Dravis, J.J. (1982). Hardened subtidal stromatolites, Bahamas. *Science*, **219**, 385–6.

Droser, M.L., Fortey, R.A., and Li, X. (1996). The Ordovician radiation. *American Scientist*, **84**, 122–31.

Droser, M.L., Bottjer, and Sheehan, P.M. (1997). Evaluating the ecological architecture of major events in the Phanerozoic history of marine invertebrate life. *Geology*, **25**, 167–70.

Drozdova, N.A. (1980). Algae in Lower Cambrian organogeneous buildups of western Mongolia. *Trans. Joint Soviet Mongolian Palaeontological Expedition*, **10**, 1–140. (In Russian.)

Dunham, R.J. (1962). Classification of carbonate rocks according to depositional texture. In *Classification of carbonate rocks* (ed. W.E. Ham), pp. 108–22. *American Association of Carbonate Petrologists Memoir*, **1**.

Dunham, R.J. (1970). Stratigraphic reefs versus ecologic reefs. *Bulletin American Association Petroleum Geologists*, **54**, 1931–2.

Dunne, R.P. (1994). *Environmental data handbook for Pari Island, Indonesia*. Overseas Development Administration, London, UK.

Durham, J.W. and Melville, R.V. (1957). A classification of echinoids. *Journal of Palaeontology*, **31**, 242–72.

Edhorn, A.-S. (1977). Early Cambrian algae croppers. *Canadian Journal of Earth Sciences*, **14**, 1014–20.

Elias, R.J. (1989). Extinctions and origins of solitary rugose corals, latest Ordovician to earliest Silurian in North America. In *Fossil Cnidaria 5* (ed. P.A. Jell, and J.W. Pickett), pp. 319–26. Memoirs of the Association of Australasian Palaeontologists, **8**.

Elliott, G.F. (1973). A Miocene solenoporoid alga showing reproductive structures. *Palaeontology*, **16**, 223–30.

Embry, A.F. and Klovan, J.E. (1971). A Late Devonian reef tract on Northeastern Banks Island, Northwest Territories. *Bulletin of Canadian Petroleum Geology*, **19**, 730–81.

Embry, A.F. and Suneby, L.B. (1994). The Triassic–Jurassic boudary in the Sverdrup Basin, Arctic Canada. *Canadian Society of Petroleum Geologists, Memoir*, **17**, 857–68.

Erwin, D.H. (1993). *The great Paleozoic crisis: life and death in the Permian.* Columbia University Press, New York.

Erwin, D.H. (1994). The Permo-Triassic extinction. *Nature*, **367**, 231–6.

Erwin, D.H. (1996). Understanding biotic recoveries: extinction, survival, and preservation during the End-Permian mass extinction. In *Evolutionary paleobiology* (ed. D. Jablonski, D.H. Erwin, and J.H. Lipps), pp. 419–33. University of Chicago Press, Chicago.

Ezaki, Y. (1993). The last representatives of Rugosa in Iran and Transcaucasus, west Tethys. *Bulletin of the Geological Survey of Japan*, **44**, 447–53.

Ezaki, Y., Kawamura, T., and Nakamura, K. (1994). Kapp Starostin Formation in Spitsbergen: a sedimentary and faunal record of Late Permian paleoenvironments in an Arctic region. *Canadian Society of Petroleum Geologists, Memoir*, **17**, 647–55.

Fagerstrom, J.A. (1987). *The evolution of reef communities.* Wiley, New York.

Fagerstrom, J.A. (1994). The history of Devonian–Carboniferous communities: extinctions, effects, recovery. *Facies*, **30**, 177–92.

Fan, J., Rigby, J.F., and Qi, J. (1990). The Upper Permian reefs in west Hubei, China. *Facies*, **6**, 1–14.

Fautin, D. G. (1991). Developmental pathways in anthozoans. *Hydrobiologia*, **216/217**, 143–9.

Faure, K., de Wit, M.J., and Willis, J.P. (1995). Late Permian global coal hiatus linked to 13C-depleted CO_2 flux into the atmosphere during the final consolidation of Pangaea. *Geology*, **23**, 507–10.

Fedorowski, J. (1989). Extinction of the Rugosa and Tabulata near the Permian/Triassic boundary. *Acta Palaeontologica Polonica*, **34**, 47–70.

Fitt, W.K. and Trench, R.K. (1981). Spawning, development, and the acquisition of zooxanthellae by *Tridacna squamosa* (Mollusca, Bivalvia). *Biological Bulletin*, **161**, 213–35.

Fitt, W.K. and Warner, M.E. (1995). Bleaching patterns of four species of Caribbean reef corals. *Biological Bulletin*, **189**, 298–307.

Fitt, W.K., Chang, S.S., and Trench, R.K. (1981). Motility patterns of different strains of *Symbiodinium* (= *Gymnodinium*) *microadriaticum* Freudenthal in culture. *Bulletin of Marine Science*, **31**, 435–43.

Flügel, E. (1981). 'Tubiphyten' aus dem fränkischen Malm. *Geoleclogisches Bulletin NO-Bayern*, **31**, 126–42.

Flügel, E. (1994). Pangean shelf carbonates: controls and paleoclimatic significance of Permian and Triassic reefs. In *Pangea: paleoclimate, tectonics, and sedimentation during accretion, zenith, and breakup of a supercontinent* (ed. G.D. Klein), pp. 247–66. *Special Paper of the Geological Society of America*, **288**.

Flügel, E. and Reinhardt, J. (1989). Uppermost Permian reefs in Skyros (Greece) and Sichuan (China): implications for the Late Permian extinction event. *Palaios*, **4**, 502–18.

Flügel, E. and Stanley, G.D. (1984). Re-organisation, development and evolution of post-Permian reefs and reef-organisms. *Paleontographica Americana*, **54**, 177–86.

Fois, E. and Gaetani, M. (1984). The recovery of reef-building communities and the role of cnidarians in carbonate sequences of the Middle Triassic (Anisian) in the Italian Dolomites. *Paleontographica Americana*, **54**, 191–200.

Fortey, R. A. (1989). There are extinctions and extinctions: examples from the Lower Palaeozoic. *Philosophical Transactions of the Royal Society London*, **B325**, 327–55.

Frakes, L.A., Francis, J.E., and Syktus, J.I. (1992). *Climatic modes of the Phanerozoic.* Cambridge University Press, Cambridge.

Freiwald, A. (1993). Coralline algal maerl frameworks—Islands within the Phaeophytic kelp belt. *Facies*, **29**, 133–48.

Freiwald, A. and Henrich, R. (1994). Reefal coralline algal buildups within the Arctic Circle: morphology and sedimentary dynamics under extreme environmental seasonality. *Sedimentology*, **41**, 963–84.

Fritz, M.A. (1977). A microbioherm. In *Essays on paleontology in honor of Loris Shano Russell* (ed. C.S. Church), pp. 18–25. Life Sciences Miscellaneous Publications, Royal Ontario Museum, Toronto.

Frost, S. H. (1986). Mid-Tertiary origin of Caribbean–Atlantic and Indo-Pacific reef provinces. Annual Meeting Coral Reef Research Society, 17. Marburg.

Fry, W. (1979). Taxonomy, the individual and the sponge. In *Biology and systematics of colonial organisms* (ed. G. Larwood and B.R. Rosen), pp. 49–80. Academic Press, London.

Fürsich, F.T. and Jablonski, D. (1984). Late Triassic natacid drillholes: carnivorous gastropods gain a major adaptation but fail to radiate. *Science*, **224**, 78–80.

Fürsich, F.T., Palmer, T.J., and Goodyear, K.L. (1994). Growth and disintegration of bivalve-dominated patch reefs in the Upper Jurassic of southern England. *Palaeontology*, **37**, 131–71.

Gaines, S.D. and Lubchenco, J. (1982). A unified approach to marine plant–herbivore interactions. II Biogeography. *Annual Review Ecology Systematics*, **13**, 111–38.

Gale, A.S. (1987). Phylogeny and classification of the Asteroidea (Echinodermata). *Zoological Journal of the Linnean Society*, **88**, 107–32.

Gao, J. and Copper, P. (1997). Growth rates of Middle Paleozoic corals and sponges: Early Silurian of Eastern Canada. *Proceedings of the 8th International Coral Reef Syposium*, Panama City, Panama, **2**, 1651–6.

Garber, R.A., Grover, G.A., and Harris, P.M. (1989). Geology of the Capitan Shelf Margin—subsurface data from the northern Delaware Basin. In *Subsurface and outcrop examination of the Capitan Shelf Margin*, *Northern Delaware Basin* (ed. P.M. Harris and G.A. Grover), pp. 3–269. Society of Economic Paleontologists and Mineralogists, Core Workshop, **13**.

Garcia-Mondéjar, J. and Fernández-Mendiola, P.A. (1995). Albian carbonate mounds: comparative study in the context of sea-level variations (Soba, northern Spain). In *Carbonate mud mounds*; *their origin and evolution* (ed. C.L.V. Monty, D.J.W. Bosence, P.H. Bridges, and B.R. Pratt), pp. 359–84. International Association of Sedimentologists, Special Publication, **23**. Blackwell Science, Oxford.

Garrett, P. (1970). Phanerozoic stromatolites: restriction by grazing and burrowing animals. *Science*, **169**, 171–3.

Gause, G.F. (1934). *The struggle for existence*. Williams and Wilkins, Baltimore, MD.

Gebelein, C.D. (1976). The effects of physical, chemical, and biological evolution of the Earth. In *Stromatolites* (ed. M.R. Walter), pp. 499–515. Elsevier, Amsterdam.

Gehling, J.G. and Rigby, J.R. (1996.) Long expected sponges from the Neoproterozoic Ediacara fauna of South Australia. *Journal of Paleontology*, **70**, 185–95.

George, A.D., Playford, P.E, and Powell, C. McA. (1994). Carbonate breccias and quartz–feldspathic sandstones of the marginal slope, Devonian reef complexes of the Canning Basin, Western Australia: implications for sea-level changes. In *Proceedings of the Australian Basins Symposium*, *Perth* (ed. P.G. Purcell and R.R. Purcell), pp. 101–12. West Petroleum Exploration Society of Australia.

George, A.D., Playford, P.E, Powell, C. McA., and Tornatora, P.M. (1997). Lithofacies and sequence development on an Upper Devonian mixed carbonate–siliclastic fore-reef slope, Canning Basin, Western Australia: *Sedimentology*, **44**, 843–87.

Gili, E., Masse, J.-P., and Skelton, P.W. (1995). Rudists as gregarious sediment-dwellers, not reef-builders, on Cretaceous carbonate platforms. *Palaeogeography*, *Palaeoclimatology*, *Palaeoecology*, **118**, 245–67.

Gill, G.A. and Coates, A. G. (1977). Mobility, growth patterns and substrate in some fossil and recent corals. *Lethaia*, **10**, 119–34.

Gischler, E. (1995). Current and wind induced facies patterns in a Devonian atoll: Iberg Reef, Harz Mts., Germany. *Palaios*, **10**, 180–9.

Given, R.K. and Wilkinson, B.H. (1985). Kinetic control of morphology, composition and mineralogy of abiotic sedimentary carbonates. *Journal of Sedimentary Petrology*, **55**, 109–19.

Given, R.K. and Wilkinson, B.H. (1987). Dolomite abundance and stratigraphic age: constraints on rates and mechanisms of Phanerozoic dolostone formation. *Journal of Sedimentary Petrology*, **57**, 1068–78.

Gladfelter, E.H., Monaham, R.K., and Gladfelder, W.D. (1978). Growth rates of five reef-building corals in the northwestern Caribbean. *Bulletin of Marine Science*, **28**, 728–34.

Gladfelter, W.D. (1982). White band disease in *Acropora palmata*: implications for the structure and growth of shallow reefs. *Bulletin of Marine Science*, **32**, 639–43.

Glaessner, M.F. (1979). Precambrian. In *Treatise on invertebrate paleontology*, Part A (ed. R.A. Robinson and C. Techert), pp. A79–118. Geological Society of America and University Kansas Press, Lawrence, KS.

Glynn, P.W. (1973). Ecology of a Caribbean coral reef. The *Porites* reef-flat biotope. II. Plankton community with evidence for depletion. *Marine Biology*, **20**, 297–318.

Glynn, P.W. (1982). *Acanthaster* population regulation by a shrimp and a worm. *Proceedings of the 4th International Coral Reef Symposium*, Manila, Phillipines, **2**, 607–12.

Glynn, P.W. (1983). Crustacean symbionts and the defence of corals: coevolution on the reef? In *Coevolution* (ed. M.H. Nitecki), pp. 111–78. University of Chicago Press, Chicago.

Glynn, P.W. (1984). Widespread coral mortality and the 1982–83 El Niño warming event. *Environmental conservation*, **11**, 133–46.

Glynn, P.W. (1988). Predation on coral reefs: some key processes, concepts and research directions. *Proceedings of the 6th International Coral Reef Symposium*, Townsville, Australia, **1**, 51–62.

Glynn, P.W. (1990). Feeding ecology of selected coral-reef macroconsumers: patterns and effects on coral community structure. In *Coral Reefs* (ed. Z. Dubinsky), pp. 365–400. *Ecosystems of the world*, **25**. Elsevier, New York.

Glynn, P.W. (1993). Coral reef bleaching: ecological perspectives. *Coral Reefs*, **12**, 1–17.

Glynn, P.W. (1996). Coral reef bleaching: facts, hypotheses and implications. *Global Change Biology*, **2**, 495–509.

Glynn, P.W. (1997*a*). Eastern Pacific reef coral biogeography and faunal flux: Durham's dilemma revisited. *Proceedings of the 8th International Coral Reef Symposium*, (ed. H.G. Lessios and I.G. Macintyre), Panama City, Panama, 1996, **1**, 371–8.

Glynn, P.W. (1997*b*). Bioerosion and coral-reef growth. In *Life and death of coral reefs*. (ed. C. Birkeland), pp. 68–95. Chapman and Hall, New York.

Glynn, P.W. and Colgan, M.W. (1992). Sporadic disturbances in fluctuating coral reef environments; El Niño and coral reef development in the eastern Pacific. *American Zoologist*, **32**, 707–18.

Glynn, P.W. and D'Croz, L. (1990). Experimental evidence for high temperature stress as the cause of El Niño-coincident coral mortality. *Coral Reefs*, **8**, 181–91.

Glynn, P.W., Stewart, R.H., and McCosker, J.E. (1972). Pacific coral reefs of Panama: structure, distribution and predators. *Geologische Rundschau*, **61**, 483–519.

Goreau, T.F. and Goreau, N.I. (1959). The physiology of skeleton formation in corals. II. Calcium deposition by hermatypic corals under various conditions of the reef. *Biological Bulletin*, **117**, 239–50.

Goreau, T.F. and Wells, J.W. (1967). The shallow water Scleractinia of Jamaica: revised list of species and their vertical distribution range. *Bulletin of Marine Sciences*, **17**, 42–53.

Goreau, T.J., Goreau, N.I., Trench, R.K., and Hayes, R.L. (1996). Technical comment. *Science*, **274**, 117.

Gould, S.J. and Vrba, E.S. (1982). Exaptation—a missing term in the science of form. *Paleobiology*, **8**, 4–15.

Gram, R. (1968). A Florida Sabellariidae reef and its effects upon sediment distribution. *Journal of Sedimentary Petrology*, **38**, 963–8.

Grammer, G.M., Ginsburg, R.N., Swart, P.K., McNeill, D.F., Jull, A.J.T., and Prezbindowski, D.R. (1993). Rate growth rates of syndepositional marine aragonite cements in steep marginal slope deposits. Bahamas and Belize. *Journal of Sedimentary Petrology*, **63**, 983–9.

Grammer, G. M., Crescini, C.M., McNeill, D.F., and Taylor, L.H. Quantifying rates of syndepositional marine cementation in deeper platform environments—new insight into a fundamental process. *Journal of Sedimentary Research*. (In press.)

Grant, R.E. (1975). Methods and conclusions in functional analysis: a reply. *Lethaia*, **8**, 31.

Grant, S.W.F. (1990). Shell structure and distribution of *Cloudina*, a potential index fossil for the terminal Proterozoic. *American Journal of Sciences*, **290A**, 261–94.

Grant, S.W.F., Knoll, A.H., and Germs, G.J.B. (1991). Probable calcified metaphytes in the latest Proterozoic Nama Group, Namibia: origin, diagenesis and implications. *Journal of Paleontology*, **65**, 1–18.

Grassle, J.F. (1973). Variety in coral reef communities. In *Biology and geology of coral reefs*, Vol. 11 (ed. O.A. Jones and R. Endean), pp. 247–70. Biology 1.

Green, J.W., Knoll, A.H., and Swett, K. (1988). Microfossils from oolites and pisolites of the Upper Proterozoic Eleonore Bay Group, central East Greenland. *Journal of Paleontology*, **62**, 835–52.

Greenstein, B.J. (1989). Mass mortality of the west Indian echinoid *Diadema antillarum* (Echinodermata: Echinoidea): a natural experiment in taphonomy. *Palaios*, **4**, 487–92.

Greenstein, B.J. and Curran, H.A. (1997). How much ecological information is preserved in fossil coral reefs and how reliable is it? *Proceedings of the 8th International Reef Symposium*, (ed. H.G. Lessios and I.G. Macintyre), Panama City, 1996, Panama, **1**, 417–22.

Grotzinger, J.P. (1989). Facies and evolution of Precambrian carbonate depositional systems: emergence of the modern platform archetype. In *Controls on carbonate platform and basin development* (ed. P.D. Crevello, J.L. Wilson, J.F. Sarg and F.F. Read), pp. 79–106. Society of Economic Paleontologists and Mineralogists Special Publication, **44**.

Grotzinger, J.P. (1990). Geochemical model for Proterozoic stromatolite decline. *American Journal of Science*, **290A**, 80–103.

Grotzinger, J.P. and Rothman, D.H. (1996). An abiotic model for stromatolite morphogenesis. *Nature*, **383**, 423–5.

Grotzinger, J.P., Bowring, S.A., Saylor, B.Z., Kaufman, A.J. (1995). Biostratigraphc and geochronological contraints on early animal evolution. *Science*, **270**, 598–604.

Gruszczyñski, M., Hoffman, A., Malkowski, K., Tatur, A. and Halas, S. (1990). Some geochemical aspects of life and burial environments of Late Jurasic scleractinian corals from northern Poland. *Neues Jahrbuch für Geologie und Paläontologie Mitteilungen Monatshrift*, **11**, 673–86.

Guida, V.G. (1976). Sponge predation in the oyster reef community as demonstrated with *Cliona celata* Grant. *Journal of Experimental Marine Biology and Ecology*, **25**, 109–22.

Håkansson, E. and Madsen, L. (1991). Symbiosis—a plausible explanation of gigantism in Permian trepostome bryozoans. In *Bryozoaires actuels et fossiles* (ed. F.P. Bigey), pp. 151–9. Bulletin de la Societé Scientifique Naturelle Ouest France Mémoir, **HS1**. Nantes, France.

Hall, V.R. and Hughes, T.P. (1996). Reproductive strategies of modular organisms: comparative studies of reef-building corals. *Ecology*, **77**, 950–63.

Hallam, A. (1981). Facies intepretation and the stratigraphic record. W.H. Freeman, Oxford.

Hallam, A. (1984). Pre-Quaternary changes in sea level. *Annual Review of Earth and Planetary Science*, **12**, 205–43.

Hallam, A. (1986). Evidence of displaced terranes from Permian to Jurassic faunas around the Pacific margins. *Journal of the Geological Society of London*, **143**, 209–16.

Hallam, A. (1995). Oxygen-restricted facies of the basal Jurassic of north west Europe. *Historical Biology*, **10**, 247–57.

Hallam, A. and O'Hara, M.J. (1962). Aragonitic fossils in the Lower Carboniferous of Scotland. *Nature*, **195**, 273–4.

Hallam, A. and Wignall, P.B. (1997). *Mass extinctions and their aftermath*. Oxford University Press, Oxford.

Hallock, P. (1988). Diversification in algal symbiont-bearing foraminifera: a response to oligotrophy? *Benthos '86. Revue de Paléobiologie*, **2**, 789–97.

Hallock, P. and Schlager, W. (1986). Nutrient excess and the demise of reefs and carbonate platforms. *Palaios*, **1**, 389–98.

Hallock, P., Hine, A.C., Vargo, G.A., Elrod, J.A., and Japp, W.C. (1988). Platforms of the Nicaraguan Rise: examples of the sensitivity of carbonate sedimentation to excess trophic reserves. *Geology*, **16**, 1104–7.

Hamdi, B., Rozanov, A.Yu., and Zhuravlev, A.Yu. (1995). Latest Middle Cambrian metazoan reef from northern Iran. *Geological Magazine*, **132**, 367–73.

Hamilton, W.D. and May, R.M. (1977). Dispersal in stable habitats. *Nature*, **269**, 378–81.

Hansen, T.A. (1988). Early Tertiary radiation of marine molluscs and the long-term effects of the Cretaceous–Tertiary extinction. *Paleobiology*, **14**, 37–51.

Haq, B.U., Hardenbol, J., and Vail, P.R. (1987). Chronology of fluctuating sea-levels since the Triassic. *Science*, **235**, 1156–67.

Harada, T., Oshima, Y., Kamiya, H., and Yasumto, T. (1982). Confirmation of paralytic shellfish toxins in the dinoflagelate *Pyrodium bahamense* var. *compressa* and bivalves in Palau. *Bulletin Japanese Society of Science and Fisheries*, **48**, 821–5.

Harland, W.B., Armstrong, R.L., Cox, A.V., Craig, L.E., Smith, A.G., and Smith, D.G. (1990). *A geologic time scale 1989*. Cambridge University Press, Cambridge.

Harmelin-Vivien, M.L. (1985). Hurricane effects on coral reefs: introduction. *Proceedings of the 5th International Coral Reef Congress*, Tahiti, French Polynesia, **3**, 315.

Harper, D.A.T., and Rong, J. (1995). Patterns of change in the brachiopod faunas through the Ordovician-Silurian interface. *Modern Geology*, **20**, 83–100.

Harper, E.M. (1991). The role of predation in the evolution of the cemented habit in bivalves. *Palaeontology*, **34**, 455–60.

Harper, E.M. (1994). Are conchilin sheets in corbulid bivalves primarily defensive? *Palaeontology*, **37**, 551–78.

Harper, E.M. and Skelton, P.W. (1993). The Mesozoic Marine Revolution and epifaunal bivalves. *Scripta Geologica, Special Issue*, **2**, 127–53.

Harper, E.M., Forsythe, G.T.W., and Palmer, T. (1998). Taphonomy and the Mesozoic Marine Revolution: Preservation state masks the importance of boring predators. *Palaios*, **13**, 352–60.

Harper, E.M., Palmer, T.J., and Alphey, J. R. (1997). Evolutionary response by bivalves to changing Phanerozoic sea-water chemistry. *Geological Magazine*, **134**, 403–7.

Harris, M.T. (1993). Reef fabrics, biotic crusts and symdepositional cements of the Latemar reef margin (Middle Triassic), northern Italy. *Sedimentology*, **40**, 383–401.

Harrison, P.L., Babcock, R.C., Bull, G.D., Oliver, J.K., Wallace, C.C., and Willis, B.L. (1984). Mass spawning in tropical reef corals. *Science*, **223**, 1186–9.

Hart, S.F. (1994). Archaeocyath paleoecology. *Fifth North American Paleontological Convention*, Abstracts and Program, Chicago. Paleontological Society Special Volume, **6**, 122.

Hartman, W.D. and Reiswig, H. (1973) The individuality of sponges. In *Animal colonies, developments and function through time* (ed. R.S. Boardman, A.H. Cheetham, and W.A. Oliver), pp. 567–84. Dowden, Hutchinson, and Ross, Stroudsburg, PA.

Harvey, P.H. and Greenwood, P.J. (1978). Anti-predator defense strategies: some evolutionary problems. In *Behavioural ecology: an evolutinary approach* (ed. J.R. Krebs and N.B. Davies), pp. 129–51. Blackwell, Oxford, UK.

Hatcher, B.G. (1981). The interaction between grazing organisms and the epilithic algal community of a coral reef: a quantitative assessment. *Proceedings of the 4th International Coral Reef Symposium*, Manila, Phillipines, **2**, 515–24.

Hatcher, B.G. (1990). Coral reef primary productivity: a heirarchy of pattern and process. *Trends in Ecology and Evolution*, **5**, 149–55.

Hay, M.E. (1981). Herbivory, algal distribution and the maintenance of between-habitat diversity on a tropical fringing reef. *American Naturalist*, **118**, 520–40.

Hay, M.E. (1984a). Predictable spatial escapes from herbivory: how do these effect the evolution of herbivore resistance in tropical marine communities? *Oecologia*, **64**, 396–407.

Hay, M.E. (1984b). Patterns of fish and urchin grazing on Caribbean coral reefs: are previous results typical? *Ecology*, **65**, 446–54.

Hay, M.E. (1991). Fish–seaweed interactions on coral reefs: effects of herbivorous fishes and adaptations of the prey. In *The ecology of coral reef fishes* (ed. P.F. Sale), pp. 96–119. Academic Press, San Diego, CA.

Hay, M.E., Colburn, T., and Downing, D. (1983). Spatial and temporal patterns in herbivory on a Caribbean fringing reef: the effects on plant distribution. *Oecologia*, **58**, 299–308.

Hay, M.E., Paul, V.J., Lewis, S.M., Tucker, J., and Trindell, R.N. (1988). Can tropical seaweeds reduce herbivory by growing at night? Diel patterns of growth, nitrogen content, herbivory, and chemical versus morpohological defenses. *Oecologia*, **75**, 233–45.

Heckel, P.H. (1972). Possible inorganic origin for stromatactis in calcilutite mounds in the Tully Limestone, Devonian of New York. *Journal Sedimentary Petrology*, **42**, 7–18.

Heckel, P.H. (1974). Carbonate buildups in the geologic record: a review. *Society of Economic Paleontologists and Mineralogists Special Publication*, **18**, 90–154.

Hembleben, L., Spindler, M., and Anderson, O.R. (1989). *Modern planktonic foraminifera*. Springer-Verlag, New York.

Hickey, L.J. (1984). Changes in the angiosperm flora across the Cretaceous–Tertiary boundary. In *Catastrophes in earth history* (ed. W.A. Berggren and J.A. van Couvering), pp. 279–313. Princeton University Press, Princeton, NJ.

Highsmith, R.C. (1980). Geographic patterns in coral bioerosion: a productivity hypothesis. *Journal of Experimental Marine Biology and Ecology*, **46**, 177–96.

Hill, D. (1981). Rugosa and Tabulata. In *Treatise on invertebrate paleontology* (ed. C. Teichert), pp. xl–762. Part F (Suppl.1). Geological Society of America and University of Kansas Press, Boulder, Colorado and Lawrence, KA.

Hine, A.C., Hallock, P., Harris, M.W., Mullins, H.T., Belknap, D., and Japp, W.C. (1988). *Halimeda* bioherms along an open seaway: Moskito Channel, Nicaraguan Rise, SW Caribbean Sea. *Coral Reefs*, **6**, 173–8.

Hobson, E.S. (1968). Predatory behaviour of some shore fishes in the Gulf of California. *Research Report—U.S. Fish Wildlife Service*, **83**, 1–92.

Hoegh-Guldberg, O. and Smith, G.J. (1987). Influence on the population density of zooxanthellae and supply of ammonia on the biomass and metabolitic characters of the reef corals, *Seriatopora hystrix* and *Stylophora pistillata*. *Marine Biology Progress Series*, **57**, 173–86.

Hoegh-Guldberg, O., McCloskey, L.R., and Muscatine, L. (1987). Expulsion of zooxanthellae by symbiotic cnidarians from the Red Sea. *Coral Reefs*, **5**, 201–4.

Hoffman, P. (1974). Shallow and deepwater stromatolites in Lower Proterozoic platform-basin facies change, Great Slave Lake, Canada. *American Association of Petroleum Geologists Bulletin*, **58**, 856–67.

Hofmann, H.J. and Grotzinger, J.P. (1985). Shelf-facies microbiotas from the Odjick and Rocknest formations (Epworth Group: 1.89 Ga), northwestern Canada. *Canadian Journal of Earth Sciences*, **22**, 1781–92.

Hofmann, H.J., Narbonne, G.M., and Aitken, J.D. (1990). Ediacaran remains from intertillite beds in northwestern Canada. *Geology*, **18**, 1199–202.

Hopley, D. (1982). *The geomorphology of the Great Barrier Reef.* Wiley, New York.

Horbury, A. (1992). A Late Dinantian peloid cementstone–palaeoberesellid buildup from North Lancashire, England: *Sedimentary Geology*, **79**, 117–37.

Howarth, M.K. (1981). Palaeogeography of the Mesozoic. In *The evolving earth* (ed. R.M. Cocks), pp. 197–200. Cambridge University Press, Cambridge.

Hubbard, D.K. (1989). Modern carbonate environmemts of St. Croix and the Caribbean: a general overveiw. In *Terrestrial and marine geology of St. Croix, US Virgin Islands* (ed. D.K. Hubbard), pp. 85–94. Special Publication, **8**. West Indies Laboratory, St. Croix, USVI.

Hubbard, D.K. (1992). Hurricane-induced sediment transport in open-shelf tropical systems—an example from St. Croix, U.S. Virgin Islands. *Journal Sedimentary Petrology*, **62**, 946–60.

Hubbard, D.K and Scaturo, D. (1985). Growth rates of seven species of scleractinian corals from Cane Bay and Salt River, St. Croix, USVI. *Bulletin of Marine Science*, **36**, 325–38.

Hubbard, D.K, Burke, R.B., and Gill, I.P. (1986). Styles of reef accretion along a steep, shelf-edge reef, St. Croix, US Virgin Islands. *Journal of Sedimentary Petrology*, **56**, 848–61.

Hubbard, D.K, Miller, A.I., and Scaturo, D. (1990). Production and cycling of calcium carbonate in a shelf-edge reef system (St. Croix, US Virgin Islands): applications to the nature of reef systems in the fossil record. *Journal of Sedimentology*, **60**, 335–60.

Hubbell, S.P. (1997). A unified theory of biogeography and relative species abundance and its application to tropical rain forests and coral reefs. *Coral Reefs*, **16** (Suppl.), 9–21.

Hughes, R.N. (1979). Coloniality in the Vermetidae (Gastropoda). In *Biology and systematics of colonial animals* (ed. G. Larwood and B.R. Rosen), pp. 243–54. Academic Press, London.

Hughes, R.N. (1989). *A functional biology of clonal animals.* Chapman and Hall, London.

Hughes, R.N. (1991). Reefs. In *Fundamentals of aquatic biology* (ed. R.S.K. Barnes and K.H. Mann), pp. 213–29. Blackwell Scientic Publications, Oxford.

Hughes, T.P. (1984). Population dynamics based in individual size rather than age. *American Naturalist*, **123**, 778–95.

Hughes, T.P. (1989). Community structure and diversity of coral reefs: the role of history. *Ecology*, **70**, 275–9.

Hughes, T.P. (1994). Catastrophies, phase shifts, and large-scale degradation of a Caribbean coral reef. *Science*, **265**, 1547–51.

Hughes, T.P. and Connell, J.H. (1987). Population dynamics based on age or size? A reef-coral analysis. *American Naturalist*, **129**, 818–29.

Hughes, T.P. and Jackson, J.B.C. (1980). Do corals lie about their age? Some demographic consequences of partial mortality, fisson, and fusion. *Science*, **209**, 713–15.

Hughes, T.P. and Jackson, J.B.C. (1985). Population dynamics and life histories of foliaceous corals. *Ecological Monographs*, **55**, 141–66.

Hughes, T.P, Reed, D.C., and Boyle, M.J. (1987). Herbivory on coral reefs: community structure following mass mortality of sea urchins. *Journal Experimental Marine Biology Ecology*, **13**, 39–59.

Humphrey, J.D., Ransom, K.L., and Matthews, R.K. (1986). Early meteroic diagenetic control of Upper Smackover production, Oaks Field, Louisiana. *American Association of Petroleum Geologists*, **70**, 70–85.

Hurley, N.F. and Lohman, K.C. (1989). Diagenesis of the Devonian reefal carbonates of the Oscar Range, Canning Basin, western Australia. *Journal of Sedimentary Petrology*, **59**, 127–46.

Insalaco, E. (1996). Upper Juarassic microsolenid biostromes of north and central Europe: fabrics and depositional envornoment. *Palaeogeography, Palaeoclimatology, Palaeoecology*, **121**, 169–94.

Insalaco, E., Hallam, A. and Rosen, B. (1997). Oxfordian (Upper Jurassic) coral reefs in Western Europe: reef types and conceptual depositional model. *Sedimentology*, **44**, 707–34.

Irving, E., North, F.K., and Coullard, R. (1976). Oil, climate and tectonics. *Canadian Journal of Earth Sciences*, **11**, 1–17.

Jablonski, D. (1989). The biology of mass extinctions: a palaeontological view. *Philosophical Transactions of the Royal Society*, **B325**, 357–68.

Jablonski, D. (1995). Extinctions in the fossil record. In *Extincton rates* (ed. J.H. Lowton and R.M. May), pp. 25–44. Oxford University Press, Oxford.

Jackson, J.B.C. (1977). Competition on marine hard substrata: the adaptive significance of solitary and colonial strategies. *American Naturalist*, **111**, 743–7.

Jackson, J.B.C. (1979). Overgrowth competition between encrusting cheilostome ecotoprocts in a Jamaican cryptic reef environment. *Journal of Animal Ecology*, **48**, 805–23.

Jackson, J.B.C. (1983). Biological determinants of present and past sessile animal distributions. In *Biotic interactions in recent and fossil benthic communities* (ed. M. Tevesz and P.W. McCall), pp. 39–120. Plenum Press, New York.

Jackson, J.B.C. (1985). Distribution and ecology of clonal and aclonal benthic invertebrates. In *Population biology and evolution of clonal organisms* (ed. J.B.C. Jackson, L.W. Buss, and R.E. Cook), pp. 297–355.Yale University Press, New Haven, CT.

Jackson, J.B.C. (1986). Modes of dispersal of clonal benthic invertebrates: consequences for species' distribution and genetic structure of local populations. *Bulletin Marine Science*, **39**, 588–606.

Jackson, J.B.C. (1988). Does ecology matter? *Paleobiology*, **14**, 307–12.

Jackson, J.B.C. (1992). Pleistocene perspectives on coral reef community structure. *American Zoologist*, **32**, 719–31.

Jackson, J.B.C. (1995). Constancy and change of life in the sea. In *Extincton rates* (ed. J.H. Lowton and R.M. May), pp. 45–54. Oxford University Press, Oxford.

Jackson, J.B.C. (1997). Reefs since Columbus. *Coral Reefs*, **16** (Suppl.), 23–32.

Jackson, J.B.C. and Buss, L.W. (1975). Allelopathy and spatial competition among coral reef invertebrates. *Proceedings of the National Academy of Sciences USA*, **72**, 5160–3.

Jackson, J.B.C. and Coates, A.G. (1986). Life cycles and evolution of clonal (modular) animals. *Philosophical Transactions of the Royal Society of London*, **B313**, 7–22.

Jackson, J.B.C. and Hughes, T.P. (1985). Adaptive strategies of coral-reef invertebrates. *American Scientist*, **75**, 265–74.

Jackson, J.B.C. and McKinney, F.K. (1991). Ecological processes and progressive macroevolution of marine clonal benthos. In *Causes of evolution* (ed. R.M. Ross and W.D. Allmon), pp. 173–209. University of Chicago Press, Chicago, IL.

Jackson, J.B.C. and Winston, J.E. (1982). Ecology of cryptic coral reef communities.1. Distrubtion and abundance of major groups of encrusting organisms. *Journal of Experimental Marine Biology and Ecology*, **57**, 135–47.

Jackson, J.B.C., Goreau, T.E., and Hartman, W.D. (1971). Recent brachiopod coralline sponge communities and their paleontological significance: *Science*, **173**, 623–5.

Jackson, J.B.C., Budd, A.F., and Pandolfi, J.M. (1996). The shifting balance of natural communities? In *Evolutionary paleobiology* (ed. D. Jablonski, D.H. Erwin, and J.H. Lipps), pp. 89–122. Chicago University Press, Chicago, IL.

James, N.P. (1983). Reefs. In *Carbonate depositional environments* (ed. P.A. Scholle, D.G. Bebout, and C.H. Moore), pp. 345–462. *Memoirs American Association of Petroleum Geologists*, **33**.

James, N.P., Ginsburg, R.N., Maszalek, D.S., and Choquette, P.W. (1976). Facies and fabric selectivity of early subsea cements in shallow Belize (British Honduras) reefs. *Journal of Sedimentary Petrology*, **46**, 523–44.

James, N.P. and Gravestock, D.I. (1990). Lower Cambrian shelf and shelf-margin buildups, Flinders Ranges, South Australia. *Sedimentology*, **37**, 455–80.

James, N.P. and Macintyre, I.G. (1985). Carbonate depositional environments. Modern and Ancient. Part 1: Reefs: zonation, depositional facies, diagenesis. *Colorado School of Mines Quarterly*, **80**, 1–70.

Jansa, J.F., Termier, G., and Termirr, H. (1982). Les bioherms á algues spongiaires et coraux des séries carbonatées de la flexure bordiére du 'paleoshelf' au large du Canada oriental. *Revue de Micropaléontologie*, **25**, 181–219.

Jeffrey, C.H. (1997). Dawn of echinoid nonplanktotrophy: coordinated shifts in development indicate environmental instability prior to the K–T boundary. *Geology*, **25**, 991–4.

Jenkyns, H.C. (1980). Cretaceous anoxic events: from continents to oceans. *Journal of the Geological Society of London*, **137**, 171–88.

Ji, Q. (1989). On the Frasnian–Famennian mass extinction event in South China. *Courier Forshunginstitut Seneckenberg*, **117**, 275–301.

Jin, Y., Glenister, B.F., Kotlyar, G.V., and Sheng, J. (1994). An operational scheme of Permian chronostratigraphy. *Palaeoworld*, **4**, 1–13.

Johnson, C.C., Barron, E.J., Kauffman, E.G., Arthur, M.A., Fawcett, P.J., and Yasuda, M.K. (1996). Middle Cretaceous reef collapse linked to ocean heat transport. *Geology*, **24**, 376–80.

Johnson, G.A.L. and Nudds, J.R. (1975). Carboniferous coral geochronometers. In *Growth rhythms and the history of the Earth's rotation* (ed. G.D. Rosenberg and S.K. Runcorn), pp. 27–42. Wiley, London.

Johnson, J.G. (1993). Extinction at the Antipodes. *Nature*, **366**, 511–12.

Johnson, J.G. and Klapper, G. (1992). North American midcontinental T–R cycles. *Oklahoma Geological Survey Bulletin*, **96**, 567–87.

Johnson, J.H. (1961). *Limestone-building algae and algal limestones*. Johnson Publishing Company, Boulder, Colorodo.

Johnson, K.G., Budd, A.F., and Stemann, T.A. (1995). Extinction selectivity and ecology of Neogene Caribbean reef corals. *Paleobiology*, **21**, 52–73.

Jones, D.S., Williams, D.F., and Spero, H.J. (1986). Life history of symbiont-bearing giant clams from stable isotope profiles. *Science*, **231**, 46–8.

Jones, D.S., Williams, D.F., and Spero, H.J. (1988). More light on photosymbiosis in fossil mollusks: the case of *Mercenaria 'tradacnoides'*. *Palaeogeography, Palaeoclimatology, Palaeoecology*, **64**, 141–52.

Jones, G.P., Sale, P.W., and Ferrell, D.J. (1988). Do large carnivorous fishes affect the ecology of macrofauna in shallow lagoon sediments?: a pilot experiment. *Proceedings of the 5th International Coral Reef Symposium*, Tahiti, French Polynesia, 1985, **2**, 77–82.

Jones, G.P., Ferrell, D.J., and Sale, P.W. (1991). Fish predation and its impact on the invertebrates of coral reefs and adjacent sediments. In *The ecology of fishes on coral reefs* (ed. P.F. Sale), pp. 156–79. Academic Press, London.

Kaiho, K. (1992). A low extinction rate of intermediate-water benthic foraminifera at the Cretaceous/Tertiary boundary. *Marine Micropaleontology*, **18**, 229–59.

Kaljo, D. (1996). Diachronous recovery patterns in Early Silurian corals, graptolites and acritarchs. In *Biotic recovery from mass extinctions* (ed. M.B. Hart), pp. 127–33. Geological Society Special Publications, **102**.

Kaljo, D. and Klaaman, E. (1973). Ordovician and Silurian corals. In *Atlas of palaeobioeography* (ed. A. Hallam), pp. 37–47. Elsevier, Amsterdam.

Kauffman, E.G. (1984). The fabric of Cretaceous marine extinctions. In *Catastrophes and Earth history* (ed. W.A. Berggren and J. van Couvering), pp. 151–246. Princeton University Press, Princeton, NJ.

Kauffman, E.G. and Erwin, D.H. (1995). Surving mass extinctions. *Geotimes*, **40**, 14–17.

Kauffman, E. G. and Fagerstrom, A.L. (1993). The Phanerozoic evolution of reef diversity. In *Species diversity in ecological communities*. (ed. R.E. Ricklefs and D. Schluter), pp. 315–29. University of Chicago Press, Chicago, IL.

Kauffman, E. G. and Johnson, C.C (1988). The morphological and ecological evolution of Middle and Upper Cretaceous reef-building rudistids. *Palaios*, **1**, 389–98.

Kauffman, E. G. and Johnson, C.C. (1992). Bioevent concepts and models: comparative examples from the Caribbean and temperate American Cretaceous. In *Abstracts of the Fifth International Conference on Bioevents* (ed. O.H. Walliser), pp. 63. Gottingen.

Kaufman, L. (1977). The three spot damselfish: effects on benthic biota of Caribbean coral reefs. *Proceedings of the 3rd International Coral Reef Symposium*, (ed. D.L. Taylor), Miami, USA, **1**, 559–64.

Kaufman, L. (1981). There was biological disturbance on Pleistocene reefs. *Paleobiology*, **7**, 527–32.

Kayanne, H., Suzuki, A., and Saito, H. (1995). Diurnal changes in the partial pressure of carbon dioxide in coral reef water. *Science*, **269**, 214–6.

Keller, G., Barrera, E., Schmitz, B., and Mattson, E. (1993). Gradual mass extinction, species survivorship, and long-term environmental changes across the Cretaceous–Tertiary boundary at high latitudes. *Bulletin of the Geological Society of America*, **105**, 979–97.

Kendall, C.G.St.C. and Schlager, W. (1981) Carbonates and relative changes in sea-level. *Marine Geology*, **44**, 181–212.

Kennard, J.M. and James, N.P. (1986). Thrombolites and stromatolites: two distinct types of microbial structure: *Palaios*, **1**, 492–503.

Kenter, J.A.M. (1990). Slope stability on carbonate platform flanks—slope angle and sediment fabric. *Sedimentology*, **37**, 777–94.

Kerans, C. (1985). Petrology of the Devonian and Carboniferous carbonates of the Canning and Bonaparte Basins. *West Australian Mining and Petroleum Research Institute Report*, **12**, 1–203.

Kerans, C., Hurley, N.F., and Playford, P.E. (1986). Marine diagenesis in Devonian reef complexes of the Canning Basin, Western Australia. In *Reef diagenesis* (ed. J.H. Schroeder and B.H. Purser), pp. 357–80. Springer-Verlag, Berlin.

Kier, P. (1974). Evolutionary trends and their functional significance in the post- Paleozoic echinoids. *Journal Paleontology*, **48**, 1–95.

Kinsey, D.W. (1991). Water quality and its effects upon reef ecology. In *Land use patterns and nutrient loading of the Great Barrier Reef Region* (ed. D. Yellowlees), pp. 192–6. James Cook University Press, Australia.

Kinsey, D.W. and Hopley, D. (1991). The significance of coral reefs as global carbon sinks— response to Greenhouse. *Palaeogeography, Palaeoclimatology Palaeoecology*, **89**, 363–77.

Kinzie, R.A. III and Buddemeier, R.W. (1996). Reefs happen. *Global Change Biology*, **2**, 479–94.

Kirkby, K.C. (1994). Growth and reservoir development in Waulsortian mounds: Pekisto Formation, west central Alberta, and Lake Valley Formation, New Mexico. Unpublished Ph.D. thesis. University of Wisconsin, Madison.

Kirkby, K.C. and Hunt, D. (1996). Episodic growth of a Waulsortian buildup: the Lower Carboniferous Muleshoe Mound, Sacramento Mountains, New Mexico, USA. In *Recent advances in Lower Carboniferous geology* (ed. P. Strogen, I.D. Sommerville, and G.L.I. Jones), pp. 97–110. Geological Society Special Publication, **107**.

Kirkland, B.L., Moore, C.H., Jr., and Dickson, J.A.D. (1993). Well-preserved, aragonitic, phylloid algae (*Eugonophyllum*, Udoteacea) from the Pennsylvanian Holder Formation, Sacramento Mountains, New Mexico. *Palaios*, **8**, 111–20.

Kirtley, D.W. (1994). *A review and taxanomic revision of the family Sabellariidae Johnston, 1865 (Annelida; Polychaeta)*. Science series 1, Sabecon Press.

Kirtley, D.W. and Tanner, W.F. (1968). Sabellariid worms: builders of a major reef type. *Journal of Sedimentary Petrology*, **38**, 73–8.

Klein, C., Beukes, N.J., and Schopf, J.W. (1987). Filamentous microfossils in the Early Proterozoic Transvaal Supergroup: their morphology, significance and palaeoenvironmental setting. *Precambrian Research*, **36**, 81–94.

Kleypas, J.A. (1995). A diagnostic model for predicted global coral reef distribution. In *Recent advances in marine science and technology '94.* (ed. O. Bellwood) pp. 211–20. PACON International and James Cook University of N. Queensland, Townsville, Australia.

Kleypas, J.A. (1997). Modeled estimates of global reef habitat and carbonate production since the last glacial maximum. *Paleoceanography*, **12**, 533–45.

Klumpp, D.W., McKinnon, A.D., and Daniel, P. (1987). Damselfish territories: zones of high productivity on coral reefs. *Marine Ecology Progress Series*, **40**, 41–51.

Klumpp, D.W., McKinnon, A.D., and Mundy, C.N. (1988). Motile cryptofauna of a coral reef: abundance, distribution and trophic potential. *Marine Ecology Progress Series*, **45**, 95–108.

Knoll, A.H. (1992). Biological and geochemical preludes to the Ediacaran radiation. In *Origin and early evolution of the Metazoa* (ed. J.H. Lipps and P.W. Signor), pp. 53–84. Plenum Press, New York.

Knoll, A.H. (1996). Daughter of time. *Paleobiology*, **22**, 1–7.

Knoll, A.H. and Walter, M.R. (1992). Latest Proterozoic stratigraphy and Earth history. *Nature*, **356**, 673–8.

Knoll, A.H., Fairchild, I.J., Swett, K. (1993). Calcified microbes in Neoproterozoic carbonates: implications for our understanding on the Proterozoic/Cambrian transition. *Palaios*, **8**, 512–25.

Knoll, A.H., Bambach, R.K., Canfield, D.E., and Grotzinger, J.P. (1996). Comparative Earth history and Late Permian mass extinction. *Science*, **273**, 452–7.

Knowlton, N. and Jackson, J.B.C. (1994). New taxonomy and niche partitioning on coral reefs: jack of all trades or master of some? *Trends in Evolution and Ecology*, **9**, 7–9.

Knowlton, N., Lang, J.C., Rooney, M.C., and Clifford, P. (1981). Evidence for delayed mortality in hurricane-damaged Jamaican staghorn corals. *Nature*, **294**, 251–2.

Knowlton, N., Lang, J.C., and Keller, B.D. (1990). Case study of natural population collapse: post-hurricane predation on Jamaican stag-horn corals. *Smithsonian Contributions to Marine Science*, **31**, 1–25.

Knowlton, N., Weil, E., Weight, L.A., and Guzmán, H.M. (1992). Sibling species in *Montastrea annularis*, coral bleaching and the coral climate record. *Science*, **255**, 330–3.

Kobluk, D.R. (1981*a*). Lower Cambrian cavity-dwelling endolithic (boring) sponges. *Canadian Journal of Earth Sciences*, **18**, 972–80.

Kobluk, D.R. (1981*b*). Cavity-dwelling biota in Middle Ordovician (Chazy) bryozoan mounds from Quebec. *Canadian Journal of Earth Sciences*, **18**, 42–54.

Kobluk, D.R. (1985). Biota reserved within cavities in Cambrian *Epiphyton* Mounds, Upper Shady Dolomite, South-western Virginia. *Journal of Paleontology*, **59**, 1158–72.

Kobluk, D.R. (1988). Cryptic fauna in reefs: ecology and geologic importance. *Palaios*, **3**, 379–90.

Kobluk, D.R. and James, N.P. (1979). Cavity-dwelling organisms in Lower Cambrian patch reefs from southern Labrador. *Lethaia*, **12**, 193–218.

Kobluk, D., James, N.P., Pemberton, S.G. (1978). Initial diversification of macroboring ichnofossils and exploitation of the macroboring niche in the Lower Paleozoic. *Paleobiology*, **4**, 163–70.

Kojis, B.L. and Quinn, N.J. (1985). Puberty in *Goniastrea favulus* age or size related? *Proceedings of the Fifth International Coral Reef Congress*, Tahiti, French Polynesia, **4**, 289–93.

Krumm, D.K. and Jones, D.S. (1993). A new coral–bivalve association (*Actinastrea–Lithophaga*) from the Eocene of Florida. *Journal of Paleontology*, **67**, 945–51.

Kruse, P.D. (1991). Cyanobacterial–archaeocyathan–radiocyathan bioherms in the Wirrealpa Limestone of South Australia. *Canadian Journal of Earth Sciences*, **28**, 601–15.

Kruse, P.D., Zhuravlev, A. Yu., and James, N.P. (1995). Primordial metazoan–calcimicrobial reefs: Tommotian (Early Cambrian) of the Siberian platform. *Palaios*, **10**, 291–321.

Kruse, P.D., Gandin, A., Debrenne, F., and Wood, R.A. (1996). Early Cambrian bioconstructions from the Zavkhan Basin, western Mongolia. *Geological Magazine*, **133**, 429–44.

Kuznetsov, V. (1990). The evolution of reef structures through time: importance of tectonic and biological controls. *Facies*, **22**, 159–68.

Ladd, H.S. (1944). Reefs and other bioherms. National Research Council, Division of Geology and Geography, Annual Report **4**, Appendix K, 26–9.

Lafuste, J., Debrenne, F., Gandin, A., and Gravestock, D.I. (1991). The oldest tabulate coral and associated Archaeocyatha. *Geobios*, **24**, 697–708.

Land, L.S., Lang, J.C., and Smith, B.N. (1975). Preliminary observations on the carbon isotopic composition of some reef coral tissues and symbiotic zooxanthellae. *Limnology and Oceanography*, **20**, 283–7.

Landing. E. (1993). *In situ* earliest Cambrian tube worms and the oldest metazoan-constructed biostome (Placentian Series, southeastern Newfoundland). *Journal of Paleontology*, **67**, 333–42.

Lang, J.C. (1973). Coral reef project—papers in memory of Dr. Thomas F. Goreau II. Interspecific aggression by scleractinian corals. 2. Why the race is not only to the swift. *Bulletin Marine Science*, **23**, 260–79.

Lang, J.C. and Chornesky, E.A. (1990). Competition between scleractinian reef corals—a review of mechanisms and effects. In *Coral reefs* (ed. Z. Dubinsky), pp. 133–208. *Ecosystems of the world*, **25**. Elsevier, New York.

Law, R. and Lewis, D.H. (1983). Biotic environments and the maintenance of sex—some evidence from mutualistic symbioses. *Biological Journal of the Linnean Society*, **20**, 249–76.

Laxton, J.H. (1974). Aspects of the ecology of the coral-eating starfish *Acanthaster planci*. *Biological Journal of the Linnean Society*, **6**, 19–45.

Lecompte, M. (1959). Certain data on the genesis and ecological character of Frasnian reefs of the Ardennes. *International Geological Review*, **1**, 1–23.

Lee, J.J. and Hallock, P. (1986). Algal symbiosis as the driving force in the evolution of large foraminfera. *Annals of the New York Academy of Science*, **503**, 330–47.

Lee, D.-J. and Noble, J.P.A. (1990). Reproduction and life strategies in the Palaeozoic tabulate coral *Paleofavosites capax* (Billings). *Lethaia*, **23**, 257–72.

Leeder, M. R. (1987). Tectonic and palaeogeographic models for Lower Carboniferous Europe. In *European Dinantian environments* (ed. J. Miller, A.E. Adams and V.P. Wright), pp. 1–20. John Wiley, London.

Lees, A. (1975). Possible influences of salinity and temperature on modern shelf carbonate sediments contrasted. *Marine Geology*, **13**, 1767–73.

Lees, A. (1988). Waulsortian 'reefs': the history of a concept. *Memoir l'Institut Geologic de Université Louvain*, **34**, 43–55.

Lees, A. and Miller, J. (1985). Facies variation in Waulsortian buildups. Part 1: A model from Belgium. *Geological Journal*, **20**, 133–58.

Lees, A. and Miller, J. (1995). Waulsortian Banks. In *Carbonate mud mounds*; *their origin and evolution* (ed. C.L.V Monty, D.J.W Bosence, P.H. Bridges, and B.R. Pratt), pp. 191–271. International Association of Sedimentologists, Special Publication, **23**. Blackwell Science, Oxford.

Leinfelder, R. (1994). Distribution of Jurassic reef types: a mirror of structural and environmental changes during the breakup of Pangea. In *Pangea*: *global environments and resources. Canadian Society of Petroleum Geologists*, Memoir **17**, 677–700.

Leinfelder, R., Nose, M., Schmid, D.U., and Werner, W. (1993). Microbial crusts of the Late Jurassic: competition, palaeoecological significance and importance in reef construction. *Facies*, **29**, 195–230.

Leroux, H., Warme, J.E., and Doukham, J.-C. (1995). Shocked quartz in the Alamo breccia, southern Nevada: evidence for a Devonian impact event. *Geology*, **23**, 1003–6.

Levy, M. and Christie-Blick, N. (1991). Tectonic subsidence of the Early Paleozoic passive continental margin in eastern California and southern Nevada. *Bulletin of the Geological Society of America*, **103**, 1590–606.

Lewis, J.B. (1977). Processes of organic production on coral reefs. *Biological Reviews*, **52**, 305–57.

Lewis, J.B. (1982). Estimates of secondary production by reef corals. *Proceedings 4th International Coral Reef Symposium*, Manila, Phillipines, 18–22 May, 1981, **1**, 369–74.

Lewis, S. M. (1985) Herbivory on coral reefs: algal susceptibility to herbivorous fishes. *Oecologia*, **65**, 370–5.

Lewis, S. M. (1986). The role of herbivorous fishes in the organisation of a Caribbean reef community. *Ecological Monographs*, **56**, 183–200.

Liddell, W.D. and Ohlhorst, S.L. (1987). Patterns of reef community structure, North Jamaica. *Bulletin Marine Science*, **40**, 311–29.

Lidgard, S. (1986). Ontogeny in animal colones: a persistent trend in the bryozoan fossil record. *Science*, **232**, 230–2.

Lidgard, S. and Jackson, J.B.C. (1982). How to be an abundant encrusting bryozoan. *Abstracts with Programs of the Geological Society of America*, **14**, 547.

Lidgard, S., McKinney, F.K., and Taylor, P.D. (1993). Competition, clade replacement, and a history of cyclostome and cheilostome bryozoan diversity. *Paleobiology*, **19**, 352–71.

Lindberg, D.R. and Dwyer, K.R. (1983). The topography, formation and mode of home depression of *Collisella scabra* (Gould) (Gastropoda: Acmaeidae). *Veliger*, **25**, 229–34.

Little, C.T.S., Herrington, R.J., Maslennikov, V.V., Morris, N.J., and Zaykov, V.V. (1997). Silurian hydrothermal-vent community from the southern Urals, Russia. *Nature*, **35**, 146–8.

Littler, M.M. and Doty, M.S. (1975). Ecological components structuring the seaward edges of tropical Pacific reefs: the distribution, communities and producivity of *Porolithon Journal of Ecology*, **63**, 117–129.

Littler, M.M. and Littler, D.S. (1984). Models of tropical reef biogenesis. *Progress in Phycological Research*, **3**, 323–64.

Loubens, G. (1980). Biologie de quelques espéces de poissons du lagoon néo-calédonien. *Cahiers Indo-Pacifique*, **2**, 101–253.

Lowenstam, H.A. (1950). Niagran reefs of the Great Lakes area. *Journal of Geology*, **58**, 430–87.

Lowenstam, H.A. (1957). Niagran reefs in the Great Lakes area. *Geological Society of America Memoir*, **67**, 215–48.

Loya, Y. (1976). Settlement, mortality and recruitment of a Red Sea coral population. In *Coelenterate ecology and behaviour* (ed. G.O. Mackie), pp. 89–100. Plenum Press, New York.

Luchinina, V.A. (1985). Vodoroslie postroiki rannego Paleozoya severa Kembriya Sibirskoi Platformy. *Trudy Inst. Geol. Geofisiki Otel*, **628**, 45–9. [In Russian.]

Lyakhin, Yu. I. (1968). Calcium carbonate saturation of Pacific water. *Oceanology*, **8**, 45–53.

MacCracken, M.C., Budyko, M.I., Hecht, A.D., Izrael. Y.A. (ed.) (1990). *Prospects for Future Climate: A Special US/USSR Report on Climate and Climate Change*. Lewis Publications, Chelsea, MI.

McCall, J., Rosen, B.R., and Darrell, J. (1994). Carbonate deposition in accretionary prism settings: Early Miocene coral limestones and corals of the Makhran Mountain Range in southern Iran. *Facies*, **31**, 141–78.

McConnaughey, T. (1989). Biomineralization mechanisms. In *Origin, evolution and modern aspects of biomineralization in plants and animals* (ed. E. Crick), pp. 57–73. Plenum Press, New York.

McGhee, G.R., Jr. (1982). The Frasnian–Famennian extinction event: a preliminary analysis of Appalachian marine ecosystems. *Geological Society of America, Special Paper* **190**, 491–500.

McGhee, G.R., Jr. (1996). *The Late Devonian mass extinction*. Columbia University Press, New York.

McIntyre, A., Ruddiman, W.F., Karling, K., and Mix, A.C. (1989). Surface water response of the equatorial Atlantic Ocean to orbital forcing. *Paleoceanography*, 4, 19–55.

Macintyre, I.G. (1983). Growth, depositional facies and diagenesis of a modern bioherm, Galeta Point, Panama. In *Carbonate buildups—a core workshop* (ed. P.M. Harris), pp. 578–93. Society of Economic Paleontologists and Mineralogists, Core Workshop, **4**. Tulsa, Oklahoma.

Macintyre, I.G. (1984). Extensive submarine lithification in a cave in the Belize barrier reef platform. *Journal of Sedimentary Petrology*, **54**, 221–35.

Macintyre, I.G. (1985). Submarine cements—the peloidal question. In *Carbonate cements* (ed. N. Schneidermann and P.M. Harris), pp. 109–16. *Special Publication of the Society of Economic Paleoetologists and Mineralogists*, **36**.

Macintyre, I.G. (1988). Modern coral reefs of the western Atlantic: new geological perspective. *American Association of Petroleum Geologists Bulletin*, **72**, 1360–9.

Macintyre, I.G. and Glynn, P.W. (1976). Evolution of a modern Caribbean fringing reef, Galeta Point, Panama. *American Association of Petroleum Geologists*, **60**, 1054–72.

Mackenzie, G., Veauvy, C.M., Swart, P.K., Rosen, B.R., and Darrell, J. (1997). Climatic variation on the Early to Middle Eocene using the stable oxygen isotopic composition of coral skeletons.

Geological Society of America Abstracts with Programs, Salt Lake City, Utah, 20–23 October, p. 395.

McKerrow, W.S. (ed.) (1978). *The ecology of fossils*. Duckworth, London.

McKinney, F.K. (1979). Some paleoenvironments of the coiled fenestrate bryozoan *Archimedes*. In *Advances in bryozoology* (ed. G.P. Larwood and M.B. Abbott), pp. 321–36. Academic Press, London.

McKinney, F.K. and Jackson, J.B.C. (1989) *Bryozoan evolution*. Unwin Hyman, Boston.

McKinney, F.K., Webb, F., and McKinney, M.J. (1986). In situ bryozoans in an intertidal–shallow subtidal sedimentary sequence (Middle Ordovician, southwestern Virginia). *Abstracts with Programs Geological Society of America*, **18**, 254

McKinney, F.K., McKinney, M.J., and Listokin, M.R.A. (1987). Erect bryozoans are more than baffling: enhanced sedimentation rate by a living unilaminate branched bryozoan and possible implications for fenestrate bryozoan mudmounds. *Palaios*, **2**, 41–7.

MacLeod, N. and Keller, G. (1994). Comparative biogeographic analysis of planktonic foraminiferal survivorship across the Cretaceous/Tertiary (K/T) boundary. *Paleobiology*, **20**, 143–77.

McMichael, D.F. (1974). Growth rate, population size and mantle coloration in the small giant clam (*Tridacna maxima*) (Röding), at One Tree Island, Capricorn Group, Queensland. *Proceedings of the 2nd Coral Reef Symposium*, (ed. A.M. Cameron, J.S. Jell *et al.*) Aboard M.V. Marco Polo, GBR Australia, 1973, **1**, 241–54.

McNamara, K. (1992). *Stromatolites*. Western Australian Museum.

McRoberts, C.A. (1994). The Triassic–Jurassic ecostratigraphic transition in the Lombardian Alps, Italy. *Palaeogeography, Palaeoclimatology, Palaeoecology*, **110**, 145–66.

Majerus, M., Amos, W., and Hurst, G. (1996). *Evolution. The four billion year war*. Longman, Harlow.

Mamet, B. (1991). Carboniferous calcareous algae. In *Calcareous algae and stromatolites* (ed. R. Riding), pp. 370–471. Springer-Verlag, Berlin.

Manten, A.A. (1971). *Silurian reefs of Gotland*. Elsevier, Amsterdam.

Maragos, J.E. (1978). Coral growth: geometrical relationships. In *Handbook of Coral Reef Research Methods*, Vol. 5 (ed. D.R. Stoddart, and R.E. Johannes), pp. 543–50. UNESCO, Paris.

Margulis, L. (1970). *Origin of eukaryotic cells*. Yale University Press, New Haven, CT.

Marshall, A.T. (1996*a*). Calcification in hermatypic and ahermatypic corals. *Science*, **271**, 637–9.

Marshall, A.T. (1996*b*). Response. *Science*, **274**, 117.

Marshall, C.R. (1994). Confidence intervals on stratigraphic ranges: partial relaxation of the assumptions of randomly distributed fossil horizons. *Paleobiology*, **20**, 459–69.

Marshall, C.R. and Ward, P. D. (1996). Sudden and gradual molluscan extinctions in the latest Cretaceous of western European Tethys. *Science*, **274**, 1360–3.

Martin, J.M., Braga, J.C. and Riding, R. (1997). Late Miocene *Halimeda* alga–microbial segment reefs in the marginal Mediterranean Sorbas Basin, Spain. *Sedimentology*, **44**, 441–56.

Marubini, F. and Spencer Davies, P (1995). An ecotoxicology approach to the study of eutrophication in Caribbean Corals: effects of nitrate. *Biology and geology of coral reefs—Abstract Volume*, *Newcastle*. European Reef Studies Group.

Masse, J.-P. and Philip, J. (1981). Cretaceous coral–rudist buildups of France. In *European fossil reef models* (ed. D.F. Toomey), pp. 399–426. Society of Economic Paleontologists and Mineralogists, Special Publication, **30**.

Mazlov, V.P. (1956). Fossil calcareous algae of the USSR. *Academy of Sciences of the USSR, Geological Sciences Institute, Moscow*, **160**, 1–302. [In Russian.]

Meesters, E.H., Wesseling, I., and Bak, R.P.M. (1996). Partial mortality in three species of reef-building corals and the relation with colony morphology. *Bulletin of Marine Science*, **58**, 838–52.

Meesters, E.H., Wesseling, I., and Bak, R.P.M. (1997). Coral colony tissue damage in six species of reef-building corals: partial mortality in relation to depth and surface area. *Journal of Sea Research*, **37**, 131–44.

Merz, M.V.E. (1992). The biology of carbonate precipitation by cyanobacteria. *Facies*, **26**, 81–102.

Milliman, D.D. (1993). Production and accumulation of calcium carbonate in the ocean: budget of a non-steady state. *Global Biogeochemical Cycles* **7**, 927–57.

Moczydlowska, M. (1991). Arcritarch biostratigraphy of the Lower Cambrian and the Precambrian–Cambrian boundary in southeastern Poland. *Fossils and Strata*, **29**, 1–127.

Moldowan, J.M., Dahl, J., Jacobsen, S.R., Huizinga, B.J., McCaffrey, M.A., and Summons, R.E. (1994). Molecular fossil evidence for Late Proterozoic–Early Paleozoic environments. *Terra Nova*, **6** (abstr. suppl.), 6.

Moldowan, J.M., Dahl, J., Jacobson, S.R., Huizinga, B.J., Fago, F.J., Shetty, R., *et al.* (1996). Chemostratigraphic reconstruction of biofacies: molecular evidence linking cyst-forming dinoflagellates with pre-Triassic ancestors. *Geology*, **24**, 159–62.

Montaggioni, L.F. (1988). Holocene reef growth history in mid-plate high volcanic islands. *Proceedings of the 6th International Coral Reef Symposium*, Townsville, Australia, 1988, **3**, 455–60.

Moore, H.B. (1972). Aspects of stress in the tropical marine environment. *Advances in Marine Biology Letters*, **1**, 69–75.

Moran, D.P. and Reaka, M.L. (1988). Bioerosion of coral rubble and availability of shelter for benthic reef organisms. *Marine Ecology Progress Series*, **44**, 249–63.

Morton, B. (1990). Corals and their bivalve borers—the evolution of a symbiosis. In *The Bivalvia* (ed. B. Morton), pp. 12–46. Proceedings of a Memorial Symposium in Honour of Sir Charles Maurice Youge, Edinburgh, 1986. Hong Kong University Press, Hong Kong.

Mountjoy, E.W. and Riding, R. (1981). Foreslope stromatoporoid–renalcid bioherm with evidence of early cementation, Devonian Ancient Wall reef complex, Rocky Mountains. *Sedimentology*, **28**, 299–319.

Multer, H.G. and Milliman, J.D. (1967) Geologic aspects of sabellarian reefs, southeastern Florida. *Bulletin Marine Science*, **17**, 257–67.

Mundy, D.J.C. (1980). Aspects of the palaeoecology of the Craven Reef Belt (Dinantian) of North Yorkshire. Unpublished Ph.D thesis. University of Manchester.

Mundy, D.J.C. (1994). Microbialite–sponge–bryozoan–coral framestones in Lower Carboniferous (Late Visean) buildups in northern England (UK). In *Pangea: global environments and resources* (ed. A.F. Embry, B. Beauchamp, and D.J. Glass), pp. 713–29. Memoir of the Canadian Society Petroleum Geologists, **17**.

Muscatine, L. (1990). The role of symbiotic algae in carbon and energy flux in reef corals. In *Coral reefs* (ed. Z. Dubinsky), pp. 75–87. Elsevier, Amsterdam.

Narbonne, G.M. and Dixon, O.A. (1984). Upper Silurian lithistid sponge reefs on Somerset Island, Arctic Canada. *Sedimentology*, **31**, 25–50.

Nee, S. and May, R.M. (1992). Dynamics of metapopulations: habitat destruction and competitive coexistence. *Journal of Animal Ecology*, **61**, 37–40.

Nelson, J.S. (1984). *Fishes of the world* (2nd edn). Wiley, New York.

Neumann, A.C. and Macintyre, I.G. (1985). Reef response to sea-level rise: keep-up, catch-up or give-up. *Proceedings of the 5th International Congress on Coral Reefs*, Tahiti, French Polynesia, 1985, **3**, 105–10.

Neumann, A.C., Kofoed, J.W., and Keller, G.H. (1977). Lithoherms in the Straits of Florida. *Geology*, **5**, 4–10.

Newell, N.D., Rigby, J. K., Fischer, A. G., Whiteman, A. J., Hickcox, J. E., and Bradey, J. S. (1953). *The Permian reef complex of the Guadalupe Mountains region, Texas and New Mexico*. W.H. Freeman, San Francisco.

Norris, R.D. (1996). Symbiosis as an evolutionary innovation in the radiation of Paleocene planktic foraminifera. *Paleobiology*, **22**, 461–80.

Norton, J.H., Shepherd, M.A., Long, H.M., and Fitt, W.K. (1992). The zooxanthellal tubular system in the giant clam. *Biological Bulletin of Marine Biology*, **155**, 105–22.

Odum, H.T. and Odum, E.P. (1955). Trophic structure and productivity of a windward coral reef community on Eniwetok Atoll. *Ecological Monographs*, **25**, 291–320.

Ohno, T., Katoh, T., and Yamasu, T. (1995). The origin of algal–bivalve symbiosis. *Palaeontology*, **38**, 1–21.

Opdyke, B.N. and Walker, J.C.G. (1992). Return of the coral reef hypothesis: basin to shelf partitioning of CaCO$_3$ and its effect on atmospheric CO$_2$. *Geology*, **20**, 733–6.

Opdyke, B.N. and Wilkinson, B.H. (1993). Carbonate mineral saturation state and cratonic limestone accumulation. *American Journal of Science*, **293**, 217–34.

Orr, A.P. (1933). Variations in some physical and chemical conditions in the sea in the neighbourhood of the Great Barrier Reef. *Scientific Reports of the Great Barrier Reef Expedition 1928–1929* **2**, 37–86.

Paine, R.T. (1966). Food web complexes and species diversity. *American Naturalist*, **100**, 65–75.

Paine, R.T. (1984). Ecological determinants in the competition for space. *The First MacArthur Lecture. Ecology*, **65**, 1339–48.

Palmer, T.J.P., Hudson, J.D., and Wilson, M.A. (1988). Palaeoecological evidence for early aragonite dissolution in ancient calcite seas. *Nature*, **335**, 809–10.

Pandolfi, J. (1996). Limited membership in Pleistocene reef coral assemblages from the Huon Peninsula, Papua New Guinea: constancy during global change. *Paleobiology*, **22**, 152–76.

Pantić, N. and Sladic-Tritunovit, M (1984). Mesozoic floral provinces of Tethys ocean and its margin with respect ot plate tectonics. 27th International Geological Congress (Ed. Bogdanov, N.A.), **27** (1),295.

Papp, A., Zapfe, H., Bachmayer, F., and Tauber, A.F. (1947). Lebensspuren mariner Krebse. *K Academie Wissenschaft Wien, Mathematische Naturwissenschaft Klass, Sitz-Ber*, **156**, 281–317.

Paris, F., Giraud, C., Feist, R., and Winchester-Seeto, T. (1996). Chitinozoan bio-event in the Frasnian–Famennian boundary beds at La Serre (Montagne Noire, southern France). *Palaegeography, Palaeoclimatology, Palaeoecology*, **121**, 131–45.

Pascal, A. (1985). Les Systems Biosedimentaires. Urgoniens (Aptien–Albien) sur la marge nord ibérique. *Université Dijon Mémoire Geologique*, **10**, 1–569.

Pearse, V.B. and Muscatine, L. (1971). Role of symbiotic algae (zooxanthellae) in coral calcification. *Biological Bulletin*, **141**, 350–63.

Pedder, A.E.H. (1982). The rugose coral record across the Frasnian–Famennian boundary. *Geological Society of America Special Paper*, **190**, 485–90.

Pentecost, A. and Riding, R. (1986). Calcification in cyanobacteria. In *Biomineralization in lower plants and animals* (ed. B.S.C. Leadbetter and R. Riding), pp. 73–90. Clarendon Press, Oxford.

Phillips, J. (1841). Figures and descriptions of the Palaeozoic fossils of Cornwall, Devon and West Somerset. Longman, Brown, Green and Longman, London.

Pickard, G.L. (1986). Effects of wind and tide on upper layer currents at Davies Reef, during MECOR (July–August 1984). *Australian Journal of Marine and Freshwater Research*, **37**, 545–65.

Pickard, N.A.H. (1996). Evidence for microbial influence on the development of Lower Carboniferous buildups. In *Recent advances in lower Carboniferous geology* (ed. P. Strogen, I.D. Sommerville, and G.L.I. Jones), pp. 65–82. Geological Society Special Publication, **107**.

Playford, P.E. (1980). Devonian 'Great Barrier Reef' of Canning Basin, Western Australia. *American Association of Petroleum Geologists Bulletin*, **6**, 814–40.

Playford, P.E. (1981). Devonian reef complexes of the Canning Basin, Western Australia. *Geological Society of Australia, Fifth Australian Geological Convention Field Excursion Guidebook*, 1–64.

Playford, P.E. (1984). Platform-margin and marginal-slope relationships in Devonian reef complexes of the Canning Basin. In *The Canning Basin, Western Australia* (ed. P.G. Purcell), pp. 189–214. Proceedings: Geological Society of Australia and Petroleum Exploration Society of Australia, Canning Basin Symposium, Perth.

Playford, P.E. and Lowry, D.C. (1966). Devonian reef complexes of the Canning Basin, Western Australia. *Geological Survey of Western Australia, Bulletin*, **118**, 1–150.

Playford, P.E., Hurley, N.F., Kerans, C and Middleton, M.F. (1989). Reefal platform development, Devonian of the Canning Basin, Western Australia. In *Controls on carbonate platform and basin development* (ed. P.D. Crevello *et al.*), pp. 187–202. Society of Economic Paleontologists and Mineralogists Special Publication, **44**.

Pleydell, S.M. and Jones, B. (1988). Boring of various faunal elements in the Oligocene–Miocene Bluff Formation of Grand Cayman, British West Indies. *Journal of Paleontology*, **62**, 348–67.

Polunin, N.V.C. (1988). Efficient uptake of algal production by a single resident herbivorous fish on a reef. *Journal Experimental Marine Biology and Ecology*, **123**, 61–76.

Potts, D.C. (1976). Growth interactions amoung morphological variants of the coral *Acropora palifera*. In *Coelenterate ecology and behaviour* (ed. G.O. Mackie), pp. 79–88. Plenum, New York.

Potts, D.C. (1983). Evolutionary disequilibrium among Indo-Pacific corals. *Bulletin of Marine Science*, **33**, 619–32.

Potts, D.C. (1984). Generation times and the Quaternary evolution of reef-building corals. *Paleobiology*, **10**, 48–58.

Potts, D.C. and Garthwaite, R.L. (1991). Evolution of reef-building corals during periods of rapid global change. In *The unity of evolutionary biology* (ed. E.C. Dudley), pp. 170–7. Dioscorides Press, Portland, Oreg.

Potts, D.C. and Swart, P.K. (1984). Water temperature as an indicator of environmental variability on a coral reef. *Limnology and Oceanography*, **29**, 504–16.

Poty, E. (1996). The Strunian Rugosa recovery after the Late Frasnian crisis in Belgium and surrounding areas. 6th North American Paleontological Convention Abstracts Volume. *Special Publication of the Paleontological Society*, Washington, DC, USA, 1996, **8**, 310.

Powell, J.H. (1980). Palaeoecology and taxonomy of some Wenlock tabulate corals and stromatoporoids. Unpublished Ph.D thesis, University of Newcastle.

Pratt, B.R. (1982). Stromatolite decline—a reconsideration. *Geology*, **10**, 512–15.

Pratt, B.R. (1984). *Epiphyton* and *Renalcis*—diagenetic microfossils from calcification of cocoid blue-green algae. *Journal of Sedimentology Petrology*, **54**, 948–71.

Pratt, B.R. (1989*a*). Small early Middle Ordovician patch reefs, Laval Formation (Chazy Group), Caughnawaga, Montreal area, Quebec. In *Reefs. Canada and adjacent area.* (ed. H.H.J. Geldsetzer, N.P. James, and G.E. Tebbutt), pp. 218–23. Canadian Society of Petroleum Geologists, **13**.

Pratt, B.R. (1989*b*). Lower Devonian stromatoporoid reefs, Formosa Reef Limestone (Detroit River Group) of southwestern Ontario. In *Reefs. Canada and adjacent area.* (ed. H.H.J. Geldsetzer, N.P. James, and G.E. Tebbutt), pp. 506–9. Canadian Society of Petroleum Geologists, **13**.

Pratt, B.R. (1990). Lower Cambrian reefs of the Mural Formation, Southern Canadian Rocky Mountains. *13th International Sedimentological Congress*, Nottingham, p. 436.

Pratt, B.R (1991). Lower Cambrian reefs of the Mural Formation, southern Canadian Rocky Mountains (abstract). *Geological Association of Canada, Annual Meeting, Programs and Abstracts*, **16**, 102.

Pratt, B.R. (1995). The origin, biota and evolution of deep-water mud-mounds. In *Carbonate mud mounds; their origin and evolution* (ed. C.L.V. Monty, D.W.J. Bosence, P.H. Bridges, and B.R. Pratt), pp. 49–123. International Association of Sedimentologists, Special Publication, **23**. Blackwell Science, Oxford.

Pratt, B. R. and James, N.P. (1982). Cryptalgal–metazoan bioherms of Early Ordovician age in the St George Group, western Newfoundland. *Sedimentology*, **29**, 543–69.

Qi, W. (1984). An Anisian coral fauna from Guizhou, South China. *Paleontographica Americana*, **54**, 187–90.

Quinn, W.H. (1992). A Study of Southern Oscillation-related climatic activity for AD622–1900 incorporating Nile River flood data. In *El Niño: historical and palaeoclimatic aspects of the Southern Oscillation* (ed. H. Diaz and V. Markgraf), pp. 119–49.

Rasmussen, C. (1988). The use of strontium as an indicator of anthropogenically-altered environmental parameters. *Proceedings of the 6th International Coral Reef Symposium, Townsville, Australia, 1988*, **2**, 325–39.

Rasmussen, K.A. and Brett, C.E. (1985). Taphonomy of Holocene cryptic biotas from St. Croix, Virgin Islands: information loss and preservational biases. *Geology*, **13**, 551–3.

Raup, D.M. (1991). *Extinction: bad luck or bad genes?* W.W. Norton, New York.

Raup, D.M. and Boyajian, G.E. (1988). Patterns of generic extinction in the fossil record. *Paleobiology*, **14**, 109–25.

Raup, D.M. and Jablonski, D. (1993). Geography of End-Cretaceous marine bivalve extinction. *Science*, **260**, 971–3.

Rees, T.A.V. and Ellard, F.M. (1989). Nitrogen conservation and the green *Hydra* symbiosis. *Proceedings of the Royal Society of London*, **B236**, 203–12.

Reid, R.P. (1987). Non-skeletal peloidal precipitates in Upper Triassic reefs, Yukon Territory (Canada). *Journal of Sedimentary Petrology*, **57**, 893–900.

Reid, R.P., Macintyre, I.G., Browne, K.M., Steneck, R.S., and Miller, T. (1995). Modern marine stromatolites in the Exuma Cays, Bahamas: uncommonly common. *Facies*, **33**, 1–18.

Reinhardt, J. (1988). Uppermost Permian reefs and Permo-Triassic sedimentary facies from the southeastern margin of the Sichuan Basin, China. *Facies*, **18**, 231–86.

Reinthal, P.N., Kensly, B., and Lewis, S.M. (1984). Dietary shifts in the queen triggerfish *Balistes vetula*, in the absence of its primary food item, *Diadema antillarum*. PSZNI *Marine Ecology*, **5**, 191–5.

Reitner, J. (1992). 'Coralline Spongien' der versuch einer phylogenetisch-taxonomischen analyse. *Berliner Geowissenschaft Abhandlungen*, **(E)1**, 1–352.

Reitner, J. (1993). Modern cryptic microbialite/metazoan facies from Lizard Island (Great Barrier Reef, Australia), formation and concepts. *Facies*, **29**, 3–40.

Reitner, J. and Keupp, H. (1991). The fossil record of the haplosclerid excavating sponge *Aka* de Laubenfels. In *Fossil and Recent sponges* (ed. J. Reitner and H. Keupp), pp. 102–120. Springer, Berlin.

Resing, J.M. and Ayre, D.J. (1985). *Proceedings of the 5th International Coral Reef Symposium, Tahiti, French Polynesia, 1985*, **6**, 75–81.

Retallack, G.J. (1995). Permian–Triassic crisis on land. *Science*, **267**, 77–80.

Retallack, G.J. (1996). Paleoenvironmental change across the Permian–Triassic boundary on land in southeastern Australia and Antarctica. *International Geological Congress Symposium, Beijing, Abstract Volume*, 109.

Richardson, C.A., Dustan, P., and Lang, J. (1979). Maintenance of living space by sweeper tentacles of *Monastrea cavernosa*, a Caribbean coral. *Marine Biology*, **55**, 181–6.

Richmond, R.H. (1993). Coral reefs: present problems and future concerns resulting from anthropogenic disturbance. *American Zoologist*, **33**, 524–36.

Riding, R. (1993). *Shamovella obscura*: the correct name for *Tubiphytes obscurus* (Fossil). *Taxon*, **42**, 71–3.

Riding, R. and Guo, L. (1992). Affinity of *Tubiphytes*. *Palaeontology*, **35**, 37–49.

Riding, R. and Zhuravlev, A.Yu. (1995). Structure and diversity of oldest sponge–microbe reefs: Lower Cambrian, Aldan River, Siberia. *Geology*, **23**, 649–52.

Rigby, J.K. and Senowbari-Daryan, B. (1995). Permian sponge biogeography and biostratigraphy. In *The Permian of northern Pangaea*, Vol. 1 (ed. P.A. Scholle, T.M. Peryt, and D.S. Ulmer-Scholle), pp. 153–66. Springer-Verlag, Berlin.

Rinkevich, B. (1996). Do reproduction and regeneration in damaged corals compete for energy allocation? *Marine Ecology Progress Series*, **143**, 297–302.

Rinkevitch, B. and Loya, Y. (1983). Intraspecific competitive networks in the Red Sea Coral *Stylophora pistillata*. *Coral Reefs*, **1**, 161–72.

Roberts, H.H. and Suhayda, J.N. (1983). Wave–current interactions on a shallow reef (Nicaragua, Central America). *Coral Reefs*, **1**, 209–14.

Roberts, H.H., Murray, S.P., and Suhayda, J.N. (1977). Physical processes on a fore-reef shelf environment. *Proceedings of the 3rd International Coral Reef Symposium*, (ed. D.L. Taylor), Miami, USA, 1977, **2**, 507–16.

Roberts, H.H., Phipps, C.V., and Effendi, L. (1987). *Halimeda* bioherms of the eastern Java sea, Indonesia. *Geology*, **15**, 371–4.

Robertson, D.R. (1982). Fish faeces as fish food on a Pacific coral reef. *Marine Ecology Progress Series*, **7**, 253–65.

Rohling, E.H., Zachariasse, W.J., and Brinkhuis, H. (1991). A terrestrial scenario for the Cretaceous–Tertiary boundary collapse of the marine pelagic ecosystem. *Terra Nova*, **3**, 41–8.

Romanek, C.S., Jones, D.S., Williams, D.F., Krantz, D.E., and Radtke, R. (1987). Stable isotope investigation of physiological and environmental changes recorded in shell carbonate from the giant clam *Tridacna maxima*. *Marine Biology*, **94**, 385–93.

Romano, S.L. and Palumbi, S.R. (1996). Evolution of scleractinian corals inferred from molecular systematics. *Science*, **271**, 640–2.

Roniewicz, E. and Morycowa, E. (1993). Evolution of the Scleractinina in the light of microstructural data. *Proceedings of the 6th International Symposium on Fossil Cnidaria and Porifera*, Münster, Germany (ed. P. Oekentorp-Küster), pp. 233–40. Courier Forschungsinstitut Senckenberg, **164**.

Rosen, B.R. (1981). The tropical high diversity enigma—the coral's eye view. In *The evolving biosphere* (ed. P.L. Forey), pp. 103–29. British Museum (Natural History) and Cambridge University Press, London.

Rosen, B.R. (1986). Modular growth and form of corals: a matter of metamers? *Philosophical Transactions of the Royal Society London*, **B313**, 115–42.

Rosen, B.R. (1988). Progress, problems and patterns in the biogeography of reef corals and other tropical marine organisms. *Helgoländer Wissenschaftliche Meereuntersuchungen*, **42**, 269–301.

Rosen, B.R. (1998). Corals, reefs, algal symbiosis and global change: the Lazarus factor. In *Biotic response to global change: the last 145 million years* (ed. S.J. Culver and P.F. Rawson). Chapman and Hall, London. (In press.)

Rosen, B.R. and Turnšek, D. (1989). Extinction patterns and biogeography of scleractinian corals across the Cretaceous/Tertiary boundary. In *Fossil Cnidaria* 5 (ed. P.A. Jell and J.W. Pickett), pp. 355–70. Memoirs of the Association of Australasian Palaeontologists, **8**.

Ross, C.A. (1974). Evolutionary and ecological significance of large, calcareous Foraminiferida (Protozoa), Great Barrier Reef. *Proceedings of the 2nd International Coral Reef Symposium*, Aboard M.V. Marco Polo GBR Australia, **1**, 327–33.

Ross, C.A. (1977). Calcium carbonate fixation by large reef-dwelling foraminifera. *American Association of Petroleum Geologists, Studies in Geology*, **4**, 219–30

Ross, C.A. (1979). Evolution of the Fusulinacea (Protozoa) in Late Palaeozoic space and time. In *Historical biogeography, plate tectonics, and the changing environment* (ed. J. Gray and A.J. Boucot), pp. 215–26. Oregon State University Press, Corvallis, Oregon.

Rowan, R. and Knowlton, N. (1995). Interspecific diversity and ecological zonation in coral–algal symbiosis. *Proceedings National Academy of Science USA*, **92**, 2850–3.

Rowan, R. and Powers, D.A. (1991). A molecular genetic classification of zooxanthellae and the evolution of animal–algal symbioses. *Science*, **251**, 1348–51.

Rowan, R., Knowlton, N., Baker, A., and Jara, J. (1997). Landscape ecology of algal symbionts creates variation in episodes of coral bleaching. *Nature*, **288**, 265–9.

Roy, K., Valentine, J.W., Jablonski, D., and Kidwell, S.M. (1996). Scales of climatic variability and time averaging in Pleistocene biotas: implications for ecology and evolution. *Trends in Ecology and Evolution*, **11**, 458–63.

Runnegar, B., Pojeta, J., Jr., Taylor, M.E., and Collins, D. (1979). New species of the Cambrian and Ordovician chitons *Mattevia* and *Chelodes* from Wisconsin and Queensland: evidence for the early history of polyplacophoran molluscs. *Journal of Paleontology*, **53**, 1374–94.

Ryan, B.F., Watterson, I.G., and Evans, J.L. (1992). Tropical cyclone frequencies inferred from Gray's yearly genesis parameter: validation of GCM tropical climates. *Geophysics Research Letters*, **19**, 1831–4.

Rylaarsdam, K.W. (1983). Life histories and abundance patterns of colonial corals on Jamaican reefs. *Marine Ecology Progress Series*, **13**, 249–60.

Saffo, M.B. (1987). New light on seaweeds. *Bioscience*, **37**, 654–64.

Sale, P.W. (1980). The ecology of fishes on coral reefs. *Oceanographic Marine Biology*, **18**, 367–421.

Saller, A.H. (1986). Radiaxial calcite in Lower Miocene strata, subsurface Enewetak Atoll. *Journal of Sedimentary Geology*, **56**, 743–62.

Sammarco, P.W. and Carleton, J.H. (1982). Damselfish territoriality and coral community structure: reduced grazing, coral recruitment, and effects on coral spat. *Proceedings of the 4th International Coral Reef Symposium*, Manila, Phillipines, 1981, **2**, 525–35.

Sammarco, P.W., Levinton, J.S., and Ogden, J.C. (1974). Grazing and control of coral reef community structure by *Diadema antillarum*. (Echinoidermata: Echinoidea): a preliminary study. *Journal of Marine Research*, **32**, 47–53.

Sandberg, C.A., Ziegler, W., Dreeson, R., and Butler, J.L. (1988). Part 3: Late Frasnian mass extinction: condodont event stratigraphy, global changes, and possible causes. *Courier Forschungsinstitut Senckenberg*, **102**, 263–307.

Sandberg, P.A. (1985*a*). Aragonite cements and their occurrence in ancient limestones. In *Carbonate cements* (ed. N. Schneidermann and P.M. Harris), pp. 33–57. Special Publication of the Society of Economic Paleontologists and Mineralogists, **36**.

Sandberg, P.A. (1985*b*). An oscillating trend in Phanerozoic non-skeletal carbonate mineralogy. *Nature*, **305**, 19–22.

Sando, W. (1989). Dynamics of Carboniferous coral distribution, western Interior USA. *Memoir of the Association of Australasian Palaeontologists*, **8**, 251–65.

Sano, H. and Kanmera, K. (1996). Microbial controls on Panthalassa Carboniferous–Permian oceanic buildups, Japan. *Facies*, **34**, 239–56.

Satterley, A. K. (1994). Sedimentology of the Upper Triassic reef complex at the Hochkönig Massif (Northern Calcareous Alps, Austria). *Facies*, **30**, 119–50.

Satterley, A. K. (1996). Cyclic carbonate sedimentation in the Upper Triassic Dachstein Limestone, Austria: the role of patterns of sediment supply and tectonics in a platform–reef–basin system. *Journal of Sedimentary Research*, **66**, 307–23.

Satterley, A., Marshall, J.D., and Fairchild, I.J. (1994). Diagenesis of an Upper Triassic reef complex, Wilde Kirche, Northern Calcareous Alps, Austria. *Sedimentology*, **41**, 935-50.

Savarese, M., Mount, J.F., Sorauf, J.E., and Bucklin, L. (1993). Paleobiologic and paleoenvironmental context of coral-bearing Early Cambrian reefs: implications for Phanerozoic reef development. *Geology*, **21**, 917–20.

Savazzi, E. (1982). Commensalism between boring mytilid bivalves and a soft bottom coral in the Upper Eocene of Northern Italy. *Paläontologisches Zeitschrift*, **56**, 165–75.

SBS (1995). *The SBS World* Guide (4th edn). Reef Reference Australia, Melbourne.

Schlager, W. (1981). The paradox of drowned reefs and carbonate platforms. *Bulletin of the Geological Society of America*, **92**, 197–211.

Schlanger, S.O. and Konishi, K. (1975). The geographic boundary between the coral–algal and the bryozoan–algal limestone fabrics—a paleolatitude indicator. *International Sedimentological Congress*, 9th, Nice, 187–91.

Schlichter, D. (1991). A perforated gastrovasular cavity in *Leptoseris fragilis*. A new improved strategy to improve heterotrophic nutrition in corals. *Naturwissenschaftern*, **78**, 467–9.

Schubert, J. K. and Bottjer, D.J. (1992). Early Triassic stromatolites as post-extinction disaster forms. *Geology*, **20**, 883–6.

Schubert, J. K. and Bottjer, D.J. (1995). Aftermath of the Permian–Triassic mass extinction event: paleoecology of Lower Triassic carbonates in the western USA. *Palaeogeography, Palaeoclimatology, Palaeoecology*, **116**, 1–40.

Schuhmacher, H. (1984). Reef-building properties of *Tubastraea micranthus* (Scleractinia; Dendrophyllidae), a coral without zooxanthellate. *Marine Ecology Progress Series*, **20**, 93–9.

Schumann, D. (1995). Upper Cretaceous rudist and stromatoporoid associations of Central Oman (Arabian Peninsula). *Facies*, **32**, 189–202.

Scoffin, T.P. (1971). The conditions of growth of the Wenlock reefs of Shropshire (England). *Sedimentology*, **17**, 173–219.

Scoffin, T.P. (1972*a*). The fossilization of Bermuda patch reefs. *Science*, **178**, 1280–2.

Scoffin, T.P. (1972*b*). Cavities in the reefs of the Wenlock Limestone (Mid-Silurian) of Shropshire, England. *Geologisches Rundschau*, **61**, 565–78.

Scoffin, T.P. (1992). Taphonomy of coral reefs: a review. *Coral Reefs*, **11**, 57–77.

Scoffin, T.P. (1993). The geological effects of hurricanes on coral reefs and the interpretation of storm deposits. *Coral Reefs*, **12**, 203–21.

Scoffin, T.P., Stearn, C.W., Boucher, D., Frydl, P., Hawkins, C.M., Hunter, I.G., *et al.* (1980). Calcium carbonate budget of a fringing reef on the west coast of Barbados. Pt. II. Erosion, sediments and internal structure. *Bulletin Marine Science*, **32**, 457–508.

Scott, R.W. (1988). Evolution of Late Jurassic and Early Cretaceous reef biotas. *Palaios*, **3**, 184–93.

Scrutton, C.T. (1965). Periodicity in Devonian coral growth. *Palaeontology*, **7**, 552–8.

Scrutton, C.T. (1984). Origin and early evolution of tabulate corals. *Paleontographica America*, **54**, 110–18.

Scrutton, C.T. (1988). Patterns of extinction and survival in Palaeozoic corals. In *Extinction and survival in the fossil record* (ed. G.P. Larwood), pp. 65–88. Clarendon Press, Oxford.

Scrutton, C.T. (1997*a*). The Palaeozoic corals, 1: origins and relationships. *Proceedings of the Yorkshire Geological Society*, **51**, 177–208.

Scrutton, C.T. (1997*b*). Growth strategies and colonial form in tabulate corals. *Boletín de la Real Sociedad Española de Historia Natural, Sección Geológica*, **91**, 179–91.

Scrutton, C.T. (1998). The Palaeozoic corals, II: Structure, variation and palaeoecology. *Proceedings of the Yorkshire Geological Society*, **52**, 1–57.

Scrutton, C.T. and Powell, J.H. (1981). Periodic development of dimetrism in some favositid corals. *Acta Palaeontologica Polonica*, **25**, 493–6.

Senowbari-Daryan, B. and Flügel, E. (1994). *Tubiphytes* Maslov, an enigmatic fossil: classification, fossil record and significance through time. Part 1: discussion of Late Palaeozoic material. *Boletin Societa Paleontologia Italica, Special volume*, **1**, 353–82.

Senowbari-Daryan, B. and Rigby, J.K. (1996). Brachiopod mounds not sponge reefs, Permian Capitan–Tansill Formations, Guadalupe Mountains, New Mexico. *Journal of Paleontology*, **70**, 697–701.

Senowbari-Daryan, B., Zühlke, R., Bechstädt, T., and Flügel, E. (1993). Anisian (Middle Triassic) buildups of the Northern Dolomites (Italy): the recovery of reef communities after the Permian/Triassic crisis. *Facies*, **28**, 181–256.

Sepkoski, J.J., Jr. (1981). A factor analytic description of the Phanerozoic marine record. *Paleobiology*, **7**, 36–53.

Sepkoski, J.J., Jr. (1986). Phanerozoic overview of mass extinctions. In *Patterns and processes in the history of life* (ed. D.M. Raup and D. Jablonski), pp. 277–95. Springer-Verlag, Berlin.

Sepkoski, J.J., Jr. (1993). Ten years in the library: new paleontological data confirm evolutionary patterns. *Paleobiology*, **19**, 43–51.

Sepkoski, J.J., Jr. (1995). Patterns of Phanerozoic extinction: a perspective from global data bases. In *Global events and event stratigraphy* (ed. O.H. Walliser), pp. 35–51. Springer-Verlag, Berlin.

Sheehan, P.M. (1985). Reefs are not so different—they follow the evolutionary pattern of the level bottom communities. *Geology*, **13**, 46–9.

Sheehan, P.W. and Hanson, T.A. (1986). Detritus feeding as a buffer to extinction at the end of the Cretaceous. *Geology*, **14**, 868–70.

Sheehan, P.W., Fastovsky, D.E., Hoffman, R.G., Berghaus, C.B., and Gabriel, D.L. (1991). Sudden extinction of the dinosaurs: latest Cretaceous, upper Great Plains, U.S.A. *Science*, **254**, 835–9.

Sheppard, C.R.C. (1985). Unoccupied substrate in the central Great Barrier Reef: role of coral interactions. *Marine Ecology Progress Series*, **25**, 259–68.

Shields, W.M. (1982). *Philopatry, interbreeding and the evolution of sex*. Albany State University, New York Press, New York.

Shinkarenko, L. (1982). The natural history of five species of octocorals (Alcyonacea), with special reference to reproduction at Heron Island, Great Barrier Reef. Ph.D Thesis, University of Brisbane.

Signor, P. W., III and Brett, C.E. (1984). The Mid-Paleozoic precursor to the Mesozoic Marine Revolution. *Paleobiology*, **10**, 229–45.

Signor, P.W., III and Lipps, J.H. (1982). Sampling bias, gradual extinction patterns, and catastrophes in the fossil record. In *Geological implications of large asteroids and comets on the Earth* (ed. L.T. Silver and P.H. Shulz), pp. 291–6. Geological Society of America, Special Paper, **190**.

Simpson, T. L. (1973), Coloniality among the Porifera. In *Animal colonies, developments and function through time* (ed. R.S. Boardman, A.H. Cheetham, and W.A. Oliver), pp. 549–65. Dowden, Hutchinson and Ross, Stroudsburg, PA.

Skelton, P.W. (1979). Preserved ligament in a radiolitid rudist bivalve and its implications for mantle marginal feeding in the group. *Paleobiology*, **5**, 90–106.

Skelton, P.W. (1985). Preadaptations and evolutionary innovations in rudist bivalves. *Special Papers in Palaeontology*, **33**, 159–73.

Skelton, P. W. (1991). Morphogenetic versus environmental cues for adaptive radiations. In *Constructional morphology and evolution* (ed. N. Schmidt-Kittler and K. Voegel), pp. 375–88. Springer-Verlag, Berlin.

Skelton, P.W. and Gili, E. (1991). Palaeoecological classification of rudist morphotypes. *1st International Conference on Rudists, Beograd, 1888. Serbian Geological Society Special Publication*, **2**, 71–86. [Issued as reprint from unpublished volume.]

Skelton, P.W. and Wright, V.P. (1987). A Caribbean rudist in Oman: Island hopping across the Pacific in the Late Cretaceous: *Palaeontology*, **30**, 375–88.

Smith, A.B. (1984). *Echinoid palaebiology*. Allen and Unwin, London.

Smith, A. B. and Jeffrey, C. H. (1997). Selectivity of extinction among sea urchins at the end of the Cretaceous. *Nature*, **392**, 69–71.

Smith, D.B. (1981). Bryozoan–algal patch reefs in the Upper Permian Lower Magnesium Limestone of Yorkshire, Northeast England. *Society of Economic Paleontologists and Mineralogists Special Publication*, **30**, 187–202.

Soja, C. M. (1994). Significance of Silurian stromatolite–sphinctozoan reefs. *Geology*, **22**, 355–8.

Soja, C.M. and Antoshkina, A.I. (1997). Coeval development of Silurian stromatolite reefs in Alaska and the Ural Mountains: implications for paleogeography of the Alexander terrane. *Geology*, **25**, 539–42.

Somerville, I.D., Pickard, N.A.H., Strogen, P., and Jones, G.L. (1992). Early to Mid Viséan shallow water platform buildups, north Co. Dublin, Ireland. *Geological Journal*, **27**, 151–72.

Sorauf, J.E. and Pedder, A.E.H. (1986). Late Devonian rugose corals and the Frasnian–Famennian crisis. *Canadian Journal of Earth Sciences*, **23**, 1265–87.

Sorauf, J.E. and Savarese, M. (1995). A Lower Cambrian coral from South Australia. *Palaeontology*, **38**, 757–70.

Sorbini, L. (1983). Littiofauna fossile di Bolca e le sue relazioni biogeografiche con I pesci attuali: Vicarianza o dispersione? *Bulletino Societa Paleontologica Italiano*, **22**, 109–18.

Speijer, R.P. (1994). Extinction and recovery patterns in benthic foraminiferal paleocommunities across the Cretaceous/Paleogene and Paleocene/Eocene boundaries. *Mededelingen van de Faculteit Aardwetenschappen Universiteit Utrecht*, **124**, 1–191.

Spincer, B. R. (1996). Paleoecology of some Upper Cambrian microbial–sponge–eocrinoid reefs, central Texas. *Paleontological Society Special Publication*, (ed. J.E. Repetski), Washington D.C., USA, **8**, 367.

Spincer, B.R. (1998). Palaeoecology of some Upper Cambrian reefs from central Texas, the Great Basin, and Colorado, USA. Unpublished Ph.D thesis. University of Cambridge.

Spjeldnaes, N. (1974). Silurian bryozoans which grew in the shade. *Documents de la Laboratoire de Geologie, Faculté de Science, Lyon. H.S.*, **3**, 415–24.

Springer, V.G. (1982). Pacific plate biogeography, with special reference to shore-fishes. *Smithsonian Contributions to Zoology*, **367**, 1–82.

Stanley, G.D. (1981). Early history of scleractinian corals and its geological consequences. *Geology*, **9**, 507–11.

Stanley, G.D. (1988). The history of Early Mesoozic reef communities: a three-step process. *Palaios*, **3**, 170–83.

Stanley, G.D. and Beauvais, L. (1994). Corals from an Early Jurassic coral reef in British Columbia: refuge on an oceanic island reef. *Lethaia*, **27**, 35–47.

Stanley, G.D. and Swart, P.W. (1995). Evolution of the coral–zooxanthellae symbiosis during the Triassic: a geochemical approach. *Paleobiology*, **21**, 179–99.

Stanley, S. M. (1977). Trends, rates and patterns of evolution in Bivalvia. In *Patterns of evolution* (ed. A. Hallam), pp. 209–50. Elsevier, Amsterdam.

Stanley, S. M. (1988). Climatic cooling and mass extinction of Paleozoic reef communities. *Palaios*, **3**, 228–32.

Stanley, S.M. and Yang, X. (1994). A double mass extinction at the end of the Paleozoic era. *Science*, **266**, 1340–4.

Stanton, R.J. and Flügel, E. (1989). Problems with reef models: Late Triassic Steinplatte 'reef' (Northern Alps, Salzburg/Tyrol, Austria). *Facies*, **20**, 1–138.

Stasek, C.R. (1962). The form, growth, and evolution of the Tridacnidae (giant clams). *Archives de Zoologie Expérimental et Générale*, **101**, 1–40.

Stearn, C.W. (1966). The microstructure of stromatoporoids. *Palaeontology*, **9**, 75–124.

Stearn, C.W. (1987). Effect of the Frasnian–Famennian extinction event on the stromatoporoids. *Geology*, **15**, 677–9.

Stearn, C.W., Scoffin, T.P., and Martindale, W. (1977). Calcium carbonate budget of a fringing reef on the west coast of Barbados. Pt. I. Zonation and productivity. *Bulletin Marine Science*, **27**, 479–510.

Stearn, C.W., Halim-Dihardja, M.K., and Nishida, D.K. (1987). An oil-producing stromatoporoid patch reef in the Famennian (Devonian) Wabamun Formation, Normandville Field, Alberta. *Palaios*, **2**, 560–70.

Stel, J.H. (1978). *Studies on the paleobiology of favositids*. Stabol/All-Round, Groningen.

Steneck, R.S. (1982*a*). A limpet-coralline alga association: adaptations and defences between a selective herbivore and its prey. *Ecology*, **63**, 502–22.

Steneck, R.S. (1982*b*). Adaptive trends in the ecology and evolution of crustose coralline algae (Rhodophyta, Corallinaceae). Ph.D Dissertation. The John Hopkins University, Baltimore, MD.

Steneck, R.S. (1983). Escalating herbivory and resulting adaptive trends in calcareous algal crusts. *Paleobiology*, **9**, 44–61.

Steneck, R.S. (1985). Adaptations of crustose coralline algae to herbivory: patterns in space and time. In *Paleobiology: contempory research and applications* (ed. D.F. Toomey and M.H. Nitecki), pp. 352–66. Springer-Verlag, Berlin.

Steneck, R.S. (1986). The ecology of coralline algal crusts: Convergent patterns and adaptive strategies. *Annual Review of Ecology and Systemics*, **17**, 273–303.

Steneck, R.S. (1988). Herbivory on coral reefs: a synthesis. *Proceedings of the 6th International Coral Reef Symposium*, Townsville, Australia, 1988, **1**, 37–49.

Steneck, R.S., Macintyre, I.G., and Reid, R.P. (1997). A unique algal ridge system in the Exuma Cays, Bahamas. *Coral Reefs*, **16**, 29–37.

Steneck, R.S. and Watling, L. (1982). Feeding capabilities and limitations of herbivorous molluscs: a functional group approach. *Marine Biology*, **68**, 299–319.

Steuber, T. (1994). Sclerochronology of rudist bivalves. In *International Society for Reef Studies, Second European Regional Meeting* (ed. J. Geister, B. Lathuliere, A. Faber, and R. Maguil), Abstracts pp. 114. Luxembourg.

Steuber, T. (1996). Stable isotope sclerochronology of rudist bivalves: growth rates and Late Cretaceous seasonality. *Geology*, **24**, 315–18.

Steuber, T. (1997). Hippuritid rudist bivalves in siliciclastic settings—functional adaptations, growth rates and strategies. *Proceedings of the 8th International Coral Reef Symposium*, Panama City, Panama, 1996, **2**, 1761–6.

Stoddart, D.J. (1984). Genetic structure within populations of the coral *Pocillopora damicornis*. *Marine Biology*, **76**, 279–84.

Suchanek, T.H. and Colin, P.L. (1986). Rates and effects of bioturbation by invertebrates and fishes at Enewetak and Bikini Atolls. *Bulletin Marine Sciences*, **38**, 25–34.

Surge, D.M., Savarese, M., Dodd, J.R., and Lohmann, K.C. (1997). Carbon isotopic evidence for photosynthesis in Early Cambrian Oceans. *Geology*, **25**, 503–6.

Sverdrup, H.U., Johnson, M.W., and Fleming, R.H. (1942). *The oceans, their physics, chemistry and general biology*. Prentice-Hall, Engelwood Cliffs, NJ.

Swart, P.K. (1983). Carbon and isotope fractionation in scleractinian corals. *Earth Science Review*, **19**, 51–80.

Szulczewski, M. and Racki, G. (1981). Early Frasnian bioherms in the Holy Cross Mts. *Acta Geologica Polonica*, **31**, 147–62.

Talbot, F.H. and Goldman, G. (1972). A preliminary report on the diversity and feeding relationships of the reef fishes on One Tree Island, Great Barrier Reef system. *Proceedings of the Symposium on Corals and Coral Reefs*, **1**, 425–42.

Talent, J. (1988). Organic reef-building: episodes of extinction and symbiosis? *Senckenbergiana Lethaea*, **69**, 315–68.

Tanner, J.E. and Hughes, T.P (1994). Species coexistence, keystone species, and succession: a sensitivity analysis. *Ecology*, **75**, 2204–19.

Tappan, H. (1982). Extinction or survival: selectivity and causes of Phanerozoic crises. *Geological Society of America Special Publication*, **190**, 265–76.

Tappan, H. and Loeblich, A.R. Jr. (1973). Evolution of oceanic plankton. *Earth Science Reviews*, **9**, 207–40.

Tappan, H. and Loeblich, A.R. Jr. (1988). Foraminiferal evolution, diversification, and extinction. *Journal of Paleontology*, **62**, 695–714.

Taylor, J.D., Cleevely, R.J., and Morris, N.J. (1983). Predatory gastropods and their activities in the Blackdown Greensand (Albian) of England. *Palaeontology*, **26**, 521–33.

Taylor, P.D. (1990). Encrusters. In *Palaeobiology: a synthesis* (ed. D.E.G. Briggs and P.W. Crowther), pp. 346–51. Blackwell, Oxford.

Taylor, P.D. and Allison, P.A. (1998). Bryozoan carbonates through time and space. *Geology*, **26**, 459–62.

Taylor, P.D. and Larwood, G. P. (1988). Mass extinctions and patterns of bryozoan evolution. In *Extinction and survival in the fossil record* (ed. G.P. Larwood), pp. 99–119. Systematics Association Special Volume, **34**.

Termier, G. and Termier, H. (1979). Temps forts de l'evolution des Spongiares. Hypothesise environmentale et symbiotique sur l'origine des Spongiaires. In *Biologie des Spongiaires* (ed. J. Vacelet), pp. 513–20. *Colloquia International du CNRS*, **291**.

Thayer, C W. (1983). Sediment-mediated biological distubance and the evolution of marine benthos. In *Biotic interactions in Recent and fossil benthic communities* (ed. M. Tevesz and P.W. McCall), pp. 480–625. Plenum Press, New York.

Thom, B.G. and Chappell, J. (1975). Holocene sea levels relative to Australia. *Search*, **6**, 90–4.

Tilman, D., May, R.M., Lehman, C.L., and Nowak, M.A. (1994). Habitat destruction and the extinction debt. *Nature*, **371**, 65–6.

Tomascik, T., van Woesik, R., and Mah, A. (1996). Rapid coral colonisation of a recent lava flow following a volcanic eruption, Banda Islands, Indonesia. *Coral Reefs*, **15**, 169–75.

Toomey, D. F. (1976). Paleosynecology of a Permian plant dominated marine community. *Neues Jahrbuch Paläontologie Abhandlungen*, **152**, 1–18.

Toomey, D.F. and Nitecki, M.H. (1979). Organic buildups in the Lower Ordovician (Canadian) of Texas and Oklahoma. *Field Geology*, **2**, 1–181.

Trench, R.K. (1987). Dinoflagellates in non-parasitic symbioses. In *The biology of dinoflagellates* (ed. F.J.R. Taylor), pp. 531–70. Botanical Monographs, **21**. Blackwell Scientific, Oxford.

Trench, R.K. (1997). Diversity of symbiotic dinoflagellates and the evolution of microalgal–invertebrate symbioses. *Proceedings of the 8th International Coral Reef Symposium*, Panama City, Panama, 1996, **2**, 1275–86.

Trench, R.K., Wethy, D.S., and Porter, J.W. (1981). Observations on the symbiosis with zooxanthellae among the Tridacnidae (Mollusca, Bivalvia). *Biological Bulletin Marine Biological Laboratories*, *Woods Hole*, **161**, 180–98.

Tsein, H.H. (1994*a*). Construction of reefs through geologic time with emphasis on the role of non-skeletal microorganisms. *Acta Geologia Taiwanica*, **32**, 1–30.

Tsein, H.H. (1994*b*). Contributions of reef-building organisms in reef carbonate construction. *Courier Forschungsinstitut Senckenberg*, **172**, 95–102.

Tucker, M.E. and Hollingworth, N.T.J. (1986). The Upper Permian reef complex (EZ) of North East England. In *Reef Diagenesis* (ed. J.H. Shroeder and B.H. Purser), pp. 270–90. Springer-Verlag, Berlin.

Tucker, M.E. and Wright, V.P. (1990). *Carbonate sedimentology*. Blackwell Scientific, Oxford.

Tunnicliffe, V. (1981). Breakage and propagation of the stony coral *Acropora cervicornis*. *Proceedings of the National Academy of Science USA*, **78**, 2427–31.

Tunnicliffe, V. (1983). Caribbean staghorn populations: pre-Hurricane Allen conditions in Discovery Bay, Jamaica. *Bulletin of Marine Science*, **33**, 132–51.

Turner, E.C., Narbonne, G.M., and James, N.P. (1993). Neoproterozoic reef microstructures from the Little Dal group, northwestern Canada. *Geology*, **21**, 259–62.

Turner, E.C., James, N.P., and Narbonne, G.M. (1997). Growth dynamics of Neoproterozoic calcimicrobial reefs, Mackenzie Mountains, Northwest Canada. *Journal of Sedimentary Research*, **67**, 437–50.

Turnsek, D., Buser, S., and Ogorelee, B. (1981). An Upper Jurassic reef complex from Slovenia, Yugoslavia. *Society of Economic Paleontologists and Mineralogists, Special Publication*, **30**, 361–9.

Tyler, J.C. (1980). Osteology, phylogeny, and higher classification of the fishes of the order Plectognathi (Tetraodontiformes). *NOAA Technical Report NMFS Circular*, **434**, 1–422.

Ulmishek, G.F. (1988). Upper Devonian - Tournasian facies and oil resources of the Russian craton's eastern margin. In *Devonian of the world* (ed. McMillan, N.J. *et al.*), pp. 527-49. Canadian Society of Petroleum Geologists, **14 (1)**.

Vacelet, J. (1983). Les éponges calcifées et les récifs anciens. *Pour la Science*, June, 14–22.

Vacelet, J. (1985). Coralline sponges and the evolution of the Porifera. In *The origins and relationships of lower Invertebrates* (ed. S. Conway Morris, *et al.*), pp. 1–13. Systematics Association Special Volume, **28**. Clarendon Press, Oxford.

Vago, R., Gill, E., and Collingwood, J.C. (1997). Laser measurements of coral growth. *Nature*, **386**, 30–1.

Valentine, J. and Jablonski, D. (1993). Fossil communities: compositional variation at many time scales. In *Species diversity in ecological communities* (ed. R.E. Ricklefs and D. Schluter), pp. 341–9. University of Chicago Press, Chicago, IL.

Vance, R.R. (1979). The effects of grazing by the sea urchin, *Centrostephanus coronatus*, on prey community composition. *Ecology*, **60**, 537–46.

Vermeij, G.J. (1977). The Mesozoic marine revolution: evidence from snails, predators and grazers. *Paleobiology*, **2**, 245–58.

Vermeij, G.J. (1982). Unsuccessful predation and evolution. *The American Naturalist*, **120**, 701–20.

Vermeij, G. J. (1983). Intimate associations and coevolution in the sea. In *Coevolution* (ed. D.J. Futuyma and M. Slatkin), pp. 311–27. Sinauer Associates, Sunderland.

Vermeij, G.J. (1987). *Evolution and escalation*. Princeton University Press, Princeton, NJ.

Veron, J.E.N. (1986). *Corals of Australia and the Indo-Pacific*. Angus and Robertson, Sidney, Australia.

Veron, J.E.N. (1995). *Corals in space and time. The biogeography and evolution of the Scleractinia*. Comstock/Cornell, Itaca and London.

Veron, J.E.N. and Kelly, R. (1988). Species stability in reef corals of Papua New Guinea and the Indo Pacifc. *Association of Australian Palaeontologists*, **6**, 1–19.

Veron, J.N. and Minchin, P.R. (1992). Scleractinia of eastern Australia. II. Families Faviidae, Trachyphyllidae. *Australian Institute of Marine Sciences Monograph, Series* 3.

Visscher, H. and Brugman, W.A. (1988). The Permian–Triassic boundary in the Southern Alps: a palynological approach. *Memoir della Societa Geologica Italiana* **34** (for 1986), 121–8.

Vogt, P.R. (1989). Volcanogenic upwelling of anoxic, nutrient-rich water: a possible factor in carbonate-bank/reef demise and benthic faunal extinctions? *Bulletin of the Geological Society of America*, **101**, 1225–45.

Voight, E. (1977). On grazing traces produced by the radula of fossil and recent gastropods and chitons. In *Trace fossils* 2 (ed. T.P. Crimes and J. Harper), pp. 335–47. House Press, Seel.

Wagenplast, P. (1972). Ökologische Untersuchungen der Fauna aus Bank- und Schwammfazies des Weissen Jura der Schwabischen Alb. Arb. *Institüt Geologie und Paläontologie Universität Stuttgard, N.F.*, **67**, 1–99.

Walter, M.R. and Heys, G.R. (1985). Links between the rise of the Metazoa and the decline of stromatolites. *Journal of Precambrian Research*, **29**, 159–74.

Ward, P.D. (1995). The K/T Trial. *Paleobiology*, **21**, 121–42.

Ware, J.R., Smith, S.V., and Reaka-Kudla, M.L. (1992). Coral reefs: sources or sinks of atmospheric CO_2? *Coral Reefs*, **11**, 127–30.

Warme, J.E. (1975). Borings as trace fossils, and the process of marine bioerosion. In *The study of trace fossils* (ed. R.W. Frey), pp. 181–227. Springer-Verlag, New York.

Watkins, R. (1993). The Silurian (Wenlockian) reef fauna of southeastern Wisconsin. *Palaios*, **8**, 325–38.

Webb, G.E. (1987). Late Mississippean thrombolite bioherms from the Pitkin Formation of northern Arkansas. *Geological Society of America Bulletin*, **99**, 686–98.

Webb, G.E. (1990). Lower Carboniferous coral fauna of the Rockhampton Group, east-central Queensland. In *Devonian and Carboniferous coral studies* (ed. P.A. Jell), pp. 1–67. Association of Australasian Palaeontologists Memoir, **10**.

Webb, G.E. (1994). Non-waulsortian Mississippian bioherms: a comparative analysis. In *Pangea: global environments and resources*. Canadian Journal of Petroleum Geologists, Memoir **17**, 701–12.

Webb, G. E. (1996). Was Phanerozoic reef history controlled by the distribution of non-enzymatically secreted reef carbonates (microbial carbonate and biologically induced cement)? *Sedimentology*, **43**, 947–71.

Webby, B.D. (1994). Evolutionary trends in Ordovician stromatoporoids. *Courier Forschungsinstitut Senckenberg*, **172**, 373–80.

Weber, J.N. and Woodhead, P.M.J. (1972). Temperature dependence of oxygen-1 concentration in reef coral carbonates. *Journal of Geophysical Research*, **77**, 463–73.

Wefer, G. and Berger, W.H. (1991). Isotope palaeontology—growth and composition of extant calcareous species. *Marine Geology*, **100**, 207–48.

Wellington, J.M. and Trench, P.S. (1985). Persistence and coevolution of a nonsymbiotic coral in open reef environments. *Proceedings of the National Academy of Science USA*, **82**, 2432–6.

Wellington, G.M. (1982). Depth zonation of corals in the Gulf of Panama: control and facilitation by resident reef fish. *Ecological Monographs*, **52**, 223–41.

Wells, J.W. (1933). Corals of the Cretaceous of the Atlantic and Gulf coastal plains and western interior of the United States. *Bulletin of American Paleontologists*, **18**, 85–288.

Wells, J.W. (1956). Scleractinia. In *Treatise on invertetbrate paleontology. Coelenterata* (ed. R.C. Moore), pp. 328–400. Geological Society of America and University of Kansas Press, Boulder, Colorado and Lawrence, Kansas, USA.

Wells, J.W. (1957). Corals. *Geological Society of America Memoir*, **67**, 773–82.

Wendt, J. (1982). The Cassian patch reefs (Lower Carnian, southern Alps). *Facies*, **6**, 185–202.

Wendt, J. (1993). Steep-sided carbonate mud-mounds in the Middle Devonian of the eastern Anti-Atlas, Morocco. *Geological Magazine*, **130**, 69–83.

Wendt, J., Wu, X., and Reinhardt, J.W. (1989). Deep-water hexactinellid sponge mounds from the Upper Triassic of northern Sichuan (China). *Palaeogeography, Palaeoclimatology, Palaeoecology*, **76**, 17–29.

Wendt, J., Belka, Z., Kaufman, B., Kostrewa, R., and Hayer, J. (1997). The world's most spectacular carbonate mud mounds (Middle Devonian, Algerian Sahara). *Journal Sedimentary Research*, **67**, 424–36.

West, R.R. and Clark, G.R. (1983). Chaetetids. In *Sponges and spongiomorphs, notes for a short course* (ed. T.W. Broadhead), pp. 130–40. University of Tenessee Studies in Geology, **7**.

Wignall, P.B. (1990). Ostracod and foraminiferal micropaleontology and its bearing on biostratigraphy: a case study from the Kimmeridgian (Late Jurassic) of north west Europe. *Palaios*, **5**, 219–26.

Wignall, P.B. and Hallam, A. (1996). Facies change and the End-Permian mass extinction in S.E. Sichuan, China. *Palaios*, **11**, 587–96.

Whittaker, R.H. (1975). *Communities and ecosystems* (2nd edn). Macmillan, New York.

Wilkinson, B.H. and Walker, J.C.G. (1989). Phanerozoic cycling of sedimentary carbonates. *American Journal of Science*, **289**, 525–48.

Wilkinson, B.H., Owen, R.M., and Carroll, A.R. (1985). Submarine hydrothermal weathering, global eustacy, and carbonate polymorphism in Phanerozoic marine oolites. *Journal of Sedimentary Petrology*, **55**, 171–83.

Wilkinson, C. (1983). Net productivity in coral reef sponges. *Science*, **219**, 411–14.

Wilkinson, C. (1984). Immunological evidence for the Precambrian origin of bacterial symbioses in marine sponges. *Proceedings of the Royal Society London*, **B 220**, 509–17.

Wilkinson, C. (1986). The nutritional spectrum of coral reef benthos. *Oceanus*, **29**, 68–75.

Wilkinson, C. (1987). Interocean differences in size and nutrition of coral reef sponge populations. *Science*, **236**, 1654–7.

Wilkinson, C. (1993). Coral reefs are facing widespread devastation: can we prevent this through sustainable management practices. *Proceedings of the 7th International Coral Reef Symposium*, Guam, **1**, 11–21.

Wilkinson, C. (1996). Global change and coral reefs: impacts on reefs, economies and human cultures. *Global Change Biology*, **2**, 547–58.

Wilkinson, C.R. and Buddemeier, R.W. (1994). Global climate change and coral reefs: Implication for people and reefs. Report of the UNEP-IOC-APEI-ICUN Global Task Team on the implications of climate change on coral reefs. ICUN, Gland, Switzerland.

Williams, G.C. (1975). *Sex and evolution*. Princeton University Press, Princeton, NJ.

Wilson, J. L. (1975). *Carbonate facies in geologic history*. Springer-Verlag, Berlin.

Wilson, P.A., Jenkyns, H.C., Elderfield, H., and Larson, R.L. (1998). The paradox of drowned carbonate platforms and the origin of Cretaceous Pacific guyots. *Nature*, **392**, 889–94.

Wolter, K. and Hastenrath, S. (1989). Annual cycles and long-term trends of circulation and climate variability over the tropical oceans. *Journal of Climatology*, **2**, 1329–51.

Wong, P.K. (1979). Sequential cementation in the Upper Devonian Kaybob Reef, Alberta. *Canadian Society of Petroleum Geologists. Symposium in Honour of A.D. Baillie*, pp. 24–30.

Wood, R.A. (1987). Biology and revised systematics of some Late Mesozoic stromatoporoids. *Special Papers in Palaeontology*, **37**, 1–89.

Wood, R. (1990). Reef-building sponges, *American Scientist*, **78**, 224–35.

Wood, R.A. (1991*a*). Problematic reef-building sponges. In *The early evolution of Metazoa and the significance of problematic taxa* (ed. A. Simonetta and S. Conway Morris), pp. 113–24. Cambridge University Press, Cambridge.

Wood, R. (1993). Nutrients, predation and the history of reefs. *Palaios*, **8**, 526–43.

Wood, R. (1995). The changing biology of reef-building. *Palaios*, **10**, 517–29.

Wood, R.A. (1997). The importance of the cryptos. *Proceedings of the 8th International Coral Reef Symposium*, Panama City, Panama, **2**, 1687–92.

Wood, R., Reitner, J. and West, R.R. (1989). Systematics and phylogenetic implications of the haplosclerid stromatoporoid *Newellia mira* n. gen. *Lethaia*, **2**, 85–93.

Wood, R., Zhuravlev, A. Yu., and Debrenne, F. (1992*a*). Functional biology and ecology of Archaeocyatha. *Palaios*, **7**, 131–56.

Wood, R., Evans, K.R., and Zhuravlev, A.Yu. (1992*b*). A new post Early-Cambrian archaeocyath from Antarctica. *Geological Magazine*, **129**, 491–5.

Wood, R.A., Zhuravlev, A.Yu., and Tseren Anaaz, C. (1993). The ecology of Lower Cambrian buildups from Zuune Arts, Mongolia: implications for early metazoan reef evolution. *Sedimentololgy*, **40**, 829–58.

Wood, R., Dickson, J.A.D., and Kirkland-George, B. (1994). Turning the Capitan reef upside down: a new appraisal of the ecology of the Permian Capitan Reef, Guadalupe Mountains, Texas and New Mexico. *Palaios*, **9**, 422–7.

Wood, R., Dickson, J.A.D., and Kirkland-George, B. (1996). New observations on the ecology of the Permian Capitan Reef, Texas and New Mexico. *Palaeontology*, **39**, 733–62.

Woodley, J.D., Chornesky, E.A., Clifford, P.A., Jackson, J.B.C., Kaufman, L.S., Knowlton, N., *et al.* (1981). Hurricane Allen's impact on Jamaican coral reefs. *Science*, **214**, 749–55.

Wörheide, G. (1998). The reef cave dwelling ultraconservative coralline demosponge *Astrosclera willeyana* Lister 1900 from the Indo-Pacific. *Facies*, **38**, 1–88.

Wray, J.L. (1964). *Archaeolithophyllum*, and abundant calcareous algae in limestones in the Lansing group (Pennsylvanian), South eastern Kansas. *State Geological Survey of Kansas Bulletin*, **170**, 1–13.

Wray, J.L. (1969). Paleocene calcareous algae from Libya. *Symposium Geological Libya University*, *Libya*, pp. 21–22.

Wray, J.L. (1972). Environmental distribution of calcareous algae in Upper Devonian reef complexes. *Geologisches Rundschau*, **61**, 578–84.

Wray, J.L. (1977). Late Paleozoic calcareous red algae. In *Fossil algae* (ed. E. Flügel), pp. 167–77. Springer-Verlag, New York.

Wray, J.L., James, N.P., and Ginsburg, R.N. (1975). The puzzling Paleozoic phylloid algae — Holocene answer in squamariacean calcareous red algae. *American Association of Petroleum Geologists, Annual Meeting Abstracts*, **2**, 82–3.

Wright, V.P. (1991). Comment on 'Probable influence of Early Carboniferous (Tournasian–Early Viséan) geography on the development of Waulsortian-like mounds'. *Geology*, **19**, 413.

Wright, V.P. and Faulkner, T.J. (1990). Sediment dynamics of Early Carboniferous ramps: a proposal. *Geological Journal*, **25**, 139–44.

Wulff, J. (1997). Mutualisms among species of coral reef sponges. *Ecology*, **78**, 146–59.

Wylie, C.R. and Paul, V.J. (1988). Feeding preferences of the surgeonfish *Zebrasoma flavecens* in relation to chemical defences of tropical algae. *Marine Ecology Progress Series*, **45**, 23–32.

Xiao, S., Zhang, Y., and Knoll, A.H. (1997). Three-dimensional preservation of algae and animal embryos in a Neoproterozoic phosphorite. *Nature*, **391**, 553–8.

Yancey, T.E. (1982). The alatoconchid bivalves: Permian analogs of modern tridacnid clams. *Proceedings of the 3rd North American Paleontological Convention*, 589–92.

Young, G.A. and Elias, R.J. (1995). Latest Ordovician to earliest Silurian colonial corals of the east-central United States. *Bulletins of American Paleontology*, **108**, 1–148.

Zachos, J.C. and Arthur, M.A. (1986). Paleoceanography of the Cretaceous/Tertiary boundary event: inferences from stable isotopic and other data. *Palaeoceanography*, **1**, 5–26.

Zachos, J.C., Arthur, M.A., and Dean, W.E. (1989). Geochemical evidence for suppression of pelagic marine productivity at the Cretaceous/Tertiary boundary. *Nature*, **337**, 61–4.

Zankl, H. (1969). Die Hohe Göll—Aufbau und Lebensbild eines Dachsteinkalk-Riffes in der Obertrias der nördlichen Kalkalpen. *Abhandlungen Senckenberg, Naturforschungs Gesellschaft*, **519**, 1–123.

Zeigler, B. and Rietschel, S. (1970). Phylogenetic relationships of fossil calcisponges. In *Biology of the Porifera* (ed. W.G. Fry), pp. 23–40. Zoological Society of London Symposium, **25**.

Zhao, Z., Xing, Y., Ma, G., and Chen, Y. (1985). *Biostratigraphy of the Yangtze Gorge Area* (1) *Sinian*. Geological Publishing House, Beijing.

Zhu, Z.D., Guo, C.X., Liu, B.L., Hu, M.Y., Hu, A.M., Xiao, C.T., *et al.* (1993). Lower Ordovician reefs at Huang-huachang, Yichang, east of the Yangtze Group. *Scientia Geologica Sinica*, **2**, 79–90.

Zhuravlev, A. Yu. (1990). Sistematika arkheotsiat (Systematics of archaeocyaths). In *Sistematika i filogeniya bespozvonochnykh* [Taxonomy and phylogeny of *Invertebrata*] (ed. V.V. Menner), pp. 28–54. Nauka, Moscow. (In Russian.)

Zhuravlev, A.Yu. (1996). Reef ecosystem recovery after the Early Cambrian extinction. In *Biotic recovery from mass extinction events* (ed. M.B. Hart), pp. 79–96. Geological Society of London Special Publication, **102**.

Zhuravlev, A. Yu. and Wood, R.A. (1995). Lower Cambrian reefal cryptic communities. *Palaeontology*, **38**, 443–70.

Zhuravlev, A.Yu. and Wood, R.A. (1996). Anoxia as the cause of the Mid–Early Cambrian (Botomian) extinction event. *Geology*, **24**, 311–14.

Glossary

acclimatization
The ability of an organism to adapt to environmental change during its lifetime

accommodation
The space available to be filled by sediment

accretionary margin
A rimmed shelf that shows the lateral migration (*progradation*) of shallow shelf-margin reefs or carbonate sand bodies over the slope sediments

adaptation
A supposed trait that benefits a specific function or effect that has been enhanced by natural selection (Gould and Vrba 1982)

aggregation
The close-packing of organisms

algal turf
Algal communities of often microfilamentous species less than 10 mm in height

algal ridge
A topographic structure constructed by coralline algae that develops in shallow subtidal to lower intertidal zones

aptations
Supposed beneficial traits (Gould and Vrba 1982)

aquiferous filtration system
The ramifying network of canals and pores of a sponge used to pump water and so gain nutrition from filtration

atoll
An isolated carbonate platform with a horseshoe or ring-shaped shallow reef rim that forms around a deeper central lagoon. True oceanic examples form on extinct, subsiding volcanoes

back-reef
The landward side of a reef, including the area behind the *reef crest*, and including the *reef flat* and shelf *lagoon*

barrier reef
An elongated reef growing some distance from, but parallel to, land separated by a lagoon of considerable depth and width

binding
The holding of loose sediment or *in-situ* organisms by organisms that show lateral, often encrusting, growth

bioeroder
An organism that scrapes, bores, etches, or otherwise excavates into calcareous substrate to either access prey or create a dwelling

bioerosion
The erosion or destruction of carbonate rock by organisms

bioherm
A relatively small, discrete, lens-shaped reef

biostrome
A bed of often *in-situ* skeletal organisms without significant relief

bioturbation	The disruption of unconsolidated sediment by organisms
bioturbators	Organisms that disturb unconsolidated sediment
bleaching	The mass expulsion or *in-situ* degradation of photosymbiotic algae
borers	Organisms with the ability to penetrate calcareous substrates by chemical solution to form borings or boreholes, in order to access prey or create dwelling places
boundstone	Carbonate rock that was biologically bound during deposition
broadcasters	Organisms that release long-lived and widely dispersed larvae—often 4 to 8 days after fertilization
brooders	Organisms that retain larvae—often for between 1 to more than 100 days—until they are almost ready to settle
browsers	Herbivores that consume plant tissues above a substrate
build-up	Any localized, thick carbonate deposit built by organisms but not always with an apparent bound fabric
bulldozing	The action of organisms displacing and pushing aside considerable amounts of sediment through their search for food, or through rapid burial as a means of escape
bypass margin	A rimmed margin that deposits little sediment on the shelf slope, because while shelf margin sedimentation is able to keep pace with rising sea level, insufficient sediment is deposited on the slope for any significant lateral accretion
carbonate	Sedimentary rock with 95% or more calcium carbonate ($CaCO_3$) or dolomite ($CaMg(CO_3)_2$)
carbonate compensation depth (CCD)	The depth in an ocean at which the rate of sedimentation equals that of dissolution
carbonate platform	Shallow marine carbonate sequence that develops on a horizontal marine shelf
cement	Chemically precipitated minerals, usually calcium carbonate, that grow between grains of a sediment or within voids between skeletal material
chasmoliths	Organisms which nestle into depressions, sometimes on living organisms
clade	All the species derived from a common ancestor
clinoform	A landform with a large-scale sloping sedimentary surface, formed mainly by the shedding of shallow-water debris from a shelf margin into the ocean

clonality	The ability to grow and asexually reproduce through a potentially unlimited production of identical modules which are all ultimately derived from the same zygote
coevolution	The simultaneous evolution of interacting populations
colony	A modular organism where the modules are also morphological individuals
competition	Fitness of individuals of species A decreased by interaction with individuals of species B; individual of species A also suffers decreased reproductive success as a result of the interaction
competitive exclusion principle	The principle that in order for species to coexist, they must differ in the way in which they use such limiting resources (Gause 1934)
competitive network	A complex network of competitive relationships where no one species is competitively dominant over all others
conspecifics	Members of the same species
constratal growth	The growth style of immobile surface-dwelling organisms which do not substantially project above a substrate
corallite	The calcareous skeleton formed by an individual coral polyp
corallivores	Predators that feed on coral polyps
cratons	The stable central part of continental crust
cross-lamination	Bedding-inclined layers (laminations) at an angle to the horizontal, formed by the migration of ripples
crustose algae	Algae that form calcareous crusts
cryptic	Organisms that occupy protected, hidden, niches
crypts	Protected, hidden niches, such as caves, grottoes, under overhangs, small cavities, or within rubble
degree of integration	The inferred level of interdependence between functional units (*modules*) of a modular organism (with the relative prefixes low, medium or high)
deposit-feeder	Organisms living on or in sediment on the sea floor that ingest sediment rich in organic matter in order to gain nutrition
determinate (growth)	Organisms or modules that have a genetically determined upper size limit
detritivore	An animal that feeds upon dead organic matter
diagenesis	All the chemical, biological, and physical changes undergone by a sediment after deposition but prior to metamorphosis

disaster taxa	Usually long-ranged species of opportunists, whose presence in vast abundance is a clear sign of considerable environmental perturbation during the survival phase after a mass extinction
discontinuity surfaces	Surfaces that record the temporary cessation of reef growth
durophagy	Predation by shell-crushing
ecological character displacement	Where the evolution of an organism is shaped or maintained by inter specific competition
ecophenotype	Different morphologies of the same species that have changed growth form in response to local environmental conditions
encrusting	The laterally expanding growth form of organisms that attach permanently to hard substrate
endocryptic	Organisms that live partially or wholly within a hard substrate
endoliths	Organisms that live partly or wholly within substrates in dwelling places of their own making
endosymbiosis	The relationship between two species of unequal size where the whole body of the smaller (the *symbiont*) is housed entirely within the larger (the *host*)
epeiric (or epicontinental) sea	An extensive shallow sea that extends far into the interior of a continent
epibenthos	Aquatic organisms that live on a substrate
epiliths	Predators which feed directly upon sessile invertebrates or algae causing skeletal or substrate damage
erect (growth)	Morphology showing growth predominantly vertical or perpendicular to a substrate
erosional margin	A rimmed margin which occurs in areas with strong tides or ocean currents such that cliff or escarpments characterize the shelf slope. Debris may often accumulate at the toe of the slope, forming characteristic *clinoform* geometries
etchers	Microscopic organisms that are able to dissolve or otherwise remove calcareous material
eustatic sea level	Global changes in sea level that affects all oceans
euxinic conditions	An environment where circulation of water is restricted, leading to reduced oxygen levels and where free H_2S is present in the lower water column
exaptations	Supposed beneficial traits whose benefits are secondary or incidental to the primary function to which they are adapted (Gould and Vrba 1982)

excavators	Herbivores or predators which are capable of deep excavation that removes large areas of substrate
eustacy	Global sea level and its fluctuations caused by absolute changes in the quantity of sea water
fabric	The texture of a limestone or dolomite
fecundity	The reproductive capacity of an organism
fenestrae	Irregular, cement-filled cavities often found in carbonate sediments
fibrous calcite (cement)	A carbonate cement where crystals have a significant length elongation, mostly parallel to the c-axis
filter-feeder	An organism that strains tiny food particles from the surrounding water
fining-upwards cycles	Sediment that is graded from coarse at the base to fine-grained at the top, reflecting the waning of hydrodynamic energy
fissures	Fractures within a reef, which may be filled by many generations of marine sediment and cement
flanking beds	Strata that accumulate around the sides of a reef
floatstone	Carbonate rock that contains more than 10% of the grains larger than 2 mm, where the grains are not in contact
foliaceous	Growth form of overlapping thin plates supported by a relatively small basal attachment
fore-reef	The seaward side of a reef, including the reef slope and talus deposits
framework	A bound, organic carbonate structure
fringing reef	An elongated reef growing in subtidal waters bordering land
functional morphology	The form and structure of an organism in relation to its mode of life
generalist	An organism that has broad environmental or ecological requirements or tolerances
genet	Collective, all genetically identical individuals or colonies
geopetal structure	Partially sediment-filled voids capped by cement that act as geological spirit levels
global sea level	World-wide sea level determined by ocean-basin and glacial ice volumes
grainstone	Carbonate rock where more than 10% of the grains are larger than 2 mm, where the grains are in contact, but where no lime mud is present
grazers	Herbivores which crop very close to a substrate, so ingesting substantial portions of living plant tissue, associated small invertebrates, as well as underlying substrate

growth form	The morphology of an organism
heterotrophs	Organisms that are unable to manufacture their own food from simple chemical compunds. Includes *suspension-* and *filter-feeders*, and *detrivores*.
highstand	The point of highest relative sea level in any cycle
holdover taxa	Taxa that outlive the majority of their clade after a mass extinction
horizontal transmission	The process whereby symbionts must be acquired anew by successive host generations from the open environment after metamorphosis to the adult form
hybrids	Organisms which have arisen through the fusion of two or more distinct genotypes
indeterminate (growth)	The ability for potentially unlimited growth
infaunal	Organisms that live mostly or wholly within sediment
intermediate disturbance hypothesis	The hypothesis that states that intermediate levels of disturbance (physical and biological) lead to the highest standing diversities within communities (Connell 1978)
internal sediment	Sediment that accumulates within reef crypts
intervallum	The area between the two walls of an archaeocyath sponge skeleton, which was filled with soft tissue
intraclasts	Fragments of reworked cement or lithified sediment
intracratonic basin	Shallow basin formed by downwarping of continental crust
intransitive (competition)	Where in competing species, no one species is competitively dominant over all others
intrazooidal budding	The episodic addition in bryozoans of one zooid at a time
isolated platforms	Shallow-water carbonate platforms that are surrounded by deep water, often with steep margins
isopachous fibrous fringes	Those where c axis of the crystals are significantly elongated and are of the same length
lagoon	Shallow water between a reef and land, or within the rim of an atoll
Lazarus taxa	Taxa that reappear some time after they had disappeared from the fossil record during an extinction event some several million years earlier
life history	The schedule of events that occurs between birth and death
light-enhanced calcification	Where photosynthetic products released by the rates of growth and symbiont to the host often result in increased hence increased skeletal calcification when compared to organisms without photosymbionts

lowstand	The point of lowest relative sea level in any cycle
macroalgae	Algae greater than 10 mm in height
macroherbivores	Herbivores easily visible to the naked eye, e.g. echinoids, fishes, and large gastropods
mass extinction	The extinction of a significant proportion of the Earth's biota over a geographically widespread area in a geological insignificant period of time, often one that appears instantaneous when viewed at the level provided by the geological record
massive (growth form)	Large domal, mounded, or head-shaped growth form
meteoric	The zone where rainfall-derived groundwater is in contact with sediment or rock
micrite	Calcium carbonate (lime) mud, i.e. grains less than 20 μm in diameter
microbialite	Organosedimentary calcareous structure formed directly or indirectly by microbial communities
microherbivores	Herbivores not easily visible to the naked eye, e.g. tanaid isopods, amphipods, and small gastropods
mineral stabilization	The process of a metastable mineral altering to a more stable and less soluble form
mixotrophs	Organisms which gain food both by photosynthesis and from organic matter
modules	The functional units of an organism which are generally capable of independent existence from the parent organism
morphological plasticity	The ability of an organisms to adopt different growth morphologies in order to adapt to local environmental conditions
mud-mound	Reefs dominated by lime mud (micrite) with relatively few sessile skeletal biota
multiserial (growth)	Growth morphology with continuous surfaces of interconnected functional units (modules)
mutualism (mutualistic)	Reproductive success of individual of species A increased byinteraction with B and vice versa. Removal of either species will result in a reduction of the reproductive success of the other
neptunian dykes	Vertical fractures within a reef body, often filled by many generations of marine sediment and cement, and sometimes the organisms that dwelt within these submarine habitats
ooid	Coated carbonate grains produced by the inorganic precipitation of laminae around a nucleus

opportunistic (organisms)	Organisms that are able to rapidly colonize newly available habitat, which often show high growth rates and short-life cycles
osculum (pl. oscula)	Exhalant pore of a sponge
ostium (pl. ostia)	Inhalant pore of a sponge
outbreak population	A population that reaches an abnormally high density
packstone	Carbonate rock where more than 10% of the grains are larger than 2 mm, where the grains are in contact and lime mud is present
parasitism (and predation)	Reproductive success of an individual of species A increased by interaction with B; reproductive success of an individual of species B decreased by interaction
partial mortality	Localized, non-lethal injury that results in local death of tissue
passive continental margin	A continental margin which is not also a plate boundary. Often characterized by thick sedimentary sequences
patch reefs	Small, isolated reefs
peloids	Structureless micritic (lime mud) grains of multiple origin
Phanerozoic	The Eon of 'visible life', from the Cambrian to the Recent
philopatry	Settlement of larvae close to the parent
photic zone	The zone of light penetration in the oceans
photoadaptation	Any adaptive response shown by an organism to the availability of light
photosymbiosis	Where protists (foraminifera) or invertebrates have entered into a symbiotic relationship with a variety of photosynthesizing microorganisms, mainly single-celled algae
photosynthate	The metabolic products of photosynthesis
phototroph	An organism that obtains energy from (sun)light, usually by photosynthesis
phylogeny	Evolutionary ancestry and the relationships between *taxa*
pinnacle reefs	Reefs that develop on topographic highs within a basin
platform	see *carbonate platform*
polymorphism	The functional or morphological specialization of functional units (*modules*) for sexual or defensive roles
polyphyletic	A morphological feature that shows multiple ancestry and has therefore arisen independent of systematic origin

post-depositional compaction	The preferential compaction of relatively weakly cemented, sedimentary strata during burial around a more dense, cement-rich structure
preadaptation	A feature which by virtue of its fortuitous suitability for a novel function becomes *co-opted* as a new adaptation (Skelton 1985)
predation (and parasitism)	Reproductive success of individual of species A increased by interaction with B; reproductive success of individual of species B decreased by interaction
predator (and herbivore)	An organism which consumes either whole or parts of other organisms
preservation potential	The potential of an organism to fossilize and so become part of the geological record
primary reef-builders	Those organisms or their calcareous products that by mutual interconnection, encrustation, or successive overgrowth form principal reef-builders
primary succession	The sequence of community changes initiated when new substrate becomes available which has never been colonized
primary cavities	Cavities that form within a reef due to the close-packing, encrustation, or successive overgrowth of reef-building organisms
progenitor taxa	Taxa which appear during the extinction, or subsequent survival phase, and subsequently radiate during the recovery phase. Progenitor taxa are often adapted to the extreme environmental conditions of the extinction phase
progradation	The lateral migration of sediment, which forms characteristic *clinoform* geometries
radiaxial (cement)	A fibrous calcite cement where crystals have undulose extinction, curved twinned planes, and with converging extinction swing in each crystal
ramets	Genetically identical individuals or colonies
ramp	A gently sloping marine surface (generally less than 1°) on which shallow-water carbonate sediments pass gradually offshore to deeper waters
recruitment	The process wherby larvae settle from the plankton onto a substrate and subsequently metamorphose into adults
reef	A discrete carbonate structure formed by *in-situ* or bound organic components that develops topographic relief upon the sea floor
reef crest	A sharp break in slope at the seaward margin of a *reef*, or edge of a *reef flat*

reef complex	A series of *reefs*, their distinct zones, and associated sediments
reef flat	Any flat area behind a *reef crest*
reef margin	The reef community itself that forms a break in slope at the seaward margin
reef mound	Reefs constructed by relatively delicate organisms, often not in life-position
reef talus	Debris derived from, and surrounding , the reef
refugia	Safe havens—where organisms have retreated in the face of new competition. These include the deep sea and crypts, especially large caves and grottoes
regeneration	Regrowth or repair in response to injury of tissue
relative sea level	Local or regional sea level determined by local basin volumes
relict topography	The influence of inherited physical morphology (topography) upon subsequent sedimentary deposition
rifted continental margin	Actively rifting margin often with variable subsidence rates
rimmed shelf (or margin)	A carbonate platform with a pronounced break of slope at shallow depth that extends into deep water
rudstone	Carbonate rock that contains more than 10% of grains larger than 2 mm, that are in mechanical contact and so support each other. Lime mud may or may not be present
scraper	Predator or herbivores capable of removing limited amounts of calcareous material
secondary cavities	Crypts formed by bioerosion, or by the biological disturbance of unconsolidated sediment (*bioturbation*) within a reef which subsequently becomes lithified
secondary encrusters	Organisms that encrust or bind the primary reef-building community
senescence	An increased rate of mortality, or decreased rate of growth, or reproductive output, with old age
sessile (mode of life)	Immobile organisms that live on the sea floor
Signor–Lipps effect	Sampling bias which will have the effect of making even sudden extinctions appear more gradual (Signor and Lipps 1982)
spatial heterogeneity	The degree of three-dimensionality of a reef framework
specificity	The taxonomic range of partners with which an organism can form a symbiosis

spicules	Small siliceous or calcareous needle- or spine-like structures found in sponges, holothurians, and some cnidarians
spongin	Organic strands of collagen found in sponges
stratigraphy	The dating, correlation, and study of geological strata
stromatolites	Finely-layered, organosedimentary structures, formed by the trapping, binding, or precipitating activities of microbial communities, usually dominated by cyanobacteria
subsidence	The progressive depression of the Earth's crust that allows sediments to accumulate
succession	The sequence of community changes initiated when new substrate becomes available
supersaturation	Where the concentrations of a given ion exceeds its thermodynamic mineral solubility product
suprastratal growth	The growth of epifaunal organisms that produces substantial growth elevation above a substrate
suspension-feeder	An organism that actively gathers organic matter from surrounding water
symbiosome	The single or multiple membranes of host origin which surround endosymbiotic algae
tabula (pl. tabulae)	A plate-like skeletal structure that serves to separate abandoned parts of the skeleton from those occupied by living tissue
talus blocks	Eroded pieces of reef rock
taphonomy	The study of all the processes of preservation
taxon (plural: taxa)	A group of organisms of any taxonomic rank
taxonomy	The scientific classification of organisms
tectonic	The forces involved in the Earth's crust
terrigenous sediments	Silicate-mineral sediments derived from the land or continents
thin-tissue syndrome	Organisms with thin tissue layers, often over substantial underlying skeletons
thrombolite	Organosedimentary carbonate structures produced by microbial communities that show a clotted fabric to the naked eye
topographic relief	Contemporary elevation upon the sea floor
transgression	The spread or extension of the sea over land
transitive (competition)	Where the ranking between competing organisms is hierarchical and fixed

trends	Non-random, directional changes in a morphological character over time
trophic structure	The feeding relationships within a community
uniserial (growth)	Morphologies with single or branching chains with one functional unit (*module*) isolated at the tip of each chain or branch
vertical transmission	Where symbionts are transferred directly from host to offspring
wackestone	A lime-mud supported carbonate rock that contains more than 10% grains
zooidal budding	The development of a series of new zooids that allows effectively continuous growth in encrusting bryozoans
zooids	The minute, clonally produced modular units of bryozoans
zooxanthellae	Gymnodinioid ('naked') dinoflagellates when occurring in symbioses

Index